Ludwig Gattermann

The Practical Methods of Organic Chemistry

Ludwig Gattermann

**The Practical Methods of Organic Chemistry**

ISBN/EAN: 9783743359857

Manufactured in Europe, USA, Canada, Australia, Japa

Cover: Foto ©berggeist007 / pixelio.de

Manufactured and distributed by brebook publishing software (www.brebook.com)

Ludwig Gattermann

**The Practical Methods of Organic Chemistry**

# THE PRACTICAL METHODS

OF

# ORGANIC CHEMISTRY

BY

LUDWIG GATTERMANN, Ph.D.
A. O. PROFESSOR IN THE UNIVERSITY OF HEIDELBERG

*WITH NUMEROUS ILLUSTRATIONS*

TRANSLATED BY

WILLIAM B. ŞHOBER, Ph.D.
INSTRUCTOR IN ORGANIC CHEMISTRY IN THE LEHIGH UNIVERSITY

*AUTHORISED TRANSLATION*

New York
THE MACMILLAN COMPANY
LONDON: MACMILLAN & CO., Ltd.
1896

*All rights reserved*

# TRANSLATOR'S PREFACE

The success of Professor Gattermann's book in the original has warranted its reproduction in English. The translation is intended for those students of chemistry who have not yet become sufficiently familiar with scientific German to be able to read it accurately without constant reference to a dictionary. To such students this translation is offered, in the hope that it will increase their interest in the science without causing a corresponding decrease in their efforts to acquire a knowledge of German, which is indispensable to every well-trained chemist.

My grateful acknowledgments are due to my colleague, Dr. H. M. Ullmann, for many valuable suggestions, and to Professor Gattermann for his courtesy in pointing out several inaccuracies which occur in the German edition.

WILLIAM B. SHOBER.

South Bethlehem, Pennsylvania,
April, 1896.

# PREFACE

THE present book has resulted primarily from the private needs of the author. If one is obliged to initiate a large number of students at the same time into organic laboratory work, it is frequently impossible, even with the best intentions, to draw the attention of each individual to the innumerable details of laboratory methods. In order that students, even in the absence of the instructor, can gain the assistance necessary for the carrying out of the common operations, a General Part, dealing with crystallisation, distillation, drying, analytical operations, etc., is given before the special directions for Preparations. In the composition of this General Part, it has been considered of more value to describe the most important operations in such a way that the beginner may be able to carry out the directions independently, rather than to give as fully as possible the numerous modifications of individual operations. In the Special Part, to each preparation are added general observations, which relate to the character and general significance of the reaction carried out in practice; and the result follows, that the student already, during the period given to laboratory work, becomes familiar with the most varied theoretical knowledge possible, which, acquired under these conditions adheres more firmly, as is well known, than if that knowledge were obtained exclusively from a purely theoretical book. And so the author hopes that his book, along with the excellent " Introductions " of E. Fischer and Levy, may here and there win some friends.

For the assistance given by his colleagues, in pointing out deficiencies of his work, the author will always be grateful.

GATTERMANN.

HEIDELBERG, August, 1894.

# CONTENTS

## GENERAL PART

| | PAGE |
|---|---|
| Crystallisation | 1 |
| Sublimation | 14 |
| Distillation | 16 |
| Distillation with Steam | 35 |
| Separation of Liquids; Extraction; Salting Out | 39 |
| Decolourising; Removal of Tarry Matter | 43 |
| Drying | 45 |
| Filtration | 49 |
| Heating under Pressure | 55 |
| Melting-point, Determination of | 62 |
| Drying and Cleaning of Vessels | 66 |

### ORGANIC ANALYTICAL METHODS

| | |
|---|---|
| Qualitative Tests for Carbon, Hydrogen, Nitrogen, Sulphur, and the Halogens | 68 |
| Quantitative Determination of the Halogens. Carius' Method | 71 |
| Quantitative Determination of Sulphur. Carius' Method | 76 |
| Quantitative Determination of Nitrogen. Dumas' Method | 80 |
| Quantitative Determination of Carbon and Hydrogen. Liebig's Method | 92 |

## SPECIAL PART

### I. ALIPHATIC SERIES

| | |
|---|---|
| 1. Reaction: Replacement of an Alcoholic Hydroxyl by a Halogen | 105 |
| 2. Reaction: Preparation of an Acid-Chloride from the Acid | 114 |
| 3. Reaction: Preparation of an Acid-Anhydride from the Acid-Chloride and the Sodium Salt of the Acid | 120 |

| | PAGE |
|---|---|
| 4. Reaction: Preparation of an Acid-Amide from the Ammonium Salt of the Acid . . . . . . . . . . | 124 |
| 5. Reaction: Preparation of an Acid-Nitrile from an Acid-Amide | 128 |
| 6. Reaction: Preparation of an Acid-Ester from the Acid and Alcohol | 130 |
| 7. Reaction: Substitution of Hydrogen by Chlorine . . . . | 132 |
| 8. Reaction: Oxidation of a Primary Alcohol to an Aldehyde . . | 136 |
| 9. Reaction: Preparation of a Primary Amine from an Acid-Amide of the next Higher Series . . . . . . . . | 144 |
| 10. Reaction: Syntheses of Ketone Acid-Esters and Polyketones with Sodium and Sodium Alcoholate . . . . . . . | 148 |
| 11. Reaction: Syntheses of the Homologues of Acetic Acid with Malonic Acid Ester . . . . . . . . . . | 154 |
| 12. Reaction: Preparation of a Hydrocarbon of the Ethylene Series by the Elimination of Water from the Alcohol. Union with Bromine | 160 |
| 13. Reaction: Preparation of an Acetylene Hydrocarbon from an Alkylene Bromide. The Addition of Bromine . . . . | 165 |

## II. AROMATIC SERIES

| | |
|---|---|
| 1. Reaction: Nitration of a Hydrocarbon . . . . . | 172 |
| 2. Reaction: Reduction of a Nitro-Compound to an Amine . | 176 |
| 3. Reaction: Reduction of a Nitro-Compound to an Azoxy-, Azo-, or a Hydrazo-Compound . . . . . . . . . | 183 |
| 4. Reaction: Preparation of a Thiourea and a Mustard Oil from Carbon Disulphide and a Primary Amine . . . | 189 |
| 5. Reaction: Sulphonation of an Amine . . . . . . | 192 |
| 6. Reaction: Replacement of the Amido- and Diazo-Group by Hydrogen . . . . . . . . . . . | 193 |
| 7. Reaction: Replacement of the Diazo-Group by Hydroxyl . | 201 |
| 8. Reaction: Replacement of a Diazo-Group by Iodine . . . | 202 |
| 9. Reaction: Replacement of a Diazo-Group by Chlorine, Bromine, or Cyanogen . . . . . . . . . . | 204 |
| 10. Reaction: (*a*) Reduction of a Diazo-Compound to a Hydrazine. (*b*) Replacement of the Hydrazine Residue by Hydrogen . . | 206 |
| 11. Reaction: (*a*) Preparation of an Azo Dye from a Diazo-Compound and an Amine. (*b*) Reduction of the Dye . . . | 211 |

|   |   | PAGE |
|---|---|---|
| 12. Reaction: Preparation of a Diazoamido-Compound | . . . | 217 |
| 13. Reaction: Transformation of a Diazoamido-Compound into an Amidoazo-Compound . . . . . . . . | | 220 |
| 14. Reaction: Oxidation of an Amine to a Quinone | . | 221 |
| 15. Reaction: Reduction of a Quinone to a Hydroquinone | . | 225 |
| 16. Reaction: Bromination of an Aromatic Compound . | | 226 |
| 17. Reaction: Fittig's Synthesis of a Hydrocarbon | . . | 231 |
| 18. Reaction: Sulphonation of an Aromatic Hydrocarbon (I.) | . . | 235 |
| 19. Reaction: Reduction of a Sulphon-Chloride to a Sulphinic Acid and to a Thiophenol . . . . . . . . | | 240 |
| 20. Reaction: Sulphonation of an Aromatic Hydrocarbon (II.) | . . | 243 |
| 21. Reaction: Conversion of a Sulphonic Acid into a Phenol | . | 246 |
| 22. Reaction: Nitration of a Phenol . . . . . . | | 249 |
| 23. Reaction: (a) Chlorination of a Hydrocarbon in the Side-Chain. (b) Conversion of a Dichloride into an Aldehyde | . . | 251 |
| 24. Reaction: Simultaneous Oxidation and Reduction of an Aldehyde under the Influence of Concentrated Caustic Potash | . . . | 256 |
| 25. Reaction: Condensation of an Aldehyde by Potassium Cyanide to a Benzoïn . . . . . . . . . . | | 258 |
| 26. Reaction: Oxidation of a Benzoïn to a Benzil . | . . . | 260 |
| 27. Reaction: Addition of Hydrogen Cyanide to an Aldehyde | | 261 |
| 28. Reaction: Perkin's Synthesis of Cinnamic Acid | . . | 265 |
| 29. Reaction: Addition of Hydrogen to an Ethylene Derivative | . | 268 |
| 30. Reaction: Preparation of an Aromatic Acid-Chloride from the Acid and Phosphorus Pentachloride . . . . . . | | 269 |
| 31. Reaction: Schotten-Baumann's Reaction for the Recognition of Compounds containing the Amido-, Imido-, or Hydroxyl-Group | | 270 |
| 32. Reaction: (a) Friedel and Crafts' Ketone Synthesis. (b) Preparation of an Oxime. (c) Beckmann's Transformation of an Oxime . . . . . . . . . | | 272 |
| 33. Reaction: Reduction of a Ketone to a Hydrocarbon | | 281 |
| 34. Reaction: Saponification of an Acid-Nitrile . . . . | | 283 |
| 35. Reaction: Oxidation of the Side-Chain of an Aromatic Compound . | | 284 |
| 36. Reaction: Kolbe's Synthesis of Oxyacids | . . | 287 |
| 37. Reaction: Preparation of a Dye of the Malachite Green Series | | 291 |

| | PAGE |
|---|---|
| 38. Reaction: Condensation of Phthalic Anhydride with a Phenol to a Phthaleïn | 294 |
| 39. Reaction: Condensation of Michler's Ketone with an Amine to a Dye of the Fuchsine Series | 301 |
| 40. Reaction: Condensation of Phthalic Anhydride with a Phenol to an Anthraquinone-Derivative | 302 |
| 41. Reaction: Alizarin from Sodium $\beta$-Anthraquinonemonosulphonate | 304 |
| 42. Reaction: Zinc Dust Distillation | 306 |

### III. PYRIDINE AND QUINOLINE SERIES

| | |
|---|---|
| 1. Reaction: Pyridine Synthesis of Hantzsch | 308 |
| 2. Reaction: Skraup's Quinoline Synthesis | 311 |

### IV. INORGANIC PART

| | |
|---|---|
| 1. Chlorine | 314 |
| 2. Hydrochloric Acid | 314 |
| 3. Hydrobromic Acid | 316 |
| 4. Hydriodic Acid | 316 |
| 5. Ammonia | 319 |
| 6. Nitrous Acid | 319 |
| 7. Phosphorus Trichloride | 319 |
| 8. Phosphorus Oxychloride | 321 |
| 9. Phosphorus Pentachloride | 321 |
| 10. Sulphurous Acid | 322 |
| 11. Sodium | 322 |
| 12. Aluminium Chloride | 323 |
| 13. Lead Peroxide | 325 |
| INDEX | 327 |
| ABBREVIATIONS | 330 |

# THE PRACTICAL METHODS OF ORGANIC CHEMISTRY

## GENERAL PART

The compounds directly obtained by means of chemical reactions are, only in rare cases, pure; they must therefore be subjected to a process of purification before they can be further utilised. For this purpose the operations most frequently employed are:

1. Crystallisation.
2. Sublimation.
3. Distillation.

## CRYSTALLISATION

**Methods of Crystallisation.** — The crude product obtained directly as the result of a reaction is, in case it is a solid, generally amorphous or not well crystallised. In order to obtain the compound in uniform, well-defined crystals, as well as to separate it from impurities like filter-fibres, inorganic substances, by-products, etc., it is dissolved, usually with the aid of heat, in a proper solvent, filtered from the impurities remaining undissolved, and allowed to cool gradually. The dissolved compound then separates out in a crystallised form, while the dissolved impurities are retained by the mother-liquor. (*Crystallisation by Cooling.*) Many compounds are so easily soluble in all solvents, even at the ordinary temperature, that they do not separate from their solutions on mere cooling. In this case, in order to obtain crystals, a portion of the solvent must be allowed to evaporate. (*Crystallisation by Evaporation.*)

**Solvents.** — As solvents for organic compounds, the following substances are principally used:

> CLASS I. Water,
> Alcohol,
> Ether,
> Ligroïn (Petroleum Ether),
> Glacial Acetic Acid,
> Benzene.

Also mixtures of these:

> CLASS II. Water + Alcohol,
> Water + Glacial Acetic Acid,
> Ether + Ligroïn,
> Benzene + Ligroïn.

Less frequently used than these are: hydrochloric acid, carbon disulphide, acetone, chloroform, ethyl acetate, methyl alcohol, amyl alcohol, toluene, xylene, solvent naphtha, etc.

But rarely used are: pyridine, naphthalene, phenol, nitrobenzene, aniline, and others.

**Choice of the Solvent.** — The choice of a suitable solvent is often of great influence upon the success of an experiment, in that a solid compound does not assume a completely characteristic appearance until it is uniformly crystallised. In order to find the most appropriate solvent, preliminary experiments are made in the following manner: successive portions, as small as possible, of the finely pulverised substance (a few milligrammes will suffice) are treated in small test-tubes of about 1 cm. diameter and 8 cm. long, with small quantities of the solvents of Class I. If solution takes place at the ordinary temperature, or on gentle heating, the solvent in question is, provisionally, left out of consideration. The remaining portions are heated to boiling, until, after the addition of more of the solvent, if necessary, solution takes place. The tubes are now cooled by contact with cold water, and an observation will show in which tube crystals have separated in the largest quantity. At times crystallisation does not occur on mere cooling; in this case the walls of the vessel are rubbed with a sharp-angled glass rod, or the solution is "seeded," *i.e.* a small crystal of the crude product is placed in the solution; by this means, crystallisation is frequently induced. If the individual solvents of Class I. are shown to be unsuitable, experiments are made with the mixtures, — Class II. Compounds which are easily soluble in alcohol or glacial acetic acid, and which consequently do not separate out on cooling, are, as a rule, difficultly soluble in water. In order to determine whether a separation of crystals will take place on cooling, the hot solutions in the pure solvents are treated with more or less water, according to the conditions. Substances easily soluble in ether, benzene, toluene, etc., often dissolve in ligroïn with difficulty. Hence mixtures of these solvents can be frequently utilised with advantage, in the manner just described. If these experiments have shown several solvents to be suitable, the portions under examination are again heated until solution takes place, and this time are allowed to cool slowly. That solvent from which the best crystals separate in the largest quantity is selected for the crystallisation of the entire quantity of the substance. If a substance is easily soluble in all solvents, recourse must be had to crystallisation by evaporation, by allowing the different solutions to stand some time in watch-glasses. That solvent from which crystals separate out first is the most suitable. Frequently a compound dissolves in a solvent only on heating and yet does not crystallise out again on cooling; compounds of this class are said to be " sluggish" (träge). In this case, the solution may be allowed to stand for some time, if necessary over night, in as cool a place as possible. If a compound is very difficultly soluble, solvents with high boiling-points are used, as toluene, xylene, nitrobenzene, aniline, phenol, and others. The crystals obtained in these preliminary experiments, especially if they are of easily soluble substances, are preserved, so that if from the main mass of the substance no crystals can be obtained, the solution may be seeded, thus inducing crystallisation. The crystallisation of substances which boil without decomposition may often be facilitated by first subjecting them to distillation.

**To dissolve the Substance.** — When water or glacial acetic acid, or a solvent which is not inflammable or not easily inflammable, is employed, the heating may be done in a beaker on a wire gauze over a free flame if the quantity is small; if large, a flask is always used. In either case care must be taken to prevent the flask from breaking, by stirring up the crystals from the bottom with a glass rod, or by frequently shaking the vessel. This precaution is especially to be observed when, on heating, the substance to be dissolved melts at the bottom of the vessel. Alcohol and benzene may also be heated in like manner directly over a moderately large flame, if the student has already had a sufficient amount of experience in laboratory work and does not use too large quantities. If the liquid becomes ignited, no attempt to extinguish the flame by blowing on it should be made, but the burner is removed and the vessel covered with a watch-glass, a glass plate, or a wet cloth. In working with large quantities of alcohol, benzene, ether, ligroïn, carbon disulphide, or other substances with low boiling-points, they are heated on a water-bath in a flask provided with a vertical glass tube (air condenser) or a reflux condenser. A substance to be crystallised from a solvent which is not miscible with water must be dried, in case it is moist, before dissolving.

An error which even advanced students too often make in crystallising substances consists in this: an excessive quantity of the solvent is poured over the substance at once. When heat is applied, it is true, solution takes place easily, but on cooling nothing crystallises out. So much of the solvent has been taken that it holds the substance in solution even at ordinary temperatures. The result is that a portion of the solvent must be evaporated or distilled off, which involves a loss of time and substance, as well as decomposition of the substance. The following rule should, therefore, always be observed: *The quantity of solvent taken at first should be insufficient to dissolve the substance completely, even on heating, then more of the solvent is gradually added, until all of the substance is just dissolved.* In this way only is it certain that on cooling an abundant crystallisation will take place. If a mixture of two solvents is used, one of which dissolves the substance easily and the other with difficulty, *e.g.* alcohol and water, the substance is first dissolved in the former with the aid of heat; the heating is continued while small amounts of the second are gradually added (if water is used it is better to add it hot) until the first turbidity appearing does not vanish on further heating. In order to remove this cloudiness, a small quantity of the first solvent is added. On the addition of the first portions of the second liquid (water or ligroïn) resinous impurities separate out at times; in this case, these are filtered off before a further addition of the solvent is made.

At times it happens that the last portions of a compound will dissolve only with difficulty. The beginner often makes the mistake here of adding more and more of the solvent to dissolve this last residue, which for the most part generally consists of difficultly soluble impurities, like inorganic salts, etc. The result of this is that on cooling nothing crystallises out. In such cases the difficultly soluble portions may be allowed to remain undissolved, and on filtering the solution are retained by the filter.

**Filtration of the Solution.** — When a substance has been dissolved, the solution must next be filtered from the insoluble impurities like by-products, filter-fibres, inorganic compounds, etc. For filtration a funnel with a very short stem is generally used, *i.e.* an ordinary funnel the stem of which has been cut off close to the conical portion (Fig. 1). The funnels used in analytical operations have the disadvantage that when a hot solution of a compound flows through the stem, it becomes cooled to such an extent that crystals frequently separate out, thus causing an obstruction of the stem. The funnel with a shortened stem or no stem is prepared with a folded filter. In case the solution contains a substance that easily crystallises out, the filter

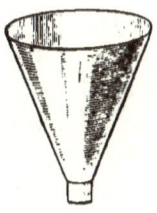

FIG. 1.

is made of rapid-filtering paper (Fig. 2). The solution to be filtered is not allowed to cool before filtering, but is poured on the filter immediately after removing it from the flame or water-bath. If inflammable solvents are used, care must be taken that the vapours are not ignited by a neighbouring flame. Under normal conditions, no crystals or only a few should separate out on the filter during filtration. If large quantities of crystals appear in a solution as soon as it is poured on the filter, it is an indication that too small an amount of the solvent has been used. In a case of this kind, the point of the filter is pierced and the crystals are washed into the unfiltered portion of the solution with a fresh quantity of the solvent; the solution is further diluted with the solvent, heated, and filtered.

FIG. 2.

Very difficultly soluble compounds crystallise during the filtration in the space between the filter and funnel, in consequence of the contact of the solution with the cold walls of the funnel.

This may be prevented when a small quantity of liquid is to be filtered, by warming the funnel previously in an air-bath, or directly over a flame. If the quantity of the liquid is large, hot-water or hot-air funnels may be used (Figs. 3 and 4), or the funnel may be surrounded by a cone of lead tubing wound around it through which steam is passed (Fig. 5). Before filtering inflammable liquids, the flame with which the hot-water or hot-air funnel has been heated is extinguished. Substances which easily crystallise out again, may also be conveniently filtered with the aid of suction and a funnel having as large a filtering surface as possible (Büchner funnel, see Fig. 33, p. 48). After filtration the solution is poured into the proper crystallisation vessel. In order to prevent the thick-walled filter-flasks from being cracked, by solvents of a high boiling-point, they are somewhat warmed before filtering by immersion in warm water.

Boiling nitrobenzene, aniline, phenol, and similar substances may be filtered in the usual way through ordinary filter-paper.

**Choice of the Crystallisation Vessel.** — The size and form of the crystallisation vessel is not without influence upon the separation of the crystals. If a compound will crystallise out on simple cooling, without the necessity of evaporating a portion of the solvent, a beaker is used for the crystallisation. The shallow dishes known as "crystallising dishes" are not recommended for this purpose, since they cannot be heated over a free flame, and further, the solution easily "creeps" over the edge, involving a loss of the substance. Moreover, the crusts collecting on the edges are very impure, since, in consequence of the complete evaporation of the solvent, they contain all the impurities which should remain dis-

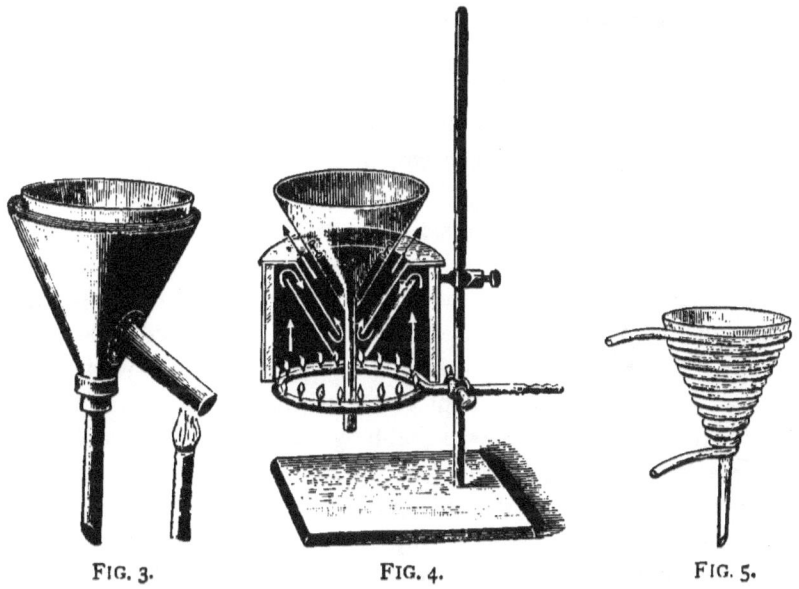

FIG. 3.      FIG. 4.      FIG. 5.

solved in the mother-liquor. The beaker is selected of such a size that the height of the solution placed in it is approximately equal to the diameter of the vessel, which is thus about one-half to two-thirds filled.

**Heating after Filtration.** — Many compounds crystallise out in the beaker during filtration. The crystals thus obtained are never well formed, in consequence of the rapid separation; therefore, after the entire solution has been filtered, it is heated again until the crystals have redissolved, and is then allowed to cool as slowly as possible without being disturbed. In order to protect the solution from dust as well as to prevent it from cooling too rapidly, the vessel is covered first with a piece of filter-paper and then with a watch-glass or glass plate. The paper is used to prevent drops of the solvent formed by the vapours condensing on the cold cover-glass from falling into the solution, by which the crystallisation would be disturbed. The paper need not be used if the vessel is covered with a watch-glass, the convex surface of which is uppermost: the condensed vapours will thus flow down the walls of the beaker.

**Crystallisation.** — In order to obtain as good crystals as possible, the solution is allowed to cool slowly without being disturbed. In exceptional cases only is it placed in cold water to hasten the separation of crystals. The vessel must not be touched until the crystallisation is ended. If a substance, on slow cooling, separates out in very coarse crystals, it is expedient, in case a sample of the substance for analysis is desired, to accelerate the crystallisation by artificial cooling, so that smaller crystals will separate out. Very coarse crystals are commonly more impure than smaller ones, in that they enclose portions of the mother-liquor. If a deposit of crystals as abundant as possible is desired, the vessel is put in a cool place — in a cellar or ice-chest if practicable. Should a compound crystallise sluggishly, the directions given on page 2, under "Choice of the Solvent," may be followed (rubbing the sides of the vessel with a glass rod; seeding the solution; allowing to stand over night). At times a compound separates out on cooling, not in crystals, but in a melted condition. This may be caused by the solution being so concentrated that crystallisation already takes place at a temperature above the fusing-point. In this case the solution is again heated until the oil which has separated out is dissolved, more of the solvent is then added, the quantity depending upon the conditions. In other cases this may be prevented by rubbing the walls of the vessel a short time with a sharp-edged glass rod, as soon as a slight turbidity shows itself, or by seeding the solution with a crystal of the same substance. This difficulty may also be avoided, in many cases, by allowing the solution to cool very slowly; *e.g.* the beaker is placed in a larger vessel filled with hot water and allowed to cool in this.

At times the separation of crystals takes place suddenly, within a few seconds, throughout the entire solution. Since the crystals thus obtained are generally not well formed, the liquid, after some of the crystals have been removed, is heated until solution again takes place. After it has partially cooled, those crystals which were taken out are now added to it, by which a gradual crystallisation is caused.

**Separation of Crystals from the Mother-Liquor.** — When crystals have been deposited, they are then to be separated from the liquid (mother-liquor). This is always done with the aid of suction, and never by merely pouring off the liquid. The filter to be used is previously moistened with the same substance which was employed as the solvent. Crusts, formed on the sides and edges of the vessel by the complete evaporation of the solvent, are not filtered with the crystals; they are removed with a spatula before the filtering, and are worked up with the mother-liquor. In order to remove the last traces of the mother-liquor adhering to the crystals, they are washed several times with fresh portions of the solvent; obviously, if the substance is easily soluble, too large quantities of the solvent must not be used. If a solvent that will not evaporate easily in the air or on the water-bath has been used, $e.g.$ glacial acetic acid, toluene, nitrobenzene, etc., it must be removed from the crystals by a more volatile substance, like alcohol or ether. This is done by first washing with a fresh quantity of the solvent, then with a mixture of the solvent and a small quantity of the more volatile liquid, the proportion of the latter in the washing mixture being gradually increased, until finally the volatile substance is used alone. Glacial acetic acid may, in this way, be displaced by water.

**Drying of Crystals.** — When crystals have been freed from the mother-liquor they must be dried. This may be effected (1) at the ordinary temperature by the gradual evaporation of the solvent in the air, and (2) at higher temperatures by heating on a water-bath or in an air-bath. In the first case the crystals are spread out in a thin layer upon several thicknesses of filter-paper and covered with a watch-glass, funnel, beaker, or similar vessel. In order that the vapours of the solvent may escape, the covering must be so placed that the air is not shut off completely from the crystals; this is conveniently done by supporting it on several corks. Crystals may also be dried in a desiccator which is partially exhausted, if necessary. In drying substances at higher temperatures the crystal form may be lost by the fusion of the substance or by the separation of the water of crystallisation. Since many substances will liquefy far below their melting-point if they contain even small quantities of the solvent, a preliminary experiment with a small portion is always made when the drying is to be effected at higher temperatures. Compounds, not easily soluble in ether, which crystallise from a solvent miscible with ether, can be very quickly dried by being washed several times with it. After a short exposure to the air they are dry.

**Treatment of the Mother-Liquor.** — The mother-liquor filtered off from crystals still contains more or less of the substance, in proportion to its solubility at the ordinary temperature; in many cases it is advantageous to extract the last portions remaining in solution. A "second crystallisation" is obtained by distilling or evaporating off a portion of the solvent. The mother-liquor may also be diluted with a second liquid, in which the dissolved substance is difficultly soluble; *e.g.* a solution in alcohol or glacial acetic acid may be diluted with water, or a solution in ether or benzene with ligroïn.

**Crystallisation by Evaporation.** — If a compound is so easily soluble in all solvents that it will only crystallise out on partial evaporation, then, in order to get good crystals, a solution, not too dilute, is made, by the aid of heat if necessary, and filtered from the impurities remaining undissolved. In this case, as a crystallisation vessel, one of the various forms of shallow dishes — the so-called crystallising dishes — is used, in which the solution is allowed partially to evaporate. In order to protect the vessel from dust, it is covered with a funnel or watch-glass, in the manner indicated under "Drying of Crystals." In crystallising by this method, it sometimes happens that the solution, owing to capillary action, will "creep" over the edge of the dish. To avoid loss of the substance from this source, the dish is placed on a watch-glass or glass plate. Under these conditions, the vessel is never covered with filter-paper, since, after standing some time, it may absorb the entire quantity of the substance. If, in order to obtain well-formed crystals, the solvent is to be evaporated as slowly as possible, the solution is placed in a beaker or test-tube, which is then covered with filter-paper. Evaporation may be hastened by placing the crystallisation vessel in a desiccator, charged, according to the nature of the solvent, with different substances; for the absorption of water or alcohol, calcium chloride or sulphuric acid is used; glacial acetic acid is absorbed by soda-lime, solid potassium hydroxide, or sodium hydroxide. The evaporation of all solvents may be hastened by exhausting the desiccator.

Since the purifying effect of crystallisation depends upon the fact that the impurities remain dissolved in the mother-liquor, and with this are filtered off, in no case must the solvent be allowed to evaporate completely, but the crystals must be filtered off while still covered with the mother-liquor. Before filtering, crusts deposited, generally on the edges of the vessel, are removed with the aid of a small piece of filter-paper or a spatula. Even though the substance is very soluble, the mother-liquor adhering to the crystals is washed away with small quantities of the solvent. If the quantity of crystals is very small, the adhering mother-liquor may be separated, in cases of necessity, by placing them on porous plates (biscuit or gypsum) and moistening with a spray of the solvent.

**Fractional Crystallisation.** — Up to this point, it has been assumed that the substance to be crystallised possessed an essentially homogeneous nature, and the object of crystallisation was only to change it to a crystallised form. Crystallisation is often employed for another purpose — that of separating a mixture of different substances into its individual constituents, — a task that is generally far more difficult than the crystallisation of an individual substance. The simplest case is one in which two substances are to be separated. If the solubilities of the two substances are very different, as is generally the case when a mixture of two different highly substituted compounds is under examination, it is frequently not difficult to find a solvent which will dissolve a considerable portion of the more easily soluble substance, and but a small portion of the less soluble. If, now, the mixture be treated with such a solvent, in not too large quantities, a solution will be obtained containing all of the easily soluble substance and a small portion of the difficultly soluble substance. This is filtered from the residue remaining undissolved. The mixture has thus been divided into two fractions. By evaporating the solution to a certain point, the more insoluble compound will crystallise out, unaccompanied by any of the other compound; the crystals are filtered off, and the solution further evaporated. If the crystallisation of the two fractions be repeated a second time, a complete separation will be effected. For separating a mixture of this kind, specially constructed apparatus — the so-called extraction apparatus — may be employed, the use of which possesses the advantage over the method of simple heating, in that much smaller quantities of the solvent are required. An apparatus of this kind is represented in Figs. 6 and 7. To a wide glass tube $d$ is fused a narrow tube which acts as a siphon, bent as in Fig. 7. This portion of the apparatus is surrounded by a glass jacket $b$, narrowed at its lower end. This is connected with the flask that is to contain the solvent. A cork bearing a reflux condenser — a ball condenser is convenient — is fitted in the opening at the upper end of the jacket. A shell of filter-paper is next prepared

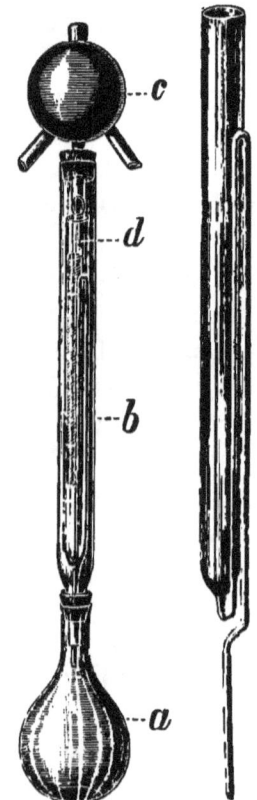

FIG. 6.   FIG. 7.

in the following manner: Three layers of filter-paper are rolled around a glass tube with half the diameter of the inner tube $d$. One end of the roll must extend somewhat beyond the edge of the glass tube; this is turned over and securely fastened with thread. To preserve the form of the roll, thread is loosely wound around its middle and upper portion. The length of the roll is such that it extends 1 cm. above the highest point of the narrow siphon-tube. In the shell is placed the mixture of the easily soluble and difficultly soluble substance, to be extracted; the upper end is closed by a loose plug of absorbent cotton. The flask $a$, containing the solvent, is now heated on a water-bath or over a free flame, according to the nature of the solvent. The condensed vapours drop from the condenser into the shell, dissolve the substance, filter through the paper, and fill the space between shell and inner glass tube. As soon as the liquid has reached the highest point of the siphon-tube, the solution siphons off and flows back into the flask $a$. This operation may be continued as long as necessary. The amount of solvent used should be one and a half or two times the volume of the inner tube up to the highest point of the siphon.

The construction of a ball condenser is represented in Fig. 8. In order to distinguish the tube by which the water enters from the outlet-tube, the former is marked with an arrow. Comparatively easy also is the separation of two substances about equally soluble, if the one is present in larger quantity than the other. If a mixture of this kind is dissolved, then, on cooling, the substance which was present in larger quantity generally crystallises out. Occasionally, after standing some time, crystals of the second substance will appear; under these conditions the crystallisation must be carefully watched, and as soon as crystals differing from those first appearing are observed, the solution is filtered with suction at once, even though it is still warm.

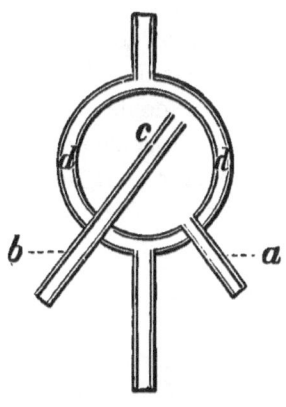

FIG. 8.

If two compounds crystallise simultaneously at the outset, as is the case when they possess approximately the same solubility and are present in almost equal quantities, they can be separated mechanically. If, *e.g.*, one of the compounds crystallises in coarse crystals, and the other in small ones, they may be separated by sifting through a suitable sieve or wire-gauze. A compound crystallising in leaflets can frequently be separated from one crystallising in needles by a sieve. If these methods fail, the separation may be effected by picking out the crystals with small pincers or a quill. In all these mechanical operations, the crystals must be as dry as possible.

In many cases, when one of the compounds is heavier than the other, it is possible to separate them by causing the lighter crystals to rise to the top of the liquid, by imparting to it a rotatory motion by rapid stirring with a glass rod. The heavier compound collects at the bottom of the vessel, and the liquid with the lighter compound floating in it can be poured off.

**Double Compounds with the Solvent.** — Many substances crystallise from certain solvents in the form of double compounds, composed of the substance and the solvent. It is well-known that many substances, in crystallising from water, combine with a certain portion of water. Alcohol, acetone, chloroform, benzene, and others also have the power of uniting with other substances to form double compounds. As a familiar example, the combination of triphenylmethane with benzene may be mentioned in this connection. If double compounds of this kind are heated, the combined solvent is generally vaporised.

## SUBLIMATION

Much less frequently than crystallisation, sublimation is used to purify a solid compound. The principle involved is this: A substance is converted by heat into the gaseous condition, and the vapours are caused to condense again on a cold surface. Under these conditions the substance frequently condenses in crystals.

The sublimation of a small quantity of a substance can be conveniently effected between two watch-glasses of the same size. The substance to be sublimed is placed on the lower one, which is then covered with a round filter perforated several times in its centre and projecting over the edges; the second watch-glass with its convex side uppermost is placed on it, and the two are held together by a watch-glass clamp. If the lower glass is now heated very slowly on a sand-bath with a free flame, the vaporised substance condenses on the cold surface of the upper watch-glass in crystals; the filter-paper prevents the very small, light crystals from falling back on the hot surface of the lower glass. To keep the upper glass cool, it is covered with several layers of wet filter-paper or with a small piece of wet cloth. If large quantities of a substance are to be sublimed, the upper watch-glass in the apparatus just described is replaced by a funnel somewhat smaller than the lower glass (Fig. 9). To prevent the escape of vapours, the stem of the funnel is closed by a plug of cotton or is covered with a small cap of filter-paper. The apparatus for sublimation designed by Brühl is admirably adapted to the purpose for which it is intended (Fig. 10). It consists of a hollow metal plate through which water flows. In

FIG. 9.

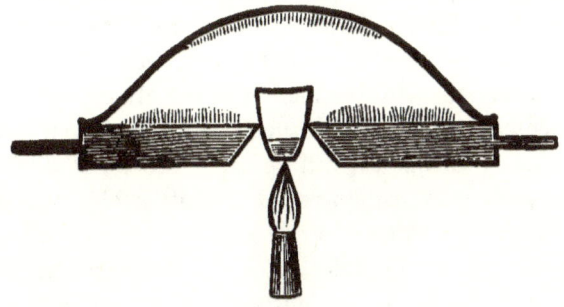

FIG. 10.

the conical opening is placed a crucible containing the substance to be sublimed. The plate is covered with a concave glass dish, the ground edges of which fit the plate tightly. The crucible is heated directly with a small flame, while cold water flows through the plate. The vapours condense in part on the glass cover, but more abundantly on the upper cold surface of the plate in crystals. The glass cover is not removed until the apparatus is completely cold.

Sublimations can also be conducted in crucibles, flasks, beakers, retorts, tubes, etc. The heating may be done in an air- or oil-bath. In order to lead off the vapours rapidly, a current of an indifferent gas is sent through the apparatus.

## DISTILLATION

**Kinds and Objects of Distillation.** — By distillation is meant the conversion by heat of a solid or liquid substance into a vapour and the subsequent condensation of this. When distillation is conducted at the atmospheric pressure, it is called *ordinary distillation;* if in a partial vacuum, *vacuum distillation.* The object of distillation is either to test the purity of an individual substance by the *determination* of its *boiling-point,* or to separate a mixture of substances boiling at different temperatures into its constituents. (*Fractional Distillation.*)

**Distillation Vessels.** — The heating of the substance to be distilled is generally effected in a fractionating flask (Figs. 11, 12, 13). These flasks differ, not only in size, but in the diameter of the condensation-tube (side-tube), as well as in the distance of the latter from the bulb. In selecting a fractionating flask the following points are to be observed. For distillation at the atmospheric pressure a flask is selected having a bulb of such a size that when it contains the substance to be distilled it will be about two-thirds filled. There are two objections to distilling small quantities of a substance from a large flask: the vapours are easily overheated, thus giving a boiling-point that is too high; a loss of the substance follows, in that, after the distillation is finished, a larger volume of vapours which condense on cooling, remains behind in the bulb, than if a smaller flask had been used. In the distillation of low boiling compounds, a flask is selected which has its condensation-tube as high as possible above the bulb, so that the entire thread of mercury of the thermometer employed is heated by the vapour of the liquid. By using a flask of this kind it is not necessary to correct the observed boiling-point, as is the case when the mercury column is not entirely surrounded by the vapour. The higher a substance boils, the nearer must the side-tube be to the bulb, in

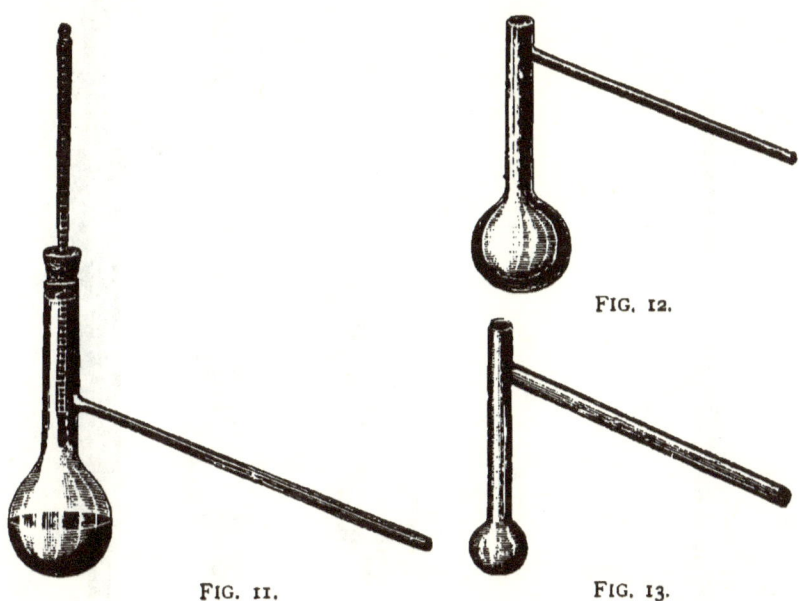

FIG. 11.   FIG. 12.   FIG. 13.

order that the vapours shall have as little opportunity as possible of condensing below the tube and flowing back into the bulb.

If large quantities of a substance are to be distilled, an ordinary flask is used. This can be converted into a fractionating flask with the aid of a cork bearing a T-tube, as illustrated in Fig. 14.

For the distillation of solid substances which solidify in the condensation-tube, a fractionating flask with a side-tube as wide as possible is used.

A fractional distillation can also be conducted in the fractionating flasks just described; but the operation can be carried out more

c

rapidly and more completely by the use of apparatus especially adapted to fractionating (Fig. 15). These can be fused directly on the bulb or they can be attached to an ordinary flask by means of a cork (Fig. 14) ; the round, short-necked flasks such as represented in Fig. 16, are well adapted to this purpose. Flasks of

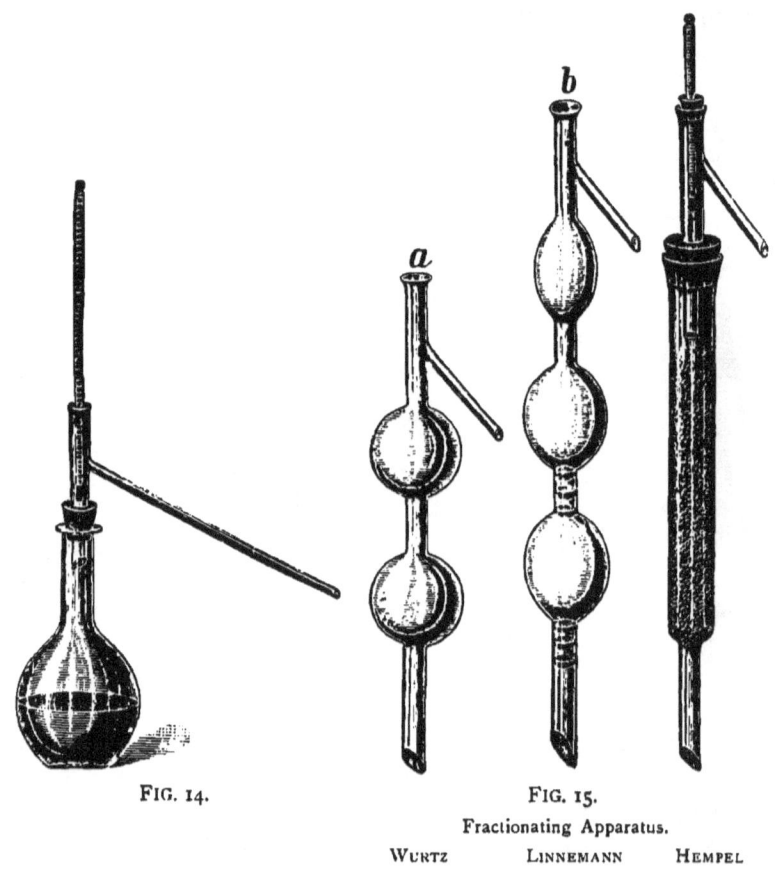

FIG. 14.      FIG. 15.
Fractionating Apparatus.
WURTZ     LINNEMANN     HEMPEL

this description can be obtained in different sizes but still possessing the same width of neck ; this enables one to use the same cork with any flask. The value of these different forms of fractionating apparatus depends upon the fact that the higher boiling portions carried along with the vapours do not pass immediately to the outlet tube, but before entering this they have an opportunity of condensing and flowing back into the flask. In the apparatus of Wurtz (*a*) the condensation takes place on the large upper surfaces of the bulbs. More complete condensation is obtained in Linnemann's apparatus (*b*), which differs from that of Wurtz in that the narrow tubes between the bulbs contain small platinum-wire sieves. Since the lower boiling portions condense to a liquid and collect in these, the ascending vapours are so far cooled by the passage through them that the accompanying portions of the higher boiling substances are likewise condensed. The apparatus of Hempel is filled with glass heads which act like the sieves in the Linnemann apparatus. For the distillation of large quantities of a liquid the Hempel apparatus is particularly well adapted; in working with it as well as the Linnemann form, the heating must be interrupted from time to time, in order that the liquid collecting in the beads or sieves may have an opportunity to flow back to the distillation flask. If the Le Bel-Henninger form is used, this precaution is unnecessary, since in this apparatus special tubes for conducting off the condensed liquid are joined to the sides of the bulb somewhat above the sieves.

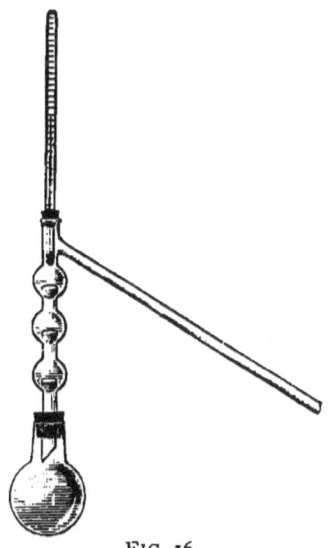

FIG. 16.

Experiments have shown that a single distillation with one of the forms of apparatus just described, effects a more complete separation than repeated fractionations in an ordinary fractionating flask.

**Supporting the Fractionating Flask.**—If it is necessary to support the fractionating flask with a clamp, it is placed as far above the outlet tube as possible, never below it; the glass expands by contact with the hot vapours, and since the expansion is impeded by the clamp, particularly if it is firmly attached, the flask frequently breaks.

**Supporting the Thermometer.** — The thermometer is passed through a cork (no rubber) which fits the neck of the flask. The most exact determinations of the boiling-point are obtained if the entire thread of mercury is surrounded by the vapour of the substance. With low boiling compounds this condition is easily obtained by the use of a fractionating flask having its outlet tube at a sufficient distance above the bulb. In this case the thermometer is so placed that the degree corresponding to the boiling-point of the liquid is opposite the outlet tube, but the bulb of the thermometer must not extend into the bulb of the flask and never into the liquid; if it does, another flask must be used, the outlet tube of which is still higher above the bulb. If in dealing with high boiling compounds such an arrangement is not possible, the thermometer is thrust so far into the neck of the flask that the thermometer-bulb is somewhat below the outlet tube. In this case, if an exact determination of the boiling-point is desired, the observed reading is corrected in the manner described below. In order to avoid making a correction a special form of thermometer is used, the graduation of the scale beginning at $100°$, $200°$, or at other convenient points. By employing an instrument of this kind the mercury column may be kept in the vapours at any temperature.

In making distillations, it occasionally happens that the mercury column ascends to that point in the scale which is hidden by the cork supporting the thermometer, thus preventing the temperature from being read. In a case of this kind the thermometer is either raised or lowered, so that the top of the mercury is visible, or if this is not possible, from that portion of the cork which projects above the flask, a section is cut which will enable the scale to be seen.

**Condensation of Vapours.** — The condensation of vapours is effected in various ways, depending upon the height of the boiling-point. If a compound boils at a relatively low temperature (up to $100°$), the outlet tube of the fractionating flask is connected with a Liebig condenser by a cork (not a rubber stopper). For very low boiling compounds a long condenser is used, and for those of high boiling-points a short one. If the boiling-point of a compound is very low, the flask in which the condensed liquid collects (the receiver) is connected with the condenser by means of a cork and a bent adapter (Fig. 54), and the receiver is cooled by ice or a freezing mixture. If the boiling-point is moderately high, between 100° and 200°, a wide glass tube 50 cm. long (air condenser) is used instead of the ordinary condenser. It is connected with the fractionating flask by a cork (Fig. 17). With still higher boiling substances even this is superfluous, since the

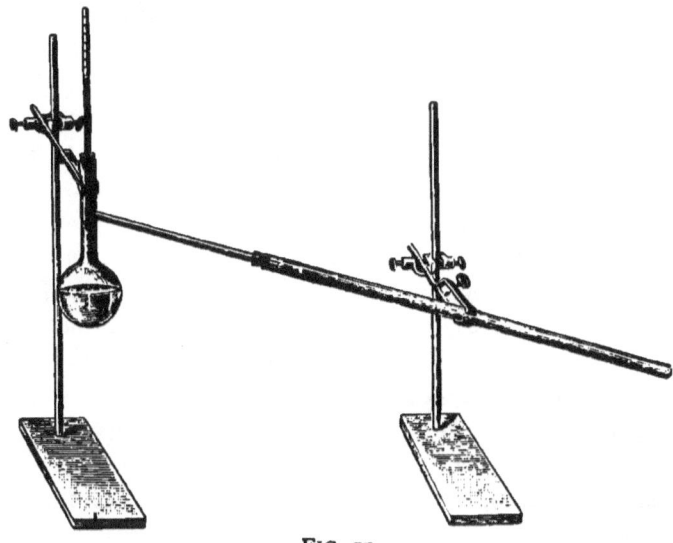

FIG. 17.

condensation tube of the fractionating flask, provided it is not too short, will suffice for the condensation.

If a small quantity of a substance is to be distilled, and it is desired to avoid the loss of substance necessarily incident to the use of a condenser, the distillation even of low boiling compounds is conducted in a small distillation flask as slowly and carefully as possible, the source of heat being a minute flame (the so-called Microburner).

If large quantities are to be distilled, a condenser is always used, since when other condensation apparatus is employed, the tube finally becomes so hot that the vapours are not completely condensed. If the vapours of a substance attack corks, the outlet-tube is inserted far enough into the condenser or extension tube so that the vapours do not come in contact with the cork. But generally a cork is not used; the outlet-tube being inserted sufficiently far into the condenser.

**Heating.** — Low boiling substances (those boiling up to about 80°) are not generally heated over the free flame, but on the water-bath heated gently or to full boiling. Frequently it is more convenient to immerse the bulb of the fractionating flask as far as the level of the liquid which it contains in a dish or beaker filled with water, which is heated gently or strongly as the case requires. Low boiling substances may also be heated by immersing the bulb of the flask from time to time in a vessel filled with warm water. If a substance is not distilled over a free flame, in order to prevent "bumping" a few pieces of platinum wire or foil, or bits of glass, are thrown into the liquid (see below). High boiling substances are heated over the free flame. In this case the flask may be protected by heating it on a wire gauze; still by working carefully the gauze need not be used. In heating, the flame is not placed under the flask at once, since the latter is likely to break easily on sudden heating; it is better to pass the flame back and forth slowly and uniformly over the bottom of the flask until the liquid is brought to incipient ebullition. Substances which have been previously dissolved, after the evaporation of the solvent in the water-bath, often stubbornly refuse to give up the last portions of the solvent, particularly when ether has been used. If now a free flame be applied, it frequently happens that in consequence of a retarded boiling during which the solution becomes overheated, a sudden active ebullition and foaming will take place. In order to prevent this the flask is shaken repeatedly during the heating, since if the liquid is kept in motion, overheating cannot easily take place. It may also be prevented frequently by heating the flask on the side. During the actual distillation the heating may be continued by slowly passing the flame over the bottom of the flask as in the preliminary heating, but in this case care must be taken not to apply the flame to the flask at any point above the liquid inside, since an overheating of the vapours would result. In order to protect the hand in case the flask should break, the burner is held obliquely and not directly under the flask; or during the distillation the burner may be placed under the flask and allowed to remain stationary. The size of the flame is so regulated that the condensed distillate flows into the receiver regularly in drops. If vapours escape from the receiver, it is an indication that the heating has been too strong. Toward the end of the distillation the burner is turned down somewhat.

**To collect the Fractions.** — If a substance which is not quite pure is being treated, and it is desired to test the purity by a determination of its boiling-point, then on distillation a small portion will generally pass over below the true boiling-point ("first runnings"); this is collected separately in a small receiver. Then follows the principal fraction, passing over at the true boiling-point, the temperature remaining constant. If there is only a small quantity of the liquid in the bulb of the flask, it is difficult, in spite of using a small flame, to prevent the vapours from being somewhat overheated; this will cause a rise of the mercury. The portion passing over a few degrees above the true boiling-point can, in preparation work, be collected with that portion which boils at correct temperature, without evil results. High boiling portions collected separately are designated as "last runnings." The operation of practical distillation is conducted in a wholly different manner. The preparation of benzoyl chloride (see page 248) will furnish a practical example of the method of procedure. This compound is obtained by treating benzoic acid with phosphorus pentachloride. The product of the reaction is a mixture of phosphorus oxychloride (b. p. $110°$) and benzoyl chloride (b. p. $200°$). If this mixture is subjected to distillation, the entire quantity of phosphorus oxychloride does not pass over at about $110°$, and afterwards the benzoyl chloride at $200°$; but the distillation will begin below $110°$, and a mixture consisting of a large quantity of the lower boiling substance and a small quantity of the higher boiling substance will pass over; the temperature then rises gradually; while the quantity of the former steadily decreases, that of the latter increases, until finally, at 200°, a mixture consisting essentially of the higher boiling substance passes over. A quantitative separation of the constituents of a mixture cannot be effected by the method of fractional distillation. However, in most cases, it is possible to obtain fractions which contain the largest part of the individual constituents, particularly when, as in the example selected, the boiling-points of the constituents lie far apart, by collecting the different fractions and repeating the distillation a number of times. It is almost impossible to give definite rules of general application for fractional distillation; the number of fractions to be collected depends upon the difference of the boiling-points, upon the number of compounds to be separated, upon the relative proportion of the compounds present, and upon other factors. If but two substances are to be separated as is generally the case in preparation work, the procedure is, very commonly, as follows: as a basis for the fractions to be collected, the interval between the boiling-points is divided into three equal parts; in the case of the example selected the temperatures would be 100°, 140°, 170°, 200°. The fraction passing over between the temperature at which the distillation first begins, up to 140°, is collected (fraction I.), then in another vessel the fraction passing over between 140°–170° (fraction II.), and finally in another receiver that passing over between 170° and 200° (fraction III.). The quantities of the three fractions thus obtained are about equal. Fraction I. is now redistilled from a smaller flask, and the portion passing over up to 140° is collected as in the first distillation in the empty receiver I., which in the meantime has been washed and dried. When the temperature reaches 140°, the distillation is stopped, and to the residue remaining in the flask is added fraction II., and the distillation continued. The portion passing over up to 140° is collected in receiver I., that from 140°–170° in the empty receiver II. When the temperature reaches 170°, the distillation is again interrupted, and to the residue in the flask is added fraction III., and the distillation is again continued: in this way the three fractions are collected. These are again distilled as in the first distillation, but now the lower and higher boiling fractions are much larger than the intermediate one; further, a larger portion of these end fractions boil nearer the true boiling-points than in the first distillation. If it is now desired to obtain the two substances in question in a still purer condition, the two end fractions are once more distilled separately, and the portion passing over a few degrees above and below the true boiling-point, for phosphorus oxychloride about 105°–115°, for benzoyl chloride, 190–205° are collected.

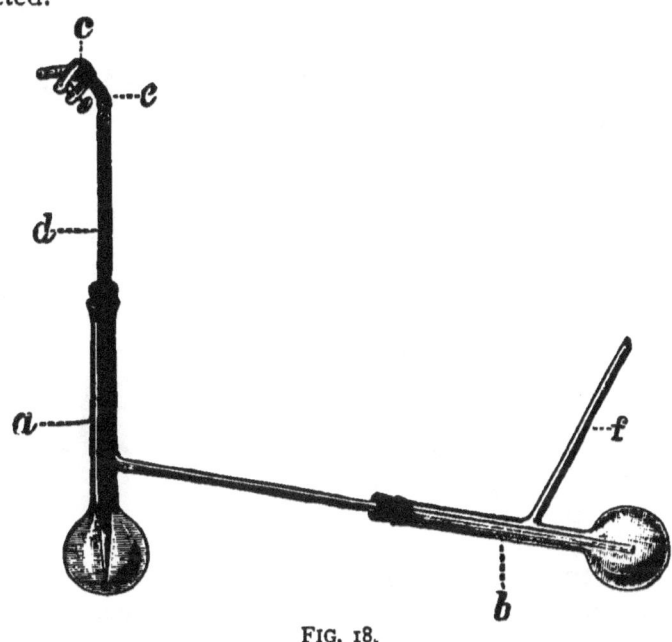

FIG. 18.

**Vacuum Distillation.** — Many compounds, not volatile at the atmospheric pressure without decomposition, may be distilled undecomposed in a partial vacuum. The vacuum distillation is used advantageously for the fractionation of small quantities of a substance, since the separation of the individual constituents can be effected more rapidly and more completely than at the atmospheric pressure.

**Vacuum Apparatus.** — The simplest form of a vacuum apparatus is represented in Fig. 18. Two fractionating flasks $a$ and $b$ are connected by a cork. The neck of *a* is closed by a tightly fitting cork bearing the glass tube *d*, reaching to the bottom of the flask, its lower end being drawn out to a fine point, the object of which will be explained below. A thermometer is placed in the tube. In place of the flask *b*, a suction-flask such as finds application in filtering under pressure, may be used (Fig. 19). But this kind of flask is used only in case low boiling substances are to be distilled, since the contact of too hot liquids with the thick walls

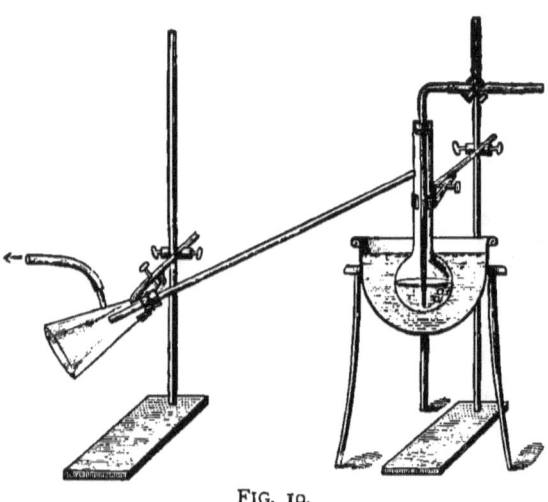

FIG. 19.

causes them to crack easily : this is likely to prove very destructive in vacuum distillation. With low boiling substances, in order to get complete condensation of the vapours, the jacket of a Liebig condenser through which water is allowed to flow is fitted over the outlet-tube of the fractionating flask. These simple forms of apparatus are used only when it is desired to collect a few fractions, since it is troublesome to be obliged to change the receiver, and thus destroy the vacuum for each new fraction.

If it is desired to collect a larger number of fractions, an apparatus is employed by means of which the receiver can be changed without destroying the vacuum.

DISTILLATION 27

Brühl's apparatus is very well adapted to this purpose (Figs. 20 and 21). By turning the axis *b*, so arranged that it supports the receivers firmly, each receiver may in turn be brought under the end of the condenser tube *c*.

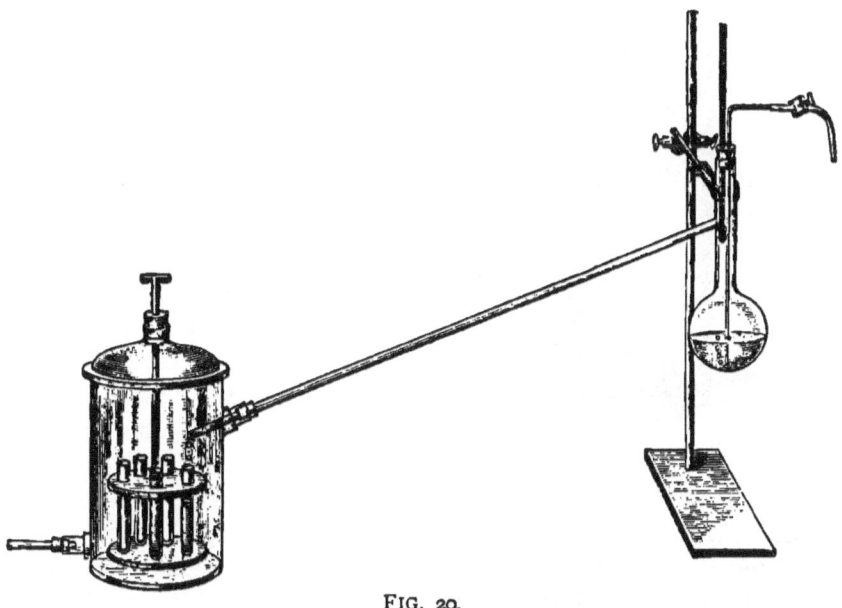

FIG. 20.

**Construction of a Vacuum Apparatus.** — In vacuum distillations the evolution of bubbles of vapour occurs to a much greater extent than under ordinary conditions. In order to prevent the liquid from foaming up and passing over, a flask of such a size is selected, that when it contains the liquid it must in no case be more than half-full; it is better to have it but one-third full. The individual parts of the apparatus are connected by rubber stoppers. Ordinary corks may also be used with almost equally good results, but only those are selected which are as free as possible from pores; they are pressed in a cork-press, and then very carefully bored. If, after the apparatus is put together, the corks are coated with a thin layer of collodion, there is no difficulty in obtaining a vacuum. The thermometer and capillary tube may be arranged as shown in Fig. 18. It is also a very excellent arrangement to use a two-hole cork, the thermometer passing through one, and the capillary tube through the other, as in Fig. 20. The capillary tube is made by drawing out a glass tube of ½ mm. diameter; the narrow

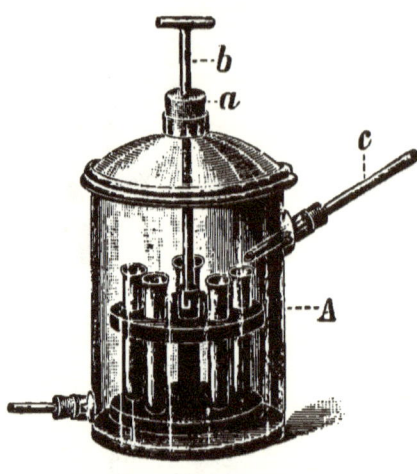

FIG. 21.

hole in the cork through which this passes is made conveniently by a hot knitting-needle. Instead of using a capillary tube to prevent "bumping," other means may be employed (see below), in which case the thermometer is supported in the fractionating flask as in ordinary distillations. When a tube drawn out to a capillary point is used, a short piece of thick-walled rubber tubing, which can be closed by a screw pinch-cock (Fig. 18, *e* and *c*), is attached to the upper end. In order to determine the efficiency of the vacuum, the lower tube of the Brühl apparatus is connected with a manometer (Fig. 22), by means of a thick-walled rubber tubing which will not collapse

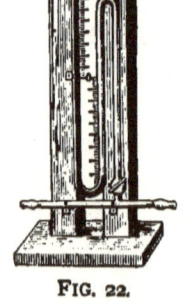

FIG. 22.

upon exhausting the apparatus. The other end of the manometer is connected with suction, by the same kind of rubber tubing.

In order that the apparatus may be perfectly tight, the corks, ends of the rubber tubing, as well as the ground surfaces of the Brühl receiver, are covered with a thin layer of grease or vaseline. If ordinary corks are used, these, as well as the ends of the tubing, are covered with collodion after the apparatus is set up. Before the distillation, the apparatus is tested to determine whether it will give the desired vacuum. For this purpose, the pinch-cock on the capillary tube is closed, the suction attached, and after some time the manometer is read; this will indicate whether the desired vacuum has been obtained. In case it is not, the corks are pressed more firmly into the tubes, greased again or covered with more collodion, and the rubber tubing is pushed farther over the ends of the glass. Frequently the suction pump will not work satisfactorily; it is then examined to see if it is stopped up, or a better pump is used. When the apparatus has been exhausted, the air must not be admitted suddenly, by removing a rubber joint, for the sudden rushing in of the air may easily destroy the apparatus. The rubber tube attached to the suction is closed by a screw pinch-cock which has been placed on it beforehand, and in case a capillary tube has been used, the pinch-cock on this is gradually opened and the air allowed to enter through it, or after disconnecting the rubber tubing from the suction, the pinch-cock which has just been closed may be opened. The same object may be accomplished most rapidly by closing the tubing leading to the suction with the fingers, detaching it and opening the tube repeatedly for an instant at a time, until the rushing sound made by the inflowing air ceases. After a test has shown that the apparatus does not leak, the liquid to be distilled is poured in the flask and the distillation begun.

**Heating.** — In vacuum distillation the flask can be heated directly with a free flame, but the flame must be applied to the side, and not to the bottom of it, as in the ordinary way. Care must be taken to keep the flame constantly moving. It is much more satisfactory and safer to use an oil- or paraffin-bath, or better a metallic air-bath (iron crucible). The latter is covered with a thick asbestos plate containing a round opening in the centre, through which the neck of the fractionating flask may pass : from the opening to the edge of the plate there is a straight narrow slit. The air-bath must not be too large ; the bottom is covered with a thin layer of asbestos, which will prevent the flask from coming in contact with the metal. The temperature of the oil- or air-bath should, in exact experiments, not be more than $20°$–$30°$ higher than the boiling-point indicated by the thermometer. A thermometer is immersed in the bath and the flame so regulated that the difference between the two thermometers is not greater than that mentioned. The heating is not begun until the apparatus is exhausted.

**To prevent Bumping.** — In vacuum distillations a troublesome bumping (a sudden, violent ebullition) frequently occurs. To prevent this a slow, continuous current of air is drawn through the liquid, thus keeping it in constant motion. The air current, controlled by a pinch-cock, must not be allowed to enter too rapidly, otherwise it will be difficult to maintain a high vacuum. The same effect may be obtained by placing certain substances in the liquid — splinters of wood the size of a match, capillary tubes, bits of glass, pieces of porcelain, powdered talc, scraps of platinum wire or foil. Small pieces of pumice-stone bound with platinum wire also act satisfactorily. For further details concerning vacuum distillation consult " Distillation under Diminished Pressure in Laboratories," R. Anschütz.

**Lowering of the Boiling-Point.** — In order that some idea may be obtained as to the approximate lowering of the boiling-point, by diminishing the pressure, the following table is given:

| Substance. | Boiling-point at 12 mm. | Boiling-point at Ordinary Pressure. | Difference. |
| --- | --- | --- | --- |
| Acetic acid . . . . . | 19° | 118° | 99° |
| Monochloracetic acid . . | 84° | 186° | 102° |
| Chlorbenzene . . . . . | 27° | 132° | 105° |
| p-Nitrotoluene . . . . | 108° | 236° | 128° |
| Acetanilide . . . . . | 167° | 295° | 128° |

**Corrections of the Boiling-Point.** — If it is not possible in making an exact determination of the boiling-point to have the mercurial column entirely surrounded by the vapour of the liquid, — a condition usually obtained by employing a flask, the side-tube of which is at a sufficient distance from the bulb, or a sectional thermometer, or both, — then a correction may be applied to the observed boiling-point in one of two ways. The portion of the mercurial column not heated by the vapours — that portion above the side-tube — is read in degrees ($L$). Another thermometer is brought as near as possible to the middle point of this column, the temperature of which is also read ($t$). If $T$ is the observed boiling temperature, then the following correction is added: $L(T-t) \cdot 0.000154$. The so-called " corrected " boiling-point may also be obtained as follows: The boiling-point is determined in the usual way; after the distillation, another substance, the corrected boiling-point of which is known, and which lies near the one in question, is placed in the same flask and distilled under the same conditions. The difference between the corrected and observed boiling-points is applied to the boiling-point of the first substance.

**Distilling off a Solvent.** — An operation frequently employed in organic work is distilling off a solvent from the substance dissolved in it. When the boiling-point of the solvent is sufficiently far away from that of the dissolved substance, a complete separation can be effected by a single distillation. The methods used depend upon the quantity of the solution, that of the dissolved substance and the boiling-point of the solvent. The methods which can be used for distilling off low boiling solvents, like ether, ligroïn, carbon disulphide, alcohol, and others, will be described first. If a small quantity of a solvent is to be evaporated, and it is not worth the trouble to recover it by condensation, then, in case the solvent is ether, ligroïn, or carbon disulphide, the solution is poured into a small flask, and this is immersed in a larger vessel filled with warm water. The vaporisation is considerably accelerated by shaking the flask. The operation is more rapidly performed by heating the flask on a water-bath. To prevent a sudden foaming, due to retarded ebullition, some small pieces of platinum wire or capillary tubes are placed in the liquid; the evaporation is also facilitated by frequent shaking. Should the vapours become ignited from the flame of the water-bath, no attempt to blow out the burning vapours should be made; but the burner is extinguished, the flask removed from the bath with a cloth, and the mouth covered with a watch-glass. Carbon disulphide, on account of its great inflammability, is never vapourised in this way, but always without a flame.

Large quantities of solvents may also be evaporated by these two methods, but the entire quantity is not treated at once. A portion is placed in a small flask, and when this has been evaporated, a second portion is added, and so on. The danger of ignition of the solvent may be avoided by inserting in the flask a glass tube, extending to within a few centimetres of the level of the liquid, supported firmly by a clamp, and attached by rubber tubing to the suction. The tube must at no time touch the liquid.

For rapid evaporation of small quantities of ether, the following method of procedure is recommended: A few cubic centimetres of the solution are placed in a sufficiently wide test-tube; this is warmed, with continuous shaking, over a small, luminous flame. After the first portion is evaporated, the second is added, and so on. Since the vapours of the ether almost regularly become ignited, this event should always be expected, and should occasion no alarm. When it happens, the heating is interrupted for a moment, and the flame is easily extinguished by blowing on it or covering the mouth of the test-tube. If the tube is held as nearly horizontal as possible during the heating, the danger of ignition is lessened.

If it is desired to distil off a larger quantity of ether, ligroïn, or carbon disulphide, and to recover it by condensation, the receiver is attached to the condenser tube by a cork, and the flask is heated by immersing it in a water-bath containing hot water. To prevent the liquid from being superheated, a silk thread as frayed as possible at the end reaching to the bottom of the flask is suspended from the neck. The entire quantity of the liquid is not placed in the flask at once, but only a portion: after the solvent has been distilled off from this, another portion is added, and so on.

By the use of the so-called safety water-bath, *i.e.* one in which the flame is surrounded by a wire gauze as in Davy's Safety Lamp, ether and ligroïn can be distilled by continuous heating with a flame. It is not safe to distil off carbon disulphide even from this apparatus, since, when it becomes sufficiently hot, it will ignite spontaneously without the intervention of a flame.

The apparatus best adapted to distilling off any desired quantity of ether is represented in Fig. 24. A fractionating flask, into the neck of which a dropping-funnel is inserted, is connected with a condenser. During the heating by means of hot water, or in special cases, the water-bath may be heated with a flame, the ethereal solution is allowed to flow gradually from the dropping-funnel into the flask in the bottom of which are a few scraps of platinum. If the flow of the solution is regulated so that the

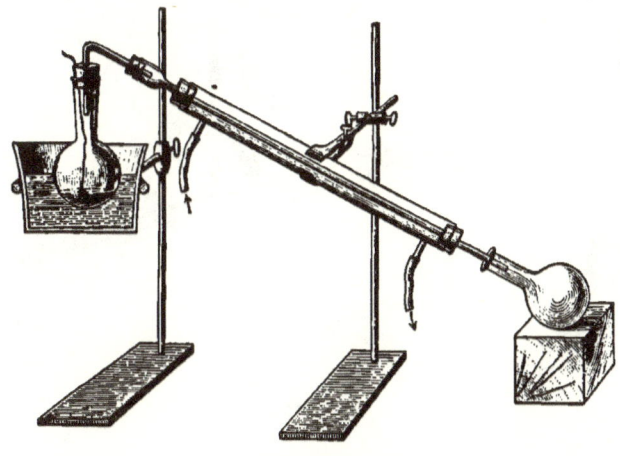

FIG. 23.

same quantity of liquid is added as that distilled, the operation may be carried on continuously for hours. The quantity of ether collected in the receiver is prevented from becoming too large, by pouring it into a larger vessel from time to time. To protect the ether from ignition, the mouth of the receiver is closed by a loose plug of cotton. Besides its convenient manipulation, this method possesses the further advantage that after the completion of the distillation the dropping-funnel may be replaced by a thermometer, and the residue can be distilled directly from the fractionating flask. This is an especially economical procedure when the quantity of the dissolved substance is small. In a case of this kind the size of the flask is selected with reference to the residue that may be expected. In distilling off alcohol, it is necessary to heat the water-bath continuously, and to always use threads. The distillation may be hastened by placing the flask not upon, but *in*, the water-bath. If a solution of common salt is employed in the bath, the temperature is raised, and the distillation proceeds still more rapidly.

If one has had sufficient experience in laboratory work, alcohol may be distilled off by heating the flask on a wire gauze or sand-

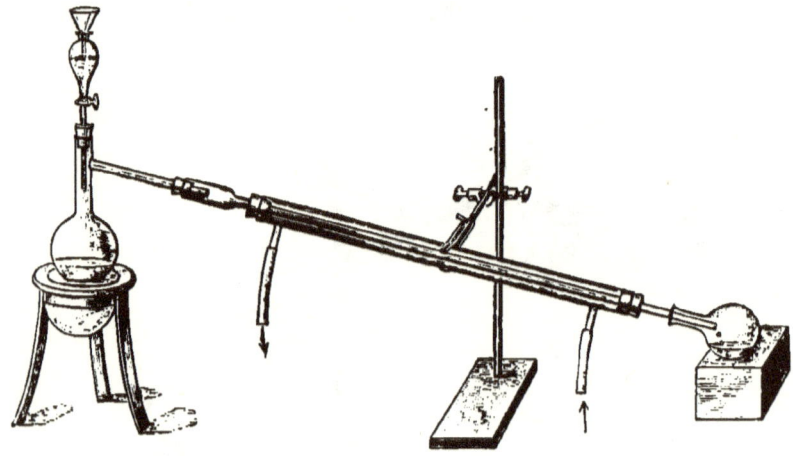

FIG. 24.

bath over a flame. In this case especial care must be taken not to use too large quantities at one time. Benzene can be distilled off under the same conditions as alcohol. The methods applicable to high boiling liquids have been given under "Distillation." (See page 15.)

In comparatively few cases the difference between the boiling-points of the solvent and the dissolved substance is a slight one; under these conditions the separation must be effected by a systematic fractional distillation with the aid of fractionating apparatus.

## DISTILLATION WITH STEAM

A particular kind of distillation, very frequently employed in organic work for the purification or separation of a mixture, is distillation with steam.  Many substances, even those distilling far above 100°, or those not volatile without decomposition, possess the property, when heated with water, or when steam is passed over or through them, of volatilising with the steam.  This phenomenon finds its explanation in the fact that the atmospheric pressure acting upon the mixture is naturally divided between the steam and the other substance, so that the partial pressure upon the latter is accordingly less than the atmospheric pressure, in consequence the volatility is increased.  The distillation with steam is therefore to be regarded as a special case of distillation in a partial vacuum.

**Apparatus.** — The apparatus used for distillation with steam is represented in Fig. 25.  A round flask inclined at an angle is closed by a two-hole cork; through one hole passes a not too narrow glass tube reaching to the bottom and serving to lead in the steam; the other hole bears a short glass tube the end of which is just below the cork, the other end is connected with a condenser as long as possible.  The distillation flask selected is of such a size that the liquid fills it not more than half-full. In order that the steam may act on an oil at the bottom of the flask, the inlet-tube is bent so that it may reach the lowest point of the flask.  Steam is generated in a tin vessel about half-filled with water, the neck being closed by a two-hole stopper; into one hole is inserted a safety-tube partially filled with mercury; the lower end of this tube does not touch the water; through the other hole passes the outlet-tube bent at a right angle.

**Method of Procedure.** — The experiment is begun by heating the steam generator and the flask simultaneously, the former conveniently by means of a low burner (Fletcher burner).  The flask may be heated on a wire gauze over a free flame; but since at times a very troublesome "bumping" will occur, it is better, if this happens, to heat on a briskly boiling water-bath. As soon as the water in the generator boils and the liquid in the flask has been heated to the proper point, the tubes of the two vessels are connected with rubber tubing. The distillation is then continued until the condensed steam passes over unaccompanied by any of the substance. Should the steam escape from the safety tube, the generator is being heated too strongly, and the flame should be lowered. To prevent the partial condensation of vapours in the upper cool part of the flask this should be covered with several layers of thick cloth to lessen the radiation of heat. If the quantity of substance to be distilled is small, so that only a small flask need be used, the preliminary and continued heating of the latter is superfluous: the steam can be passed at once into the cold liquid. If a compound is very easily volatile with steam, the introduction of the latter may be omitted; in this case it is only necessary to mix the compound with several times its volume of water and distil directly from the flask. If the substance to be distilled is solid, and its vapour forms crystals in the condenser, these may be removed provided the substance melts below 100°, by drawing off the water in the condenser for a short time. The substance is melted by the hot steam and flows into the receiver. If after this operation the water is to be turned into the condenser again, it must be done slowly at first, otherwise the cold water coming in contact with the hot condenser may easily crack it. When the melting-point of the substance is above 100°, in order to keep the condenser free from crystals, the distillation is interrupted for a short time, and the crystals are pushed out of the tube by a long glass rod.

 The end of the operation is indicated, if the substance is difficultly soluble in water, by the fact that the water passing over carries no drops of oil or crystals with it. But when the substance is soluble, even though the condensed water is apparently pure, it may still contain considerable quantities of the dissolved substance. In this case, to determine when the end of the operation has been reached, a small quantity, about 10 c.c., is collected in a test-tube,

# DISTILLATION WITH STEAM

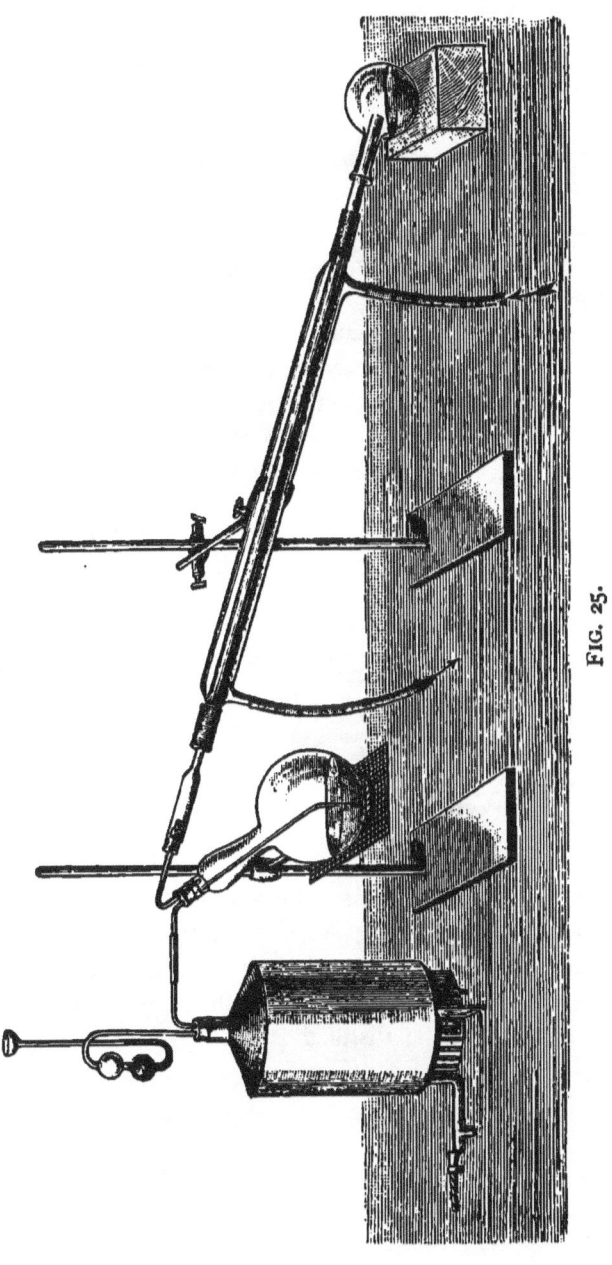

FIG. 25.

shaken up with ether, the ether decanted and evaporated. If no residue remains, the distillation is finished. When a substance shows a colour reaction, *e.g.* aniline with bleaching powder, advantage is taken of this to decide the question. After the distillation is ended the rubber tubing is first removed from the distilling flask, and then, *after this has been done*, the flame under the generator is extinguished. This point is also carefully observed when the distillation is interrupted; otherwise it may happen that the contents of the flask will be drawn back into the generator.

**Superheated Steam.** — In dealing with very difficultly volatile compounds, it is frequently necessary to conduct the distillation with the aid of superheated steam. A conically wound copper tube is interposed between the steam generator and the distilling flask (Fig. 26). In order to superheat the steam, a burner as

FIG. 26.

large as possible is placed under the spiral in such a position that the flame comes in contact with the interior of the spiral. The distillation is still further facilitated by heating the flask to a high temperature in an oil- or water-bath. Under these conditions the substance to be distilled is not covered with a layer of water.

## SEPARATION OF LIQUID MIXTURES. SEPARATION BY EXTRACTION. SALTING OUT

**Separation of Liquids.** — If the separation of large quantities of two non-miscible liquids, one of which is for the most part water or a water solution, is to be effected, it can be done with a separating funnel (Fig. 27). If the liquid desired is of a greater specific gravity than that of water, it is allowed to flow off through the stem by opening the cock. If, however, it floats upon the water, the latter is first allowed to flow off, and the liquid remaining is poured out of the *top of the funnel*. By this manipulation the liquid is prevented from coming in contact with the portion of water remaining in the cock, and adhering to the sides of the stem. For the separation of small quantities of liquids a small separating funnel, the so-called dropping funnel, is employed. If the quantity of liquid is so small that even a dropping funnel is too large, a capillary pipette is used (Figs. 28 and 29). The mixture to be separated is placed in a test-tube as narrow as possible, the pipette is immersed in the mixture almost to the surface of contact of the two liquids in case the upper layer is to be removed, the test-tube is brought to the level of the eyes, and the upper layer drawn off. The tubing is thus closed by pressure with the teeth or fingers and the pipette removed from the tube. If the lower layer is the one desired, the pipette is immersed to the bottom of the test-tube, and the operation conducted as before. Pipettes of this kind are very easily made from glass tubing; and if one is accustomed to work with them, are indispensable.

FIG. 27.

**Separation by Extraction.** — When a substance is held in suspension or dissolved in a liquid, generally water, the removal of the dissolved substance may be effected by agitating the solution with another solvent which will more readily dissolve the substance, but which is not miscible with the first liquid, drawing off this and distilling it. For extraction, ether is generally used; in special cases carbon disulphide, ligroïn, chloroform, benzene, amyl alcohol, etc., may be used. In the discussion following it will be assumed that the extraction is made with ether.

If the liquid to be separated is insoluble in water and is present in such a small quantity that a direct separation would cause a loss owing to the adhesion of the liquid to the walls of the vessel, or if it is held in suspension by the water in the form of individual

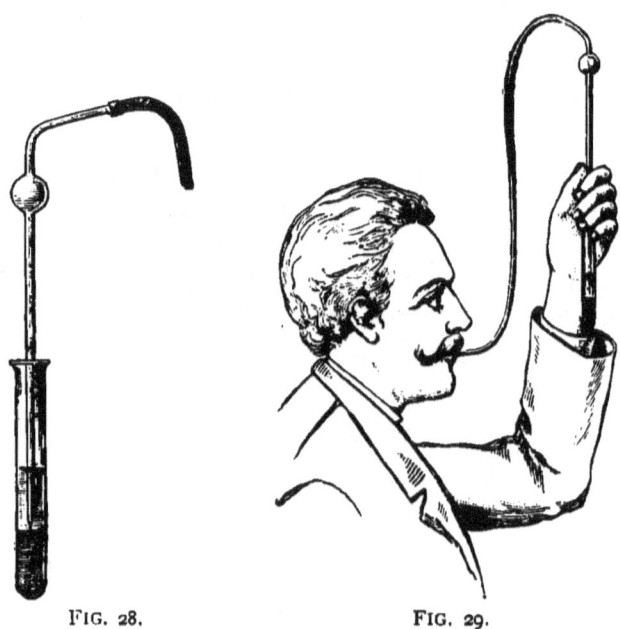

FIG. 28.  FIG. 29.

drops, ether is added to the mixture; it is then shaken, allowed to stand, the two layers separated, and the ether evaporated.

A similar method is followed if the substance is completely soluble in water, or if a portion of it is soluble and the remainder held in suspension. If from an aqueous solution a considerable portion of a solid or liquid separates out, it is not immediately extracted with ether, but if the substance separating out is a solid, it is first filtered off; if a liquid, the oily layer is separated in a dropping funnel, and then the filtrate is extracted with ether. In many cases, *e.g.* when the layer of oil is very turbid, or the solution strongly acid or alkaline, thus rendering filtration difficult, the entire solution may be treated at once with ether. With substances soluble in water the extraction must be repeated one or more times in proportion to the greater or less solubility of the substance. If it is desirable to avoid using large quantities of ether, after the first extraction it is distilled off and used for the second extraction, and so on. The extraction is repeated until a test portion of the ethereal solution, evaporated on a watch-glass, leaves no residue. If the test portion is evaporated by blowing the breath on it, the beginner will probably often be deceived by the appearance of an abundant quantity of crystals of *ice*, formed by the condensation of the moisture of the breath, due to the cold produced by the evaporation of the ether. Another error into which the beginner often falls when extracting with ether is this: in most chemical processes small quantities of coloured impurities are formed; on extraction, these impart a colour to the ether. The tendency is to continue the extraction until the ether is no longer coloured; but the proper method of procedure is just the reverse of this. On extracting colourless compounds, the colouring of the ether does not show that it still contains some of the substance dissolved; the above-mentioned test gives the only safe indication of the presence of the substance.

When using ether as an extraction agent the following points are to be observed: It frequently happens that, after long standing, two layers of liquids will not separate, in consequence of the formation of a flocculent precipitate floating in the liquids at the surface of contact. This difficulty may be obviated either by stirring the liquids with a glass rod, or by giving the separating funnel a circular motion in a horizontal plane, in such a way as not to cause the two layers to be mixed by the shaking. Under these conditions the neck of the funnel may be closed with a cork bearing a glass tube attached to the suction; by this means the space over the liquid is exhausted. The bubbles of gas which now rise through the liquid frequently destroy the troublesome emulsion. The same object may also be attained by adding a few drops of alcohol to the ether. If the separation of the layers is very imperfect, the cause is often due to the fact that an insufficient quantity of ether has been used; in this case more ether is added. If all of these methods prove ineffectual, then a complete separation may be obtained by first filtering the mixture, best with the aid of a Büchner funnel, which will retain the precipitate causing the emulsion, and then allowing it to stand.

Occasionally when extracting an aqueous solution containing inorganic salts with ether, they will separate out as solids. In this case water is added until the salts are redissolved, or the solution is filtered, with suction.

If the specific gravity of an ethereal solution is approximately the same as that of the water solution, the separation often takes place only with difficulty. Some common salt is then added, by which the specific gravity of the aqueous solution is increased.

Under certain conditions both the ethereal and aqueous solutions are so coloured that they cannot be distinguished. In this case the separating funnel is held toward the light; in the evening a luminous flame is placed behind it, and the eye is directed to the liquid at a point just above the cock. On opening the cock, the eye will readily detect the layer separating the two liquids, as it approaches the opening.

**Salting Out.** — A very valuable method to induce substances dissolved in water to separate out is known as "salting out." Many substances soluble in pure water are insoluble or difficultly soluble in an aqueous solution of certain salts; if, therefore, sodium chloride, potassium chloride, potash, calcium chloride, ammonium chloride, Glauber's salt, sodium acetate, or other salts is added to the solution, this is dissolved, and the substance previously in solution separates out. By this method many compounds like alcohol, acetone, etc., which are so easily soluble in water that they cannot be removed from it by extraction with ether, can be separated out with ease. The method of procedure is this: One of the above-mentioned salts, usually solid potash, is added to the solution until no more will dissolve. The substance thus forced out of solution collects above the heavier salt-solution and is removed by decantation or suction.

A combination of extraction and salting out also presents many advantages. If to the solution of a compound in water one of the salts mentioned is added — it is best to use finely pulverised sodium chloride — before the extraction with ether, this latter is greatly facilitated for several reasons. In the first place, a portion of the dissolved substance will separate out, due to the "salting out" action; furthermore, the solubility of the substance in the the new solvent — sodium chloride solution — will be diminished so that on extracting, a larger portion is dissolved by the ether than on treating the solution directly with it, and finally, ether does not dissolve so readily in a sodium chloride solution as in water, so that the volume of the ethereal solution is larger. The amount of sodium chloride to be added is about 25–30 grammes of the finely pulverised salt to 100 c.c. of the aqueous solution. Unfortunately the method of "salting out" has not been so generally adopted in scientific laboratories as it deserves, while in the laboratories of technical chemists it has long been in daily use. Among the reagents constantly used, a bottle of solid sodium chloride should not be wanting. In many cases, instead of the salt, a concentrated aqueous solution may also be used.

## DECOLOURISING. REMOVAL OF TARRY MATTER

As is well known, animal charcoal possesses the property of being able to remove the colour from certain solutions; for this reason it is frequently employed in the laboratory to free a colourless substance from coloured impurities. If it is to be used to remove the colour of a solid substance, the latter is first dissolved in a suitable solvent, then boiled with the animal charcoal and filtered. Before treating a hot solution, it is allowed to cool somewhat, since when animal charcoal comes in contact with liquids heated nearly to the boiling-point, a violent ebullition is frequently caused, and an overflowing of the liquid may easily take place. When a solvent not miscible with water is used, the animal charcoal, which is generally moist, is previously dried on the water-bath. The solvent selected is such that upon cooling the decolourised solution, the substance will crystallise out. In carrying out this operation, the general rule that no animal charcoal is added until the substance to be decolourised has completely dissolved, should be followed. Under these conditions only, is it certain that a portion of the substance does not remain undissolved mixed with the charcoal. The quantity of animal charcoal to be added to a solution depends upon the intensity of the colour of the latter. To a solution very slightly coloured, a small quantity is added; to a deeply coloured solution, a larger quantity. Very finely divided precipitates in water which pass through the filter may also be removed by the use of animal charcoal. When, *e.g.* tin is precipitated with hydrogen sulphide, the tin sulphide is often so finely divided that it runs through the filter. If the liquid is boiled with animal charcoal, the filtration presents no difficulty.

The use of animal charcoal, especially when it is in a very finely divided condition, has the disadvantage that at times it passes through the filter and contaminates the filtrate. This may be prevented frequently, by filtering again, or by boiling the filtrate a few minutes before the second filtration. When substances to be analysed have been decolourised with animal charcoal, care must always be taken to prevent the contamination of the substance. In such cases it is again crystallised without the use of animal charcoal. This difficulty may also be prevented or essentially lessened, by washing the charcoal with water several times before using; the portion suspended in the water is decanted, and only the coarser residue which easily settles at the bottom is used.

Recently the use of animal charcoal in the sugar industry has been replaced in part by a mixture of fine wood meal and floated infusorial earth (kieselguhr). This mixture ought to be of great advantage in the laboratory, for decolourising purposes, if used in the same way on a small scale. To the mixture is ascribed very superior purifying properties, so that by using much smaller quantities the same effect is obtained as with far larger quantities of animal charcoal. In order to prevent an easily oxidisable liquid from decomposing when it is heated in the air — this action being generally attended with more or less colouration — a gaseous reducing or protecting agent is passed through it; *e.g.* sulphur dioxide, hydrogen sulphide, or carbon dioxide. Very easily oxidisable substances are not evaporated in a dish, but in a flask, since in this the liquid is better protected from the action of the air.

Not only coloured impurities, but those of a tarry character, may also be removed by boiling with animal charcoal as above described. A mixture of wood meal and infusorial earth with which the solution may likewise be boiled is said to be of great value.

For the absorption of tarry impurities, in so far as they are liquid or oily, unglazed, porous plates (drying plates) may be used with advantage, the substances being firmly pressed out with a spatula in a thin layer. If one pressing out is insufficient, the substance is spread out again upon a fresh, unused portion of the plate. The absorption of an oil may often be facilitated by moistening the substance on the plate with alcohol, ether, or ligroïn, which at times will dissolve the impurities without causing a solution of the substance. Oily by-products may also be removed by pressing the substance between a number of layers of filter-paper. For this purpose either a screw-press is used, or the substance is placed in layers of filter-paper between two wooden blocks, the upper one of which bears a heavy object.

## DRYING

**Drying Solid Compounds.** — Under the chapter on "Crystallisation," page 9, the method of drying moist crystals has already been given. This method is naturally applicable to all solids, even if they are not crystallised, or only imperfectly crystallised, so that it will be unnecessary to repeat the directions already given. But a few methods, not so refined, and generally employed in dealing with crude products, will be referred to here. Before a substance is dried by allowing it to lie in the air or in a desiccator, the greatest portion of the moisture is removed by pressure, as follows: The substance lying between a number of layers of filter-paper is placed in a screw-press, and pressure applied. The operation is repeated, and the paper renewed, until it is no longer moistened. If a solid is not contaminated by water or other solvent, but by a liquid by-product, which one desires to obtain, the paper, after it has absorbed this liquid substance, can be extracted with a solvent, like ether. Large masses of a compound not too finely granulated can be tied up in a piece of filter-cloth of small mesh, placed in the screw-press, and pressure applied. Smaller quantities of a substance may be pressed out between two wooden blocks, the upper of which bears a heavy object. Generally one places the blocks on the floor and stands on them.

Very often solids may be dried by making use of the power of unglazed porcelain to absorb liquids with avidity. The substance to be dried is spread out in a thin layer upon a suitable piece of an unglazed porcelain plate, with a spatula, and is allowed to stand for some time, longer or shorter, as may be necessary. If one pressing out is not sufficient, the operation is repeated, using a fresh plate. Oily and tarry impurities may also be removed in this way, as mentioned above.

Compounds which fuse without decomposition may be dried either upon the water-bath or in an air-bath, or by heating over a free flame until they melt, allowing them to solidify, and then pouring off the water.

In order to dry a substance at a high temperature in a vacuum, two glass hemispheres, the edges of which are ground and fitted together, are used. The upper vessel is supplied with a tubulure, the opening of which is closed by a cork bearing a glass tube bent at a right angle connected with suction. The sphere may be heated by immersion in a large quantity of hot water or on a boiling water-bath. The upper hemisphere is enveloped in a cloth to prevent the condensation of the vapours.

**Drying Agents for Liquids.** — Liquids are dried (deprived of water) either by placing in them, or in a solution of them, drying agents. The most frequently employed drying agents are:

Calcium chloride,
  (*a*) granulated,
  (*b*) fused,
Potassium hydroxide,
Sodium hydroxide,
Ignited potash,
Fused sodium sulphate.

Less frequently used are: lime, barium oxide, anhydrous sodium carbonate, anhydrous copper sulphate, phosphorus pentoxide, sodium, and others.

In the choice of a drying agent care must be taken to select one which will not react with the substance to be dried. For example, calcium chloride unites to form double compounds with alcohols as well as with bases. Consequently, for drying these two classes of compounds, calcium chloride is never used. Caustic potash and caustic soda, as is well known, react with acids and phenols to form salts, upon alcohols to form alcoholates, and upon esters, saponifying them. These drying agents are never used with these substances. Further, acids are never dried with carbonates, owing to the salt formation taking place.

Calcium chloride is employed in two forms, — granulated and fused. The former acts more energetically, since it contains less water of crystallisation and possesses a larger acting surface. Still, it has the disadvantage of being more porous than the fused variety and in consequence of this porosity the loss of the substance being dried is greater. For drying small quantities of a substance, or liquids containing very little moisture, it is better, therefore, to use fused calcium chloride.

On drying bases with caustic potash or caustic soda, it must be borne in mind that these drying agents may be contaminated, at times, with potassium nitrite or sodium nitrite. Since these latter act upon bases, decomposing them, it is necessary to use the pure alkalies, or in place of them potassium carbonate or Glauber's salt.

**Methods of Drying.** — As already mentioned, liquids may be dried either in the undiluted form or in solution. The first method is followed when the quantity of the liquid is considerable, so that the loss of the substance necessarily incident to the adhesion of the liquid to the drying agent cannot amount to a large percentage of the whole. Low boiling liquids are always dried directly, without the use of a solvent. If a solution of a higher boiling compound is to be dried, it is done before the solvent is distilled off. A small quantity of a substance or a viscous substance is designedly treated with a diluting agent, generally ether, and is then dried.

The drying is accomplished by placing the drying agent in the liquid and allowing the two substances to remain in contact for a longer or shorter time, according to circumstances. So long as a liquid appears turbid, it has not been deprived of its moisture. A liquid about to be dried must never contain drops of water which are visible; in case it does, it must be treated in a separating funnel or the water drawn off with a capillary pipette: it is then dried. If only a few small drops of water are present, the liquid is first filtered through a small folded filter, or it is poured carefully into another vessel, and the water drops will remain in the first vessel, adhering to the walls. When a separation of an ethereal from an aqueous solution is to be made, to prevent a portion of the water from being carried along with the ethereal solution, the former is not drawn off through the cock of the vessel, but is poured out of the mouth, as has already been mentioned.

If a liquid contains very much moisture, and this is the case especially in turbid, milky liquids, it frequently happens that the drying agent will absorb enough water to dissolve itself and thus form an aqueous solution. In this case a fresh quantity of the drying agent is not added at once, but the separation of the two layers is effected by a separating funnel, pipette, or by decanting one of the layers.

A similar rule obtains for undiluted liquids. The drying of high boiling substances, if they are not volatile at the temperature of the water-bath, may be greatly facilitated by heating them with the drying agent on the water-bath.

# FILTRATION

Before distilling a liquid which has been dried, or before distilling off the solvent from such a liquid, it is poured off from the drying agent. To obtain the small portions which adhere to the latter, it may be washed with a small quantity of the dried solvent. Low boiling individual liquids (boiling on water-bath) can, under certain conditions, be distilled without a previous separation from the drying agent. If the liquid to be dried is of such a specific gravity that the drying agent will float in it, then, in order to effect the separation, it is poured through a funnel containing a small quantity of glass wool, or asbestos. In some cases, which, however, are rare, a liquid not easily volatile may be dried by exposing it in a dish as shallow as possible in a partially exhausted desiccator.

## FILTRATION

While in analytical operations it is much more desirable to conduct filtrations without employing pressure, the precipitates obtained in organic preparation work are filtered with pressure whenever it is possible. The method presents a number of advantages: the filtration may be made in a much shorter time; the liquid may be much more completely separated from the precipitate, in consequence of which the latter will dry more rapidly, etc.

**Filtration with Suction.** — For filtering under pressure (suction), a filtering flask $a$ (suction flask) with a side-tube $b$ (Fig. 30) is used. An ordinary flask may be converted into a suction flask, by fitting to it a two-hole rubber stopper; through one hole is passed the stem of a funnel, through the other, a glass tube, bent at a right angle, one end of which passes just through the cork, while the other is attached to the suction. For this purpose, flasks with walls as thick as possible are selected, in order that they may not be crushed by the atmospheric

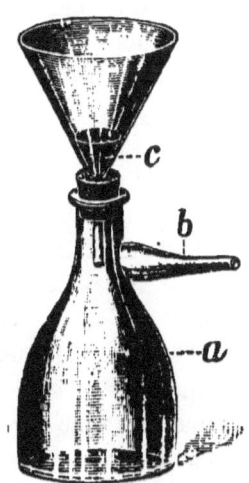

FIG. 30.

E

pressure on exhaustion; if a thin-walled flask is used, it must be exhausted but slightly.

The funnel used is, in many cases, the ordinary conical glass form, in which is placed the filter. If the funnel is imperfect in construction, and does not possess the correct angle (60°), the filter is made narrower or wider, as the case may be, to accommodate it to the angle of the funnel. In order that the point of the filter, not supported by the glass walls of the funnel, may not tear on exhaustion, a platinum cone $c$ is previously placed in the funnel.

If a platinum cone is not at hand, it may be replaced by a conically folded piece of parchment paper or filter-cloth. The filter is moistened with the same liquid which is to be filtered, otherwise it may happen that the filtration is prevented, or, at least, rendered difficult; *e.g.* if the filter has been moistened with water, and an alcoholic solution is to be filtered through it, the substance dissolved in the alcohol may be precipitated in the pores of the filter by the water. If a liquid foams excessively on filtering, as happens at times with alkaline liquids, the rubber tubing is removed suddenly from time to time from the filter-flask. The pressure of the in-rushing air destroys the bubbles. The foaming may also be prevented at times by treating the filtrate with a few drops of alcohol or ether. This is one of the common methods of preventing foaming in general. When filtering very small quantities of a liquid, a test-tube is placed in the filter-flask, as is represented in Fig. 31.

The suction surface may be increased by placing a so-called filter-plate of glass or porcelain in the funnel (Fig. 32). If a filter-plate is used, the filter-paper should be of two thicknesses. Upon the plate is first placed a round filter of exactly the same size as the plate, and upon this another round filter, the edge of which projects about 2–3 mm. beyond that of the plate.

The Büchner funnel is indispensable in working with organic substances. In consequence of its large suction surface, a very rapid filtration is possible. In the filtration of large quantities of substance it should always be used (Fig. 33).

A double filter described above may be used in this, but in

## FILTRATION

most cases, a single filter is sufficient. Since the Büchner funnels are made of porcelain, and consequently are opaque, they must be carefully cleaned immediately after using.

Similar to the Büchner funnel in its construction and action is the so-called "Nutsch" filter. This consists of a shallow dish with a perforated bottom, which is fitted to the cover of a tubulated cylinder by means of a rubber ring, the joint being air-tight. This apparatus is especially adapted to the filtration of larger quantities than can be conveniently treated in the Büchner funnel.

If the solution to be filtered acts on the filter-paper, filter-cloth may be used in its place; a fine or coarse meshed cloth is selected

FIG. 31.    FIG. 32.    FIG. 33.

according to the nature of the precipitate; it is moistened before the filtration.

If this is also attacked, the precipitate is retained by using glass wool, or better, long-fibrous asbestos, with which the bottom of the funnel, containing in this case a platinum cone, is filled, or it is spread out in thin layers over a filtering plate, or on the surface of a Büchner funnel. Under these conditions, the suction is applied gently at the beginning of the filtration; as soon as a large quantity of the precipitate has accumulated, the suction is increased. Very coarse-grained precipitates can be filtered without the use of a filter by placing in the point of an ordinary glass funnel a sphere of glass (a marble), this is surrounded by glass wool, or asbestos, if necessary.

**Pukall Cells.** — For the filtration of precipitates, like calcium sulphate, barium sulphate, of strongly acid liquids, etc., Pukall's cells (porous cells), made of unglazed clay, are very useful. They may be procured in different sizes in the market, and possess either the form of a cylinder or a mortar pestle. The operation is performed as follows: In the mouth of the cell is placed a closely fitting stopper bearing a glass tube bent twice at right angles, connected with a filter-flask. The tube at each end projects slightly below the stopper (Fig. 34). The cell is now immersed in the liquid to be filtered, contained in a beaker, not too wide, until it almost touches the bottom. When the suction is applied, the liquid filters through the porous walls until the cell is filled, and is then drawn into the flask; the precipitate remains behind in the vessel, and for the most part is deposited on the exterior walls of the cell.

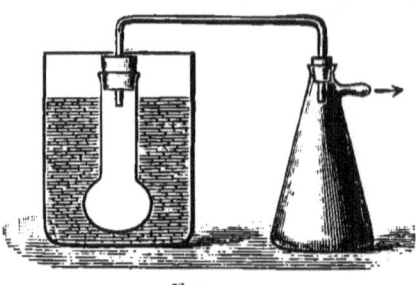

Fig. 34.

**Filter-Press.** — For the filtration of large quantities of substance which filter with difficulty, especially dye-stuffs, barium sulphate, calcium sulphate, etc., filter-presses are often used, of which the Hempel form will be described (Fig. 35). The separation of the liquid from the precipitate is effected in the cell $c$, which consists of two perforated porcelain plates between which is a rubber ring. The first operation in working with the press is the preparation of the cells. Two circular pieces of filter-cloth and two of filter-paper the same size as the plates are cut; after the cloth (linen or muslin) has been thoroughly moistened with water, the cells are made as follows: At the bottom comes the perforated plate upon which is placed one layer of the filter-paper, and upon this the cloth. After a wide glass tube $g$, which extends almost to the opposite side of the cell, has been inserted into the opening of the rubber ring, this is placed upon the cloth, then follows the other piece of cloth, filter-paper, and finally the second plate.

# FILTRATION

The cell is now secured by three clamps, one of which is attached near the glass tube, and the others equally distant from this. The cell is now ready for the filtration and is placed between the two corrugated glass plates *d*. Before it is connected with the vertical tube *b*, the pinch-cock on this is closed, water is poured into the funnel *a*, and the cock is now opened until the vertical tube is filled. The cock is again closed, and the tube is connected with the cells, the liquid to be filtered poured into the funnel and the cock opened. During the resulting filtration care is taken to keep the funnel partially filled so that the vertical tube is constantly full. If the first portions run through turbid, they are returned to the funnel. In order to wash the precipitate collecting in the cell, the glass tube passing through the rubber ring is partly withdrawn, so that it projects into the cell but a few centimetres. This causes a canal to be formed

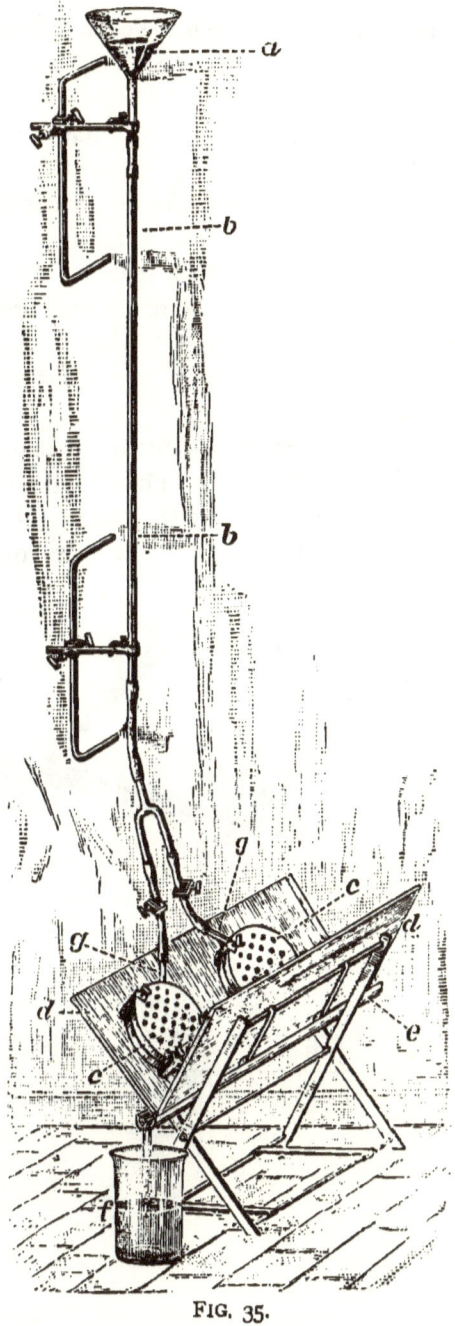

FIG. 35.

in the cell from which the wash water can permeate the precipitate in all directions. If the precipitate is large enough to completely fill the interior space of the cell, it forms a solid cake that can be removed without difficulty. But if the precipitate is small, and it is desired to obtain it, the glass tube is withdrawn from the rubber ring, the contents of the cell, generally half-fluid, are poured into a beaker, the cell taken apart, and the precipitate adhering to the sides scraped off with a spatula. By filtering with suction a complete separation of the liquid and precipitate is effected. If it is desired to filter larger quantities of a precipitate than can conveniently be done in a single cell, two cells connected by a Y-tube may be used.

**Filtering through Muslin.** — Precipitates which are not too finely divided may be filtered off through a filter-cloth (muslin) stretched over a wooden frame (filter-frame). A square piece of muslin or linen, after being thoroughly moistened, is fastened on the four nails of the frame in such a way as to cause a shallow bag in the middle. The frame is placed over a dish of the proper size and the liquid to be filtered is poured on the cloth and generally filters rapidly through it. If it is desired after washing the precipitate to press it out, the cloth is taken from the four corners, folded together, and squeezed with the hands. The precipitate may be further dried, by tying up the opening of the bag with twine, and then pressing it out carefully under a screw-press.

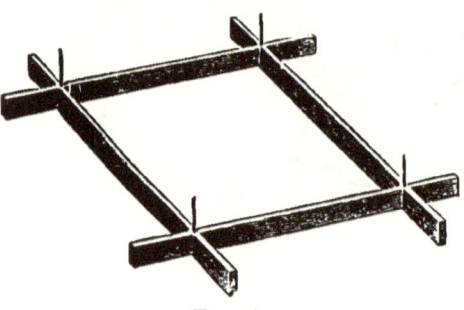

FIG. 36.

# HEATING UNDER PRESSURE

**Sealed Tubes. Method of Filling.** — If it is desired to induce a reaction between two substances at a temperature above their boiling-points, they are generally heated in sealed tubes. If a quantitative determination is not to be made, if the substances to be heated do not attack the glass or generate no gases, and if the heating is not to be high, soft glass tubes may be used. But generally, and in quantitative determinations always, difficultly fusible tubes of potash glass are used, since they are not so easily acted upon and do not crack so readily as the former. In filling the tubes the following points are observed. The tube is dried before placing the substance in it. Never put solid or liquid substances directly in the tube, but with the aid of a funnel-tube which should be as wide as possible when the substance is a solid. In proportion to the temperature of the heating a greater or less pressure is developed; therefore more or less of the substance is placed in the tube, depending on the conditions. The tube is never more than half-filled. Easily volatile substances as well as those giving off vapours, like hydrochloric and hydriodic acids, which render the sealing of the tube difficult, are transferred to the tube just before the sealing is to be done. In withdrawing the funnel-tube care is taken to avoid bringing it in contact with the walls of the tube.

**Sealing.** — To seal the open end of a tube charged with the substance, it is warmed by holding it at an angle of about $45°$, with constant turning, in the small luminous flame of a blast-lamp, and then heated strongly in a larger non-luminous flame; when the glass becomes soft, a previously somewhat warmed glass rod is fused to it (Fig. 37, I). The flame is then applied to the tube at a short distance from the opening, and as soon as the glass has become soft the tube is narrowed by drawing it out suddenly (II). After breaking off or cutting off the end of the capillary tube at $a$, to allow the air to escape on further heating, it is heated at $b$, when the tube is softened at this point it is drawn out slightly, the heat is applied just below *b*, it is drawn out again, and so on; the result is that the form of the end of the tube gradually changes from a cylinder to a sharp-pointed cone. The narrowest part of the latter is then heated with a not too large flame without drawing it further. The soft glass melts together, and there is thus obtained a thick-walled capillary tube which is melted off at the proper place (III). Figure 38 shows the sealed portion of a tube in its natural size. In the formation of

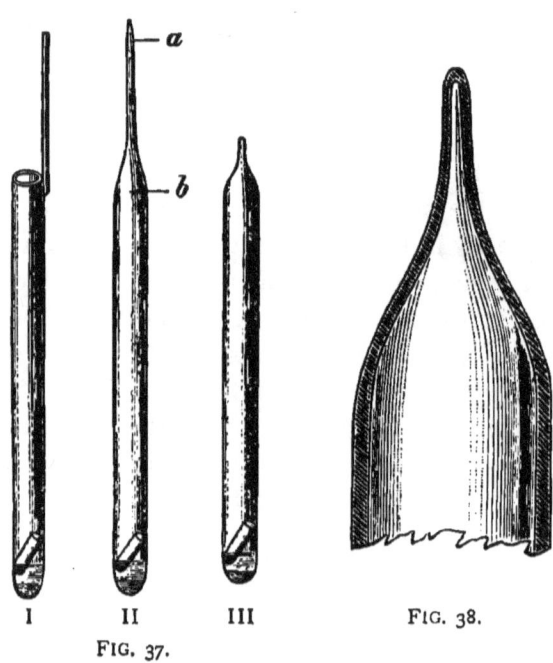

FIG. 37.   FIG. 38.

the capillary portion, it is desirable not to turn the tube in the manner previously directed, but to give it a few turns in one direction and then to reverse the motion, otherwise a spiral would be formed owing to the smallness of the glass at that point. After sealing, the heated portion is cooled gradually by holding it in the luminous flame until it is blackened. The sealing of hard glass tubes may be facilitated by placing a brick or tile near the flame in such a position that the heat will be reflected. If one is in possession of a cylinder of oxygen, it may be attached to the blast-lamp in place of the blast. At the high temperature of the illuminating gas-oxygen flame, the sealing may be effected with great ease.

In many cases the operation is rendered difficult by the vapours of the substance attacking the glass, or by the decomposition of the substance with the evolution of troublesome products like carbon, iodine, etc. Under these conditions, the tube is not drawn out first to a narrow tube, as above, but the glass rod fused on is allowed to remain, and this is used to draw out the tube. The sealing is rendered less difficult by allowing the air to have free access to the tube, in order that the evolved vapours may pass out unimpeded. The separation of carbon may be avoided by having an assistant direct a continuous current of air, during the heating, through a narrow tube into the upper part of the tube being sealed; this will cause the oxidation of the carbon. When dealing with very volatile substances, during the sealing the lower part of the tube is cooled by water, ice, or a freezing mixture. In this case, the services of an assistant will be needed to give to the vessel containing the cooling agent a circular motion corresponding to that of the tube. Under these conditions, it is often advisable to narrow the tube before charging it with the substance, so that it will just admit a funnel-tube as narrow as possible.

**Heating.** — The heating of sealed tubes (bombs) is conducted in the so-called "bomb-furnace," of which a convenient form is represented in Fig. 39. To be able to carry out the operation of heating at a definite temperature, a cork, covered with asbestos paper, bearing a thermometer, is fitted into the opening at the top of the furnace. The bulb of the thermometer must be about 1 cm. above the bottom of the iron tube. The sealed tube is not heated directly, but in a thick-walled protecting case of iron closed at one end, in which the glass tube is so placed that the capillary portion is at the open end. In transferring the glass tube to the iron casing the latter is not held vertically, but is slightly inclined from the horizontal, so that the glass tube may not be broken by suddenly striking the bottom. The iron case is pushed into the furnace open end first, so that in case of an explosion the fragments of glass are not thrown out of the forward end but from the rear of the furnace, directed toward a wall. A "fragment cage" renders the flying pieces of glass harmless. After the tube is in position the front opening is closed by a "drop-slide." The tubes are not heated at once up to the desired temperature, but are warmed gradually. If it is desired to heat a furnace similar to the one represented to a low temperature, the gas tubes are raised and small flames used, rather than a lowering of the gas-tube and the corresponding increase in size

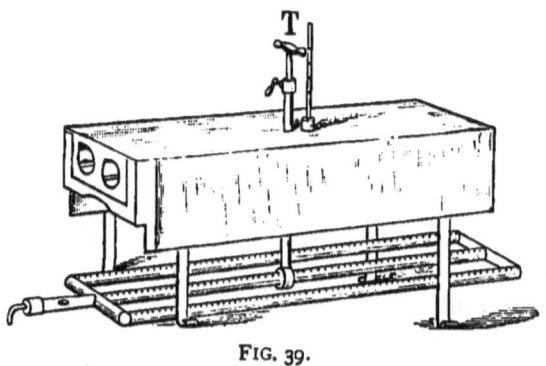

FIG. 39.

of flame. The danger of the bursting of the glass tubes may be diminished in many cases, particularly in those in which a very high pressure is developed, by interrupting the heating after a certain length of time, opening the capillary after the tube has completely cooled, and allowing the gases which have been generated to escape. The tube is then resealed and heated again.

If tubes are to be heated not higher than 100°, the convenient so-called "water-bath cannon" is used, in which the case enclosing the tube is heated by steam at ordinary pressure; in this case overheating is impossible.

**Opening the Tubes.** — *Sealed tubes must not be opened until after they are completely cold. The protecting case of iron,* containing the tube, is removed from the furnace and held in a slightly inclined position, the end of the capillary being higher than the rear end. By means of a slight jerk the capillary end of the glass tube is caused to project from the iron case. The extreme end of the capillary is now held in the flame of a Bunsen burner. In case there is an internal pressure in the tube, the glass on becoming soft will be blown out and the gases will escape from the opening thus made, often with such force as to extinguish the flame. If on the softening of the glass the capillary is not blown out, it may be due to the absence of internal pressure or the tube may be stopped up by some of the substance. In the latter case the substance is removed by heating. To show that there is an internal pressure the capillary is held after it has been opened near a small luminous flame; if the latter is blown out in a long thin flame sidewise, obviously there is pressure. If great pressure exists in a tube to be opened, before blowing the capillary the hand holding the iron casing is protected by a thick glove or a cloth is wrapped around the casing several times at the point where it is held, so that if the tube bursts, in consequence of the sudden diminution of pressure, and the seam of the case should be torn open, the hand is protected from injury. *In handling an unopened tube the greatest care possible must be observed. It is never removed from the iron casing to look at it or for any other purpose. On opening, it is held in such a position that neither the operator nor any one else can be injured in case of bursting.*

On heating substances with hydriodic acid and phosphorus, it sometimes happens, that the tube, on being opened by a flame, explodes. In this case the explosion is due to the fact that the phosphine as well as the hydrogen evolved in the reaction have formed an explosive mixture with the oxygen of the air present in the tube. Under these conditions, the capillary is opened by snipping off the end with pincers or tongs, but in doing so the greatest care must be observed. To remove the end of the cone, it is not necessary to proceed as described below, but the end of the tube is broken directly with a blow of a hammer.

In order to break off the end of a tube after it has been opened, so that the contents may be emptied out, the procedure is as follows: At that point of the tube where the cone begins, a well-defined file mark is made, not extending completely around the tube; this is touched lightly with the hot end of a glass rod, previously heated to fusion in the blast-flame. If the crack caused by this does not extend entirely around the tube, the extreme end of it is again touched with a hot glass rod, by which it is extended, so that the conical end may be lifted off. Instead of a glass rod, a thick iron wire, the end of which has been bent around the iron casing to a semicircle, may be used. If this is heated to redness, the file mark touched with it, and the wire turned, the end of the tube breaks off smoothly. To prevent the fragments of glass from falling into the tube (when a quantitative determination is being made), the method of procedure is this: As before, a deep file mark is made, and on each side of it, at a distance of $\frac{1}{2}$ cm., a strip of moistened filter-paper 1 cm. wide is wrapped around the tube several times. That portion of the tube between the strips is heated by a small flame, the tube being constantly turned, this causes the end to split off smoothly without splintering. If the glass does not crack at once, the heated portion is moistened with a few drops of water, and the breaking off will follow with certainty.

**Pressure Flasks. Autoclaves.** — In order to heat substances under pressure at a moderate temperature which on reacting with each other evolve no gaseous products, so that no pressure due to the reaction is developed, they are sometimes enclosed in strong-walled flasks (pressure flasks), wrapped up in a cloth and heated in a water-bath.

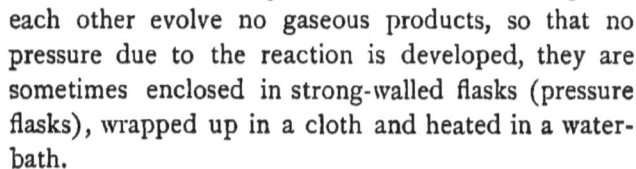

FIG. 40.

Very well adapted to this purpose are the common soda-water or beer bottles, of the kind represented in Fig. 40. In using them they are not immersed in water already heated, but are slowly heated with the water. The water-bath is closed by a loosely fitting cover, so that in case the bottle bursts, one may not be burned by the hot water. The flasks are not opened until after they are completely cold.

# HEATING UNDER PRESSURE

Large quantities of substances which do not act on metals may be heated under pressure in closed vessels, generally made of iron, bronze or copper (autoclaves). Such vessels are not suited for heating acid substances, but may be used for neutral or alkaline substances. In this laboratory Mannesmann tubes (without seams) are in use, one end being welded together, and the other is supplied with a screw-thread and cap. The open end is cone-shaped. The tube is closed by a threaded cap, which in section shows a cone. The cap is partially filled with lead. After the substance has been put in, the cover is screwed on as far as possible with the hand, the tube is then clamped in a vise, and the cap made fast with a wrench. The conical end of the tube is pressed into the soft lead, thus giving an excellent joint. The heating may be conducted in an oil-bath, or directly in the bomb-furnace. If the heating is to be carried beyond the point at which lead softens, a short metallic condenser about 10 cm. in length may be screwed on the threaded portion of the tube. A slow current of water is passed through the condenser. Tubes of this kind have proved of excellent service in many cases, *e.g.* for the preparation of the phenol ethers.

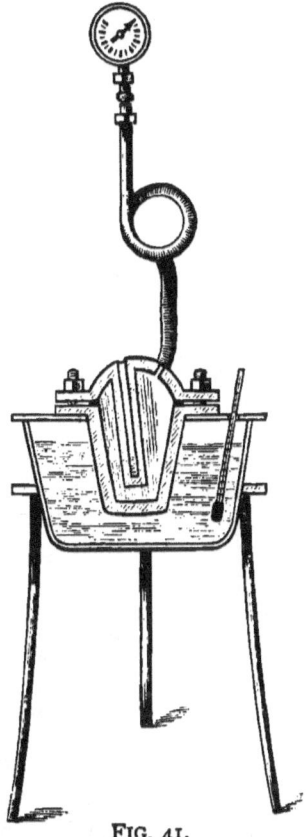

FIG. 41.

Another form of autoclave is represented in Fig. 41. For the packing a ring of lead or asbestos is used. The tube leading to the interior is designed for a thermometer. The lower portion contains oil in which the thermometer is placed.

## MELTING-POINT

In organic work the most common method of testing the purity, of characterising and of recognising a solid compound, is the determination of its melting-point. The apparatus most generally used for this purpose is represented in Figs. 42 and 43. A long-necked flask is closed by a cork provided with several canals cut in the sides, through which the heated air and vapours may escape, bearing a thermometer. The bulb of the flask is two-thirds filled with pure concentrated sulphuric acid, into which is dropped a crystal of potassium nitrate the size of a pinhead, to prevent it from becoming dark in colour. The substance is placed in a small narrow tube (melting-point tube), made in the following way: A glass tube 4–5 mm. wide is heated at one point while constantly turned, in a small, blast-lamp flame, until it becomes soft and is then drawn out, with constant turning, from both ends to a tube 1 mm. wide. The narrow tube thus produced is then fused off at its middle point; the portion lying next to that part of the glass tube which has not been drawn out is heated as before and is again drawn out, and so on. There is thus produced a tube having the form represented in Fig. 44, *a*. In order to prepare the melting-point tube from this a file-mark is made at the points indicated, the tube broken off and fused at the narrow end by holding it nearly vertical in a Bunsen flame. Fig. 44, *b* represents the melting-point tube in its natural size. A supply of several dozen of these is made and preserved in a closed bottle. To transfer to the tube the substance the melt-

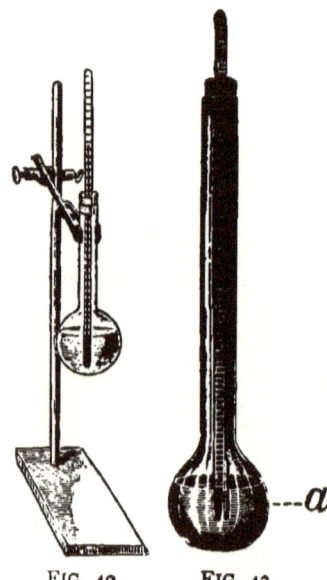

FIG. 42.   FIG. 43.

ing-point of which is to be determined, a small portion of it is pulverised, the end of the tube dipped into it; by gentle tapping the substance is caused to fall from the upper end to the bottom of the tube. In order that it may not form a too loose layer, it is packed by a thin glass rod or platinum wire. The height of the layer should be 1 mm. and in no case more than 2 mm. To attach the tube different methods may be used. The upper end of the tube may be touched with a drop of sulphuric acid; this, when brought in contact with the thermometer, will cause it to adhere. It is safer to fasten the tube, just below the mouth, to the thermometer with a thin platinum wire or a rubber ring 1 mm. wide. The substance is placed at the middle point of the thermometer bulb. The thermometer is now immersed in the sulphuric acid until the bulb is at about the centre of the liquid; the flask is heated with a free flame which is given a continuous, uniform motion as in distillation. The burner is inclined at a convenient angle, so that, if the flask should break, the hand would not be directly under it. When the melting temperature is reached, it is observed that the previously opaque, unfused substance suddenly becomes transparent and a meniscus is formed on its upper surface. If it is known at about what point the substance will melt, it may be heated rapidly to within 10° of this point, and then slowly with a small flame so that the behaviour of the substance from degree to degree can be easily observed. If the melting-point is not known, it can be readily ascertained on heating it rapidly to a high temperature. In this case the determination is repeated, heating rapidly until the temperature approaches the melting-point, and then slowly. In many cases when the temperature nears the melting-point this is shown by a softening of the substance before melt-

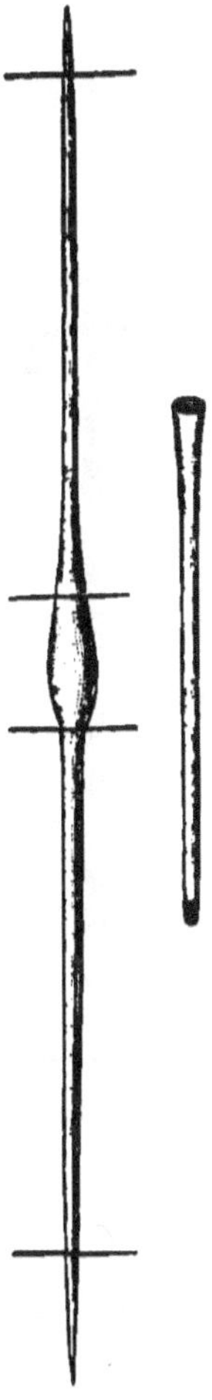

*a b*
FIG. 44.

ing; it loosens from the walls of the tube and collects in the middle. If this phenomenon occurs the heating is conducted very slowly from degree to degree. At times proximity to the melting-point may also be recognised by the fact that the particles of the substance which adhered to the upper portion of the tube during the filling, melt before the mass of the substance; since the hotter and therefore lighter layers of the acid rise to the top, the upper layers of the bath are heated somewhat higher than the lower.

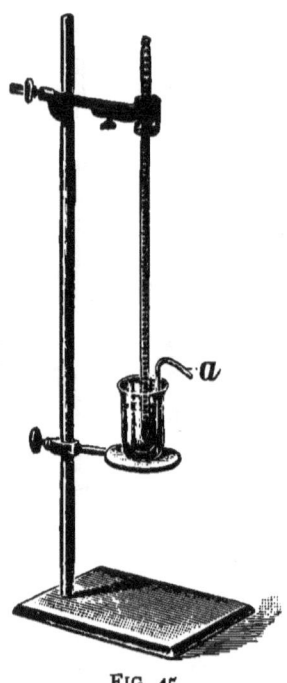

FIG. 45.

Instead of the apparatus just described the one represented in Fig. 45 serves very well for the same purpose. The liquid used may be water or sulphuric acid, depending on the melting-point of the substance to be examined, or in case of a substance with a high melting-point paraffin is placed in a beaker supported on a wire gauze. In order to keep the liquid at a uniform temperature, it is stirred by an up-and-down motion of the glass stirrer $a$.

A substance is regarded as pure in most cases, if it melts sharply within one-half or a whole degree, and if after repeated crystallisation the melting-point does not change. In determining the melting-point of a newly discovered substance, one determination is not sufficient even if it is very sharp; a small portion is recrystallised and the melting-point again determined. Many substances decompose on fusing, if this takes place suddenly at a definite temperature; this may also be regarded as a characteristic of the substance.

Since many compounds on heating decompose explosively, and since in the last few years it has happened that the explosion of minute quantities of a compound has shattered the melting-point apparatus, and serious wounds have been caused by the hot sulphuric acid, it is safer before the melting-point of a hitherto unknown substance is determined in the apparatus described above, to take the slight trouble of making a preliminary test by heating a small tube containing the substance directly in a small flame to the melting temperature, and by this means ascertaining if the substance will explode.

**Testing the Thermometer.**— At this point a few observations concerning the testing and correcting of the thermometer will be added. Since the ordinary thermometers, at least the cheaper varieties, are never exact, they must be corrected before using. If a normal thermometer is at hand, the correction to be applied may be determined by slowly heating the thermometer to be tested by the side of the normal instrument in a bath of sulphuric acid, glycerol, or vaseline, and noticing the reading of both thermometers for every 10°. There is thus obtained a table from which the corrections may be read directly. For many purposes it is sufficient to determine the deviation at only a few points; the corrections for the degrees lying between these may be calculated by interpolation. Thus, *e.g.*, the point to be considered as the true zero point may be determined as follows: A thick-walled test-tube of about $2\frac{1}{2}$ cm. in diameter and 12 cm. in length is one-third filled with distilled water. The mouth is closed by a cork bearing a thermometer dipping into the water. Through an opening cut out of the side of the cork is introduced a thick copper wire, the end of which is bent into a circle at a right angle to its length. The test-tube is surrounded by a freezing mixture of ice and salt. The water is frequently agitated with the stirrer; the temperature at which crystals first begin to form is carefully noted.

The true 100° point is found by placing distilled water in a not too small fractionating flask and determining the boiling-point of it, the entire column of mercury being in the vapour. In an analogous manner, the boiling-point of naphthalene (218° at 760 mm. pressure) and of benzophenone (306° at 760 mm. pressure) may serve for the correction of the higher degrees. Since the boiling-point is influenced by the pressure, the barometer must be read at the same time with the thermometer and a correction, taken from the table given below, applied.

| Pressure. | Water. | Naphthalene. | Benzophenone. |
|---|---|---|---|
| 720 mm. | 98.5° | 215.7° | 303.5° |
| 725 | 98.7 | 216.0 | 303.8 |
| 730 | 98.9 | 216.3 | 304.2 |
| 735 | 99.1 | 216.6 | 304.5 |
| 740 | 99.3 | 216.9 | 304.8 |
| 745 | 99.4 | 217.2 | 305.2 |
| 750 | 99.6 | 217.5 | 305.5 |
| 755 | 99.8 | 217.8 | 305.8 |
| 760 | 100.0 | 218.1 | 306.1 |
| 765 | 100.2 | 218.4 | 306.4 |
| 770 | 100.4 | 218.7 | 306.7 |

## DRYING AND CLEANING OF VESSELS

While in analytical operations, since one generally deals with aqueous solutions, the cleaned vessels may be used even if wet, it frequently happens in organic work, in experimenting with liquids not miscible with water, that dry vessels must be employed. In order to dry small pieces of apparatus rapidly, they should be rinsed first with alcohol and then with ether. To remove the last portions of the easily volatile ether, air from a blast is blown through the vessel for a short time, or the ether vapours are removed by suction. The alcohol and ether used for rinsing can frequently be used again; it is convenient to keep two separate bottles for the wash alcohol and wash ether, into which the substances, after being used, may be poured.

For rapid drying of large vessels this method is costly. In this case the procedure is as follows: The wet vessel is first drained as thoroughly as possible, and then heated with constant turning in a large luminous blast-flame, while, by means of a blast of air from bellows or other source, the water vapour is driven out. It may also be removed by careful heating and simultaneous suction. Thick-walled vessels like suction flasks must not be heated over a flame, but are dried by the first method.

Vessels may be cleaned in part by rinsing them out with water with the use of a feather or flask-cleaner. If the last portions of the solution of a solid, *e.g.* in alcohol, are to be removed from a flask, it is not washed out at once with water, but first with a small quantity of the solvent, and then afterwards with water. Resinous or tarry impurities adhering firmly to the walls can be removed by crude concentrated sulphuric acid. The action of this latter may be strengthened by adding a little water to it, by which heat is generated; also by the addition of some crystals of potassium dichromate. At times the impurities adhere so firmly that the vessel must be allowed to stand in contact with sulphuric acid for a long time. Crude concentrated nitric acid, or a mixture of this with sulphuric acid, is also used at times for cleaning purposes. Impurities of an acid character can, under certain conditions, be removed by caustic soda or caustic potash.

Finally, a method for cleaning the hands may be mentioned, if they are discoloured by dyes which cannot be removed by water. If the dye, *e.g.* fuchsine, contains an amido ($NH_2$) group, the hands are dipped into a dilute, weakly acid solution of sodium nitrite. The dye is diazotised, and may be removed by washing in water. The two methods following are applicable to all dyes; the hands are immersed into a dilute solution of potassium permanganate to which some sulphuric acid has been added, and are allowed to remain for some time; the dye is oxidised, and thereby destroyed. After the permanganate has been washed off with water, the hands, especially the nails, are coloured brown by manganese dioxide. This is removed by washing the hands with a little sulphurous acid. The second method is this: A thick paste of bleaching powder and a sodium carbonate solution is rubbed on the hands. This causes the oxidation and destruction of the dye, as above. In order to take away the unpleasant odour of the bleaching powder, the hands are scrubbed with a brush, care being taken to remove the particles adhering to the upper and under surface of the nails, and are then washed, as just described, with sulphurous acid.

## ORGANIC ANALYTICAL METHODS

### DETECTION OF CARBON, HYDROGEN, NITROGEN, SULPHUR, AND THE HALOGENS

**Tests for Carbon and Hydrogen.** — If on heating a substance on platinum foil, it decomposes with charring, it is an organic substance. Carbon and hydrogen may be detected in one operation, by mixing a small portion of the dried substance with several times its volume of ignited fine cupric oxide, placing the mixture in a small test-tube, adding more cupric oxide to the top of the mixture, and heating slowly, the tube being closed by a cork bearing a delivery tube bent twice at right angles. If the gas evolved (carbon dioxide) will cause a clear solution of barium hydroxide to become turbid, the original substance contained carbon; if it also contained hydrogen, small drops of water will collect in the upper cold part of the tube.

**Test for Nitrogen.** — To test an organic substance for nitrogen, it is heated in a small test-tube of difficultly fusible glass, about 5 mm. wide and 6 cm. long, with a piece of bright potassium the size of a lentil, which has been pressed between layers of filter-paper, in a Bunsen flame until decomposition, generally accompanied by slight detonations and dark colouration, takes place. The tube is finally heated to redness; while still hot it is dipped into a small beaker containing 10 c.c. of water; by this the tube is shattered, and any potassium unacted upon becomes ignited. The aqueous solution containing potassium cyanide, if nitrogen was present in the substance, is filtered from the carbon and glass fragments, the filtrate treated with a few drops of caustic potash or caustic soda until it shows an alkaline reaction; to this solution is then added a small quantity of ferrous sulphate solution and ferric chloride solution; it is boiled 1-2 minutes, and if potassium cyanide was present, potassium ferrocyanide will be formed. After cooling, the alkaline liquid is acidified with hydrochloric acid, the precipitated ferric and ferrous hydroxides will be dissolved, and being acted upon by the potassium ferrocyanide, will form Berlin blue. Accordingly, if nitrogen was present, a blue precipitate is obtained, otherwise only a yellow solution will be formed. If the substance contains only a small proportion of nitrogen, at times no precipitate is obtained at first, but only a bluish-green solution. If this is allowed to stand some time, under certain conditions, over night, the precipitate will separate out. In testing easily volatile substances for nitrogen, a longer tube is used and the portions of substance condensing in the upper cold part of the tube flow back a number of times on the potassium. In place of potassium, sodium may also be used in most cases, but the former acts more certainly. In testing for nitrogen, in a substance containing sulphur, a larger quantity of potassium or sodium than that given above is used. Substances which evolve nitrogen at moderate temperatures, *e.g.* diazo-compounds, cannot be tested in the manner described. In dealing with a substance of this kind it must be determined whether on heating the substance with cupric oxide in a tube filled with carbon dioxide, a gas is given off which is not absorbed by a solution of caustic potash. (See quantitative determination of nitrogen.)

In a limited number of substances containing nitrogen, the presence of the latter may be proved by heating the substance with an excess of pulverised soda-lime in a test-tube with a Bunsen flame; this causes decomposition with evolution of ammonia, which is detected by its odour or by means of a black colour imparted to a piece of filter-paper moistened with a solution of mercurous nitrate. Nitro-compounds, *e.g.*, do not give this reaction.

**Test for Sulphur.** — The qualitative test for sulphur is made in the same manner as that for nitrogen. The substance is heated in a small tube with sodium. After the mass has cooled it is treated with water, and to one-half of the solution is added a small quantity (a few drops) of a solution of sodium nitroprussiate, just prepared by shaking a few crystals with water at the ordinary temperature. A violet colouration indicates the presence of sulphur. Since the nitroprussiate reaction is very delicate, no conclusion as to the amount of sulphur can be drawn from the test, therefore the second half of the solution is treated with a lead acetate solution and acidified with acetic acid. In proportion to the amount of lead sulphide formed, the liquid will assume a dark colour, or a more or less heavy precipitate will appear, in this way indicating the original quantity of sulphur.

Easily volatile substances cannot usually be tested by this method. They are heated with fuming nitric acid in a bomb-tube to about 200 or 300°. After diluting with water the solution is tested with barium chloride for sulphuric acid. (See method for the quantitative determination of sulphur.)

**Test for the Halogens.** — The presence of chlorine, bromine, and iodine in organic compounds can only in rare cases be shown by precipitation with silver nitrate. This is explained by the fact that most organic compounds are non-electrolytes; *i.e.* that the solutions of the same do not contain free halogen ions, as is the case in solutions of the inorganic salts of the halogen hydracids.

In order to detect the halogens, the substance to be tested is heated in a not too narrow test-tube with a Bunsen flame with an excess of chemically pure lime, the tube while still hot is dipped into a little water, chemically pure nitric acid is added to acid reaction, the solution is then filtered and treated with silver nitrate.

In compounds containing no nitrogen, a test for the halogens may be made by the same method given for nitrogen — heating with sodium. In this case the solution, filtered from the decomposition products and fragments of glass, is acidified with nitric acid and silver nitrate added. Substances containing nitrogen cannot be tested in this way for the halogens, since, as shown above, these on fusion with sodium give sodium cyanide, which, like the sodium halides, reacts with silver nitrate.

The presence of halogens may be recognised very quickly and conveniently by Beilstein's test. A piece of cupric oxide the size of a lentil, or a small rod of the oxide $\frac{1}{2}$ cm. long, is wrapped around with a thin platinum wire, the other end of which is fused to a glass handle, and heated in the Bunsen flame until it becomes colourless. If after cooling, a minute particle of the substance containing a halogen is placed on this and then heated in the outer part of the flame, the carbon burns first and a luminous flame is noticed. This soon vanishes, and there appears a green or bluish-green colour due to the vaporisation of the copper halide. From the length of time the colour is visible, conclusions may be drawn concerning the presence of a trace or more of the halogen in the original substance.

### QUANTITATIVE DETERMINATION OF THE HALOGENS
### CARIUS' METHOD

The method consists in heating a weighed amount of the substance to be analysed in a sealed glass tube with silver nitrate and fuming nitric acid, by which it is completely decomposed (oxidised) and weighing the quantity of the silver halide thus formed.

Requisites for the analysis:

1. A tube of difficultly fusible glass sealed at one end, length about 50 cm.; outside diameter, 18–20 mm.; thickness of walls, about 2 mm. (Sealing-tubes, bomb-tubes.)
2. A funnel-tube about 40 cm. long, for transferring the silver nitrate and nitric acid to the glass tube.
3. A weighing-tube of hard glass (length, 7 cm.; outside diameter, about 6–8 mm.).
4. Solid silver nitrate and pure fuming nitric acid. The purity of the latter is tested by diluting 2 c.c. of it with 50 c.c. of distilled water, and adding a few drops of a silver nitrate solution. Neither an opalescence nor a precipitate should appear.

**Filling and Sealing the Tube.** — After the bomb-tube, weighing-tube, and funnel-tube have been cleaned with distilled water, they are dried, not with alcohol and ether, but by heating over a flame. (See page 66, Drying.) The exact weight of the weighing-tube is next determined. Into this, with the help of a spatula, is placed

0.15 to 0.2 gramme of the substance to be analysed, finely powdered. The open end of the tube is wiped off with a cloth, and the exact weight of the tube plus the substance is found. With the aid of the funnel-tube, about 0.5 gramme of finely powdered silver nitrate is transferred to the bomb-tube (a correspondingly larger amount up to 1 gramme is used for substances containing a high percentage of halogen) and 2 c.c. of fuming nitric acid. If a number of halogen determinations are to be made, it is advisable to measure off 2 c.c. of water in a narrow test-tube, mark the volume with a file on the outside, and then use this to measure the acid for the different determinations. After removing the funnel-tube, care being taken not to touch the walls of the bomb-tube with it, the weighing-tube is inserted in the bomb held at a slight angle, and is allowed to slide down to the bottom, but the substance must not come in contact with the acid. The tube is now sealed in the manner described on page 55. During the sealing, the substance must be prevented from coming in contact with the acid. Even after the tube is closed, this is not brought about purposely, as by violently shaking the tube.

If the substance to be analysed is liquid, it is placed in the weighing-tube with a capillary pipette, otherwise the procedure is just as described. In dealing with easily volatile substances, the weighing-tube is closed by a glass stopper, made by heating a piece of glass rod in the blast-flame until it softens, and then pressing it on a metal surface until a head is formed (Fig. 46).

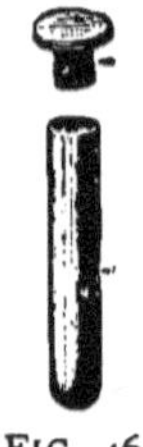

**Heating the Tube.** — After cooling, the tube is transferred to the iron protecting case, and heated in the bomb-furnace, in accordance with the directions on page 56.

FIG. 46

The temperature and time of heating depend upon the greater or less ease with which the compound is decomposed. In many cases, it is necessary to heat aliphatic compounds 2–4 hours at a temperature of 150–200°, while aromatic compounds must be heated 8–10 hours, and finally up to 250–300°. It is convenient to so plan the analysis that the bombs may be sealed in the evening, so that the heating may be begun the first thing the next day. The sealed tube is kept under the hood in the bomb-room, in the iron case, over night, which is clamped with its open end directed vertically upwards. The tube is never allowed to remain at the working table. If the furnace is not loaded, naturally it is most convenient to place the bomb at once in that. Since in many cases the oxidation begins even at the ordinary temperature, pressure is developed in the tube; therefore, after it has been standing over night, it must not be removed from the iron case to be examined. The heating is done gradually; at first, with a small flame, the gas-tubes being lowered from the furnace. Gradually these are raised, and the flames increased in size. The following table will show how the heating of a moderately refractory substance should be regulated.

The heating is begun at 9 o'clock A.M.

From 9–10 the temperature is raised to about 100°,
    10–11     ,,     ,,     ,,     150°,
    11–12     ,,     ,,     ,,     200°,
    12–3     ,,     ,,     ,,     250°,
    3–6     ,,     ,,     ,,     300°.

If an especially high pressure is generated by the decomposition of a substance, the danger of the bursting of the tube may be lessened by turning off the gas before leaving the laboratory at noon, and then in the afternoon opening the capillary, sealing and heating again to a higher temperature. The same method is followed in working with a substance so refractory that several days' heating is required; in this case at the beginning of the second day the pressure is reduced by opening the tube.

If two bombs are heated in the furnace at the same time, an entry is made in the note-book to show which tube lies to the right and which to the left. If this has been neglected, and the identity of the two tubes is in doubt, the neglect may be corrected by again weighing the two weighing tubes.

**To Open and Empty the Tube.** — The perfectly cooled tube is opened according to the directions given on page 58. Especial care must be taken before heating the capillary to softening in a large flame, to drive back into the tube by gentle heating over a small flame, any of the liquid which may have collected in the capillary. Before the conical end is broken off the tube is examined to see whether it still contains crystals or oily drops of the undecomposed substance. In case it does, the capillary is again sealed and the tube reheated; but if it does not, the conical end is removed according to the directions given on page 59. The part broken off is first washed free from any liquid or any of the precipitate which may have adhered to it, with distilled water, into a beaker; the portion in the tube is diluted with distilled water, upon which there is generally obtained a bluish-green solution, coloured by nitrous acid; this is poured, together with the weighing-tube into a beaker by inverting the tube, care being taken that the sudden falling of the weighing-tube does not break the bottom of the beaker. After the outer open portion of the tube has been washed with distilled water the tube is revolved and the precipitate in the interior is washed out; this is repeated as often as may be necessary. If a portion of the silver halide adheres firmly to the glass, it may be removed by loosening it with a long glass rod over the end of which has been drawn a piece of rubber tubing (such as is used in the quantitative analysis of inorganic substances) and then washing it out with distilled water. The weighing-tube is removed from the bottom of the beaker with a glass rod or thick platinum wire, held against the walls above the liquid, washed thoroughly inside and out with distilled water, and then raised with the fingers and washed several times again.

**To filter off and weigh the Silver Halide.** — The beaker is now heated on a wire gauze until the silver halide has settled to the bottom and the supernatant liquid is clear. Since the excess of silver nitrate at times packs together with the silver halide to form thick, solid lumps, the precipitate is from time to time crushed with a glass rod, the end of which has been flattened out to a broad head. After cooling, the silver halide is collected on a filter, the weight of the ash of which is known, and washed with hot water until, on testing the filtrate with hydrochloric acid, no turbidity follows; the filter, together with the funnel, is then dried in an air bath at 100–110°, the funnel being covered with a piece of filter-paper. In order to weigh the dry halide, as large an amount as possible is separated from the paper carefully, and transferred to a watch-glass placed on a piece of black, glazed paper. The portions which fall on the paper are swept into the watch-glass with a small feather. The filter is rolled up tightly, wrapped with a platinum wire, and ignited in the usual way over a weighed porcelain crucible; the heating is done only with the outer part of the flame, and not with the inner, reducing part. The folded filter may also be incinerated directly in the crucible, which is first heated over a small flame, and the temperature increased later; the heating is continued until the filter ash appears uniformly light. In order to convert the silver which has been reduced in the incineration back to the silver halide, the fused residue is moistened, by the aid of a glass rod, with a few drops of nitric acid, — if the latter method of incineration has been employed, only after complete cooling of the crucible: it is now evaporated to dryness on the water-bath. It is then treated with a few drops of the corresponding halogen acid, and again evaporated to dryness on the water-bath. The principal mass of the silver halide on the watch-glass is transferred to the crucible with the aid of a feather, and heated directly over a small flame until it just begins to fuse: the crucible is then placed in a desiccator, and allowed to cool. If the analysis is intended to be very exact, the principal mass of the silver halide may be moistened before fusion, with a few drops of nitric acid, and then evaporated on the water-bath with the halogen acid.

Even after taking the usual precautions, it sometimes happens that the silver halide is mixed with fragments of glass, which will, of course, cause the percentage of halogen to be too high. If the substance under examination is silver chloride, and the presence of glass is noticed in the beaker or on filtering, an error may be avoided, by pouring over the completely washed, moist silver chloride on the filter, slightly warmed dilute ammonium hydroxide several times, then washing the filter with water, and precipitating the pure silver chloride in the filtrate by acidifying with hydrochloric acid. If the compound under examination is silver bromide or iodide, and glass fragments have been noticed, the analysis is carried out to the end in the usual way. To determine the amount of glass present, the silver halide in the crucible is treated with very dilute pure sulphuric acid, and a small piece of chemically pure zinc is added. In the course of several hours, the silver halide is reduced to spongy, metallic silver. By careful decantation, the liquid is separated from the silver, water is added and decanted; this is repeated several times. It is then treated with dilute nitric acid, and heated on the water-bath until all the silver is dissolved. After dilution with water, it is filtered through a quantitative filter, the undissolved glass fragments are also well washed on the filter, and the latter incinerated. The weight of the glass is to be subtracted from the weight of the halide obtained. It is obvious that the purity of the fused silver chloride may also be tested in this way.

In conclusion, the atomic weights of the halogens as well as the molecular weights of the corresponding silver compounds are here given:

$$Cl = 35.5, \qquad AgCl = 143.5,$$
$$Br = 80.0, \qquad AgBr = 188.0,$$
$$I = 127.0, \qquad AgI = 235.0.$$

## QUANTITATIVE DETERMINATION OF SULPHUR
### CARIUS' METHOD

This method, like the preceding one, depends upon the complete oxidation of the weighed substance, by heating it with fuming nitric acid in a sealed tube. The sulphuric acid thus formed is weighed as barium sulphate. The charging, sealing, heating, opening, and emptying of the tube are performed in exactly the same way as in the halogen determinations; but in this case it is evident that the use of silver nitrate is superfluous. Before breaking off the conical end, the tube is examined to see that no undecomposed portions of the substance are present; if there should be, the capillary is again sealed and the tube reheated. Before the sulphuric acid is precipitated with barium chloride, the bottom of the beaker must be examined for any fragments of glass which may be present; if there are any, they are filtered off through a small filter.

**Precipitation of the Barium Sulphate.** — The liquid in the bomb, diluted with water up to 400 c.c., is heated almost to boiling on a wire gauze and acidified with hydrochloric acid; a solution of barium chloride heated to boiling in a test-tube is gradually added until a precipitate is no longer formed. This can be easily observed by allowing the precipitate to settle somewhat before adding more of the solution. The liquid is then heated over a small flame until the barium sulphate settles at the bottom of the beaker and the supernatant liquid is perfectly clear: at times from one to two hours' heating may be necessary. After cooling, the liquid is filtered, without disturbing the precipitate at the bottom, through a small filter the weight of the ash of which is known; the precipitate remaining in the beaker is boiled several minutes with 100 c.c. water and filtered through the same filter. The precipitate occasionally at first goes through the paper; in case it does, another beaker is placed under the funnel so that the entire quantity of liquid need not be refiltered. The precipitate is washed with hot water until a portion of the filtrate tested with dilute sulphuric acid shows no turbidity. Before throwing away the filtrate, barium chloride is added in order to be sure that a sufficient quantity was used in the first instance. If a precipitate is formed, the above process is repeated and the second precipitate collected on the filter containing the first.

The method just described has the disadvantage that if a smaller quantity of water be used for diluting the contents of the tube than that given above, the barium sulphate may easily carry along with it some barium nitrate, which is only removed with difficulty on washing with water. Since, in consequence of this, the percentage of sulphur is too high, it is for many reasons preferable to wash the contents of the bomb into a porcelain dish instead of a beaker, and to evaporate the liquid on the water-bath until the acid vapours vanish, before adding the barium chloride; by this operation the nitric acid is removed. After evaporating, the residue is diluted with water, filtered if necessary, from any glass fragments, and the operation just described above repeated. Under these conditions, too much of an excess of barium chloride is to be avoided.

**Ignition and Weighing of the Barium Sulphate.** — In order to prepare the barium sulphate for weighing, it is not necessary to dry it before incineration; if Bunsen's method is followed, it may be incinerated while still moist. With the aid of a small spatula or knife the moist filter is removed from the funnel and folded in the form of a quadrant. Should any barium sulphate adhere to the funnel, it is removed with a small piece of filter-paper, which is incinerated with the main mass. After the filter has been carefully folded toward the centre, it is pressed into the bottom of a weighed platinum crucible, placed on a platinum triangle in such a position that its axis is inclined 20–30° from a vertical position. The cover, also inclined at an angle of 20–30°, in the opposite direction, however, is supported before the crucible, so that the upper half of the opening of the latter is uncovered. The burner under the crucible is placed in such a position that the flame, which must not be too large at first, is directly under the angle formed by the crucible and cover. This will allow the ignition of the filter to take place at so low a temperature that reduction of the barium sulphate need not be feared. It sometimes happens that on heating the filter, the gases formed take fire at the mouth of the crucible, which, however, does no harm. After some time the burner is placed under the bottom of the crucible, the flame increased, and the heating continued until the residue has become white. The crucible is now placed in an upright position, heated a short time with the full flame, and then allowed to cool in a desiccator. It is entirely superfluous to treat the barium sulphate with sulphuric acid and then evaporate it off. If the percentage of sulphur found is too high, this may have been caused, under certain conditions, by the fact that in the precipitation too great an excess of barium chloride has been used, and that the barium sulphate has carried along some of it. This source of error may be rectified by treating the ignited barium sulphate with water until the crucible is half full, then adding a few drops of concentrated hydrochloric acid, and heating on the water-bath for fifteen minutes. The liquid is filtered from the precipitate through a quantitative filter; the greatest portion of the precipitate remaining in the crucible is again treated with water and hydrochloric acid, and the contents of the crucible poured on the filter already used; after washing repeatedly with water, the filter and precipitate are again ignited as before. This process is obviously only employed when the barium sulphate has not been evaporated down with sulphuric acid. For the calculation of the analysis the following data are given:

Atomic weight of sulphur = 32.
Molecular weight of barium sulphate = 233.
Percentage of sulphur in barium sulphate = 13.73.

**Simultaneous Determination of the Halogens and Sulphur.** — If a substance contains both a halogen and sulphur, they may be determined in a single operation by the following method: As in the determination of the halogens the bomb is charged with silver nitrate and nitric acid, and the silver halide filtered off after the heating as above described. The filtrate thus obtained contains, besides the excess of silver nitrate, the sulphuric acid formed by oxidation. This latter cannot be precipitated as before with barium chloride, since the silver as silver chloride would also be thrown down. In its place is used a solution of barium nitrate, the purity of which has been tested by adding silver nitrate to it. The precipitation is made hot as above directed, the solution used being as dilute as possible — the volume of which must be at least 500 c.c. A large excess of barium nitrate is particularly to be avoided. If the barium nitrate solution contains halogen salts as impurities, it is heated, and silver nitrate added so long as a precipitate is formed, the precipitate filtered off, and the solution, which is now free from halogens, is used for the precipitation.

## QUANTITATIVE DETERMINATION OF NITROGEN
### DUMAS' METHOD

In scientific laboratories, the method almost exclusively used for determining nitrogen quantitatively is that of Dumas. The principle involved is that the substance is completely burned by cupric oxide in a tube filled with carbon dioxide, the nitrogen is evolved as such, and its volume measured, while the carbon and hydrogen are completely oxidised to carbon dioxide and water.

Requisites for the analysis:

1. A combustion tube of difficultly fusible glass, 80–85 cm. long; outside diameter, about 15 mm.
2. A glass funnel-tube with wide stem (at least 10 mm. in diameter).
3. 400 grammes of coarse and 100 grammes of fine cupric oxide. The former is kept in a large flask, the latter in a small one, both of which are closed by a cork covered with tinfoil.
4. 500 grammes of magnesite, in pieces the size of a pea. The fine powder, which cannot be used, is sifted out in a wire sieve. The dark grains which have become discoloured by impurities are thrown out.
5. A small flask of pure methyl alcohol (50 grammes) for reducing the copper spiral.
6. A copper spiral, 10–12 cm. long. This is made by winding an oblong piece of copper wire gauze spirally around a thin glass rod. It is made of such a width that when in position it will touch the walls of the combustion tube; a space between the walls and spiral is disadvantageous. Also a short copper spiral from 1–2 cm. long.
7. A solution of 150 grammes of potassium hydroxide in 150 grammes of water. It is prepared in a porcelain dish, and not in a glass beaker or flask, since these are frequently broken by the heat generated by the solution. After cooling, it is preserved in a well-closed bottle.

8. A nickel crucible 6 cm. high; diameter of top 7 cm., for the ignition of the coarse cupric oxide.
9. A moderately large porcelain crucible for the ignition of the fine cupric oxide.
10. A small mortar with a glazed bottom.

Besides these, a weighing-tube, a one-hole rubber stopper for closing one end of the combustion tube, a sieve to sift the copper oxide, a small feather, thermometer, absorption apparatus, and a eudiometer.

**Preparations for the Analysis.** — The analysis is conveniently begun by heating the entire quantity of coarse copper oxide in the nickel crucible over as large a flame as possible (Fletcher burner), and the fine copper oxide in the porcelain crucible over a Bunsen flame for a long time, the crucibles being supported on wire triangles. The covers are placed on the crucibles loosely, and the copper oxide occasionally stirred with a thick wire. While the copper is being heated, one end of the combustion tube is sealed to a solid head, the narrower end being selected for this purpose, if the tube is not perfectly cylindrical. The sealing is done as follows: The end of the tube is first warmed in a luminous flame, with constant turning; it is then heated to softening, in the blast-flame, a glass rod fused on it, and the heated portion suddenly drawn out to a narrow tube. The glass rod is now fused off, and the conical part of the tube just produced is heated and drawn out. The cone is then heated in the hottest flame until it falls together; it is finally allowed to cool gradually over a small luminous flame. When this operation is finished, the open end of the tube is warmed in a luminous flame, and, with constant turning, the sharp edges are rounded by the blast-flame: it is then allowed to cool in the luminous flame again. After complete cooling, the soot is removed, the tube rinsed out several times with water, the water allowed to drain off as completely as possible, and the tube finally dried in one of the two following ways: The tube, with constant turning, is repeatedly passed through the large luminous flame of a blast-lamp, while a current of air is blown from a blast into the bottom of it by means of a narrower, longer (10 cm.) tube inserted in the larger tube; this operation is continued until all moisture is removed. Or the combustion tube is clamped in a horizontal position, a narrower tube extending to the sealed end, attached to suction, is inserted, and the combustion tube equally heated with a Bunsen burner throughout its entire length; the water vapour is drawn off by the suction. To reduce the long copper spiral which is to be used for the reduction of oxides of nitrogen which may be formed, the method of procedure is as follows: Into a test-tube large enough to admit the spiral, 1 c.c. of methyl alcohol is placed; the spiral, held by crucible tongs, is then heated to glowing in a large, somewhat roaring blast-flame, and dropped as quickly as possible into the test-tube; since this becomes strongly heated at its upper end, it is clamped in a test-tube holder, or wrapped in a cloth or strips of paper. The dark spiral soon assumes a bright metallic lustre, while vapours, having a sharp, pungent odour (oxidation products of methyl alcohol like formic aldehyde and formic acid), which frequently become ignited, are formed; after a few minutes, the tube may be loosely corked, and the spiral allowed to cool. When this operation is ended, the copper oxide will have been sufficiently heated, and the flames may be removed. During the cooling, the substance to be analysed is weighed. A convenient method is this: The weight of the weighing-flask is determined with exactness to centigrammes, this weight is entered in the note-book at a convenient place for future use. The substance to be analysed is now placed in the weighing-tube, and the weight of the tube, plus substance, is determined exactly to the tenth of a milligramme. In the mean time, the copper oxide has cooled sufficiently to be transferred to the appropriate flask. The combustion tube is next filled.

**Filling the Tube.** — At the edge of the working table is placed a stand; fastened firmly near the bottom of this is a clamp projecting over the edge of the table supporting the combustion tube in a vertical position, the mouth being at about the level of the table. The tube is now directly filled with the magnesite until the layer has a height of 10-12 cm. (Fig. 47). A small roll of copper gauze, 1-2 cm. long held with pincers or tongs is heated for a short time in a Bunsen flame (it need not be reduced) and dropped on the magnesite. The funnel-tube is then placed in the tube, and from the flask coarse copper oxide is poured in until the layer measures 8 cm., and upon this is poured a layer of 2 cm. of the fine oxide. To the operation following — the mixing of the substance with copper oxide and the transference of the mixture to the tube — especial care must be given. In the bottom of a small mortar, standing on black, glazed paper a ½ cm. layer of the fine, perfectly cooled copper oxide is placed; to this is added from the weighing-tube the substance to be analysed, of which 0.15–0.20 gramme is taken. Since the weight of the empty tube is known as well as that of the substance contained therein, one can easily decide, by measuring with the eye, how much of the substance to take. Fine copper oxide is now added until the substance is completely covered, and the two are carefully mixed by stirring with the pestle, without pressure; during the mixing care must be taken

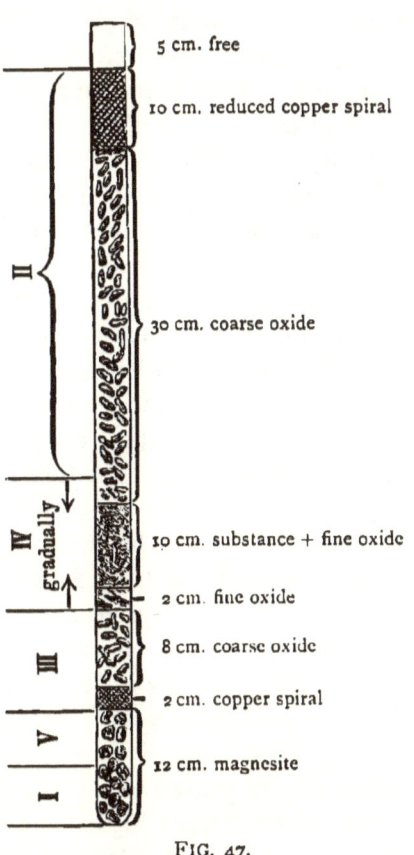

FIG. 47.

not to stir so rapidly as to cause dust-like particles of the mixture to leave the mortar. With the aid of a clean, clipped feather, such as is used in quantitative operations, or a small brush, the contents of the mortar are transferred through the funnel-tube into the combustion tube. The operation must be done cautiously to prevent the light, dusty particles from being blown away. The mortar, as well as the pestle, is now rinsed with a fresh portion of the fine copper oxide, and this is likewise transferred to the tube with the aid of the feather. The layer of substance plus copper oxide should be about 10 cm. long. Then follows a layer of 30 cm. of coarse copper oxide, and finally the reduced copper spiral.

The length of the tube, as well as that of the single layers, is regulated in accordance with the size of the combustion furnace; the figures given above refer to a furnace possessing a flame surface of 75 cm. Generally the tubes are 5 cm. longer than the furnace; the tube contents are of the same length as the flame surface.

**Heating the Tube.** — After the tube is filled it is held in a horizontal position and tapped gently on the table in order that a canal may be formed in the upper portion of the fine copper oxide; it is then connected with a rubber stopper to the absorption apparatus which has been charged with caustic potash solution, and placed in the combustion furnace, the rear end of which (that under the magnesite) has been raised on a block (Fig. 48). The following points are to be observed: In the lower part of the absorption apparatus there must be a sufficient amount of mercury to extend almost to the side-tube; if this is not the case, more mercury is added: the end of the glass tube passing through the rubber stopper must be flush with the end of the stopper. In order to protect the latter from the heat, there is placed over the portion of the tube projecting beyond the furnace, an asbestos plate having a circular opening in the centre. After opening the pinch-cock of the absorption apparatus, the burners under the last half of the magnesite are lighted; the flames, being small at first, are increased in size, as soon as the tube becomes warmed, but not sufficiently to cause them to meet above the tube. In order to raise the temperature higher when it becomes necessary, the tube is covered from both sides with the tiles. After about ten minutes a rapid current of carbon dioxide is evolved, the magnesite being decomposed by heat as represented in the following equation:

$$MgCO_3 = MgO + CO_2.$$

# ORGANIC ANALYTICAL METHODS

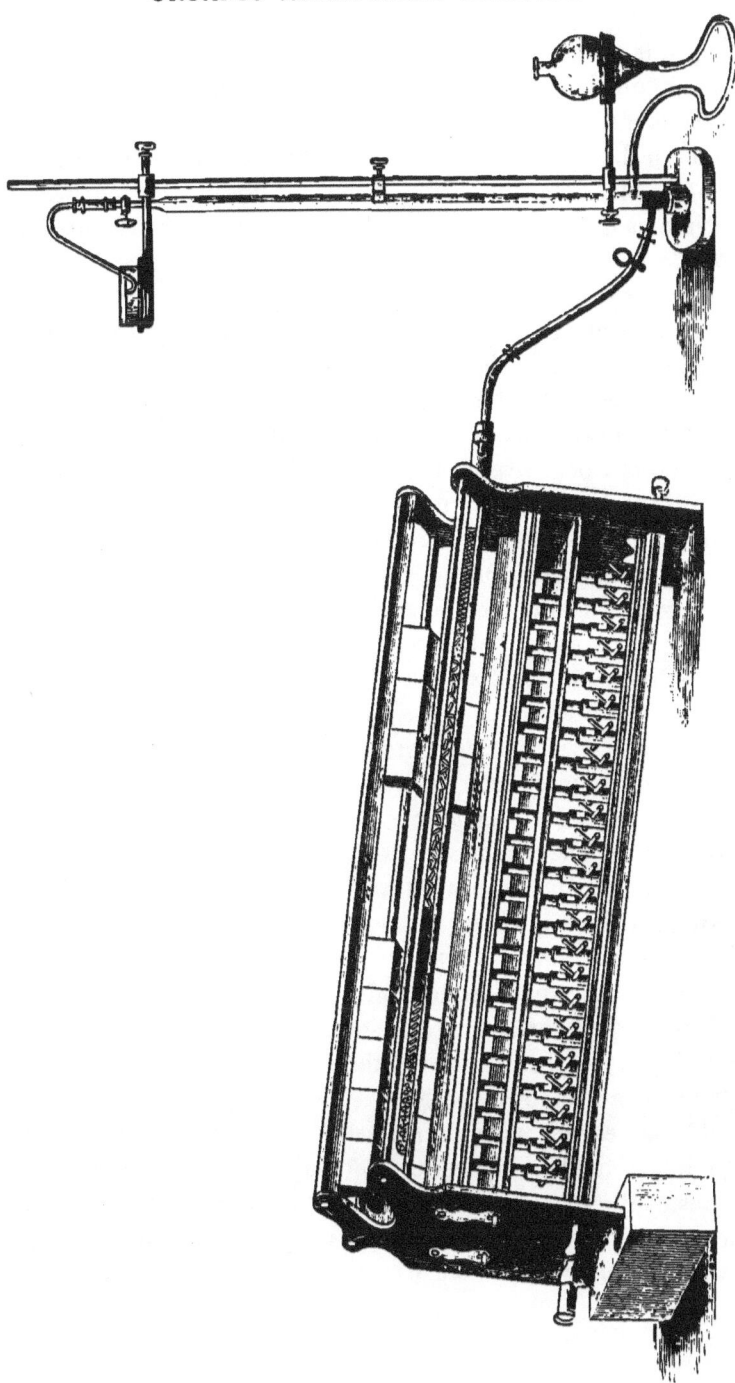

Fig. 48.

During this operation the glass stop-cock of the absorption apparatus is opened, and the pear-shaped vessel placed as low as possible, so that it contains the principal portion of the caustic potash. After a rapid current of carbon dioxide has been evolved for about fifteen minutes, the pear-shaped vessel is raised high enough to cause the caustic potash to ascend somewhat above the tubulure in the glass cock, the latter is closed, and the pear-shaped vessel again lowered as far as possible. When the air in the tube has been completely replaced by carbon dioxide, only a minimum quantity of light foam should collect over the potash in the course of two minutes. If this is not the case, and a large air volume collects, the glass cock is opened, upon which the potash flows in to the lowered pear vessel, and carbon dioxide is caused to pass through the tube for five minutes longer. The pear vessel is then raised as high as at first, the glass cock closed, and the former lowered. An observation will show whether the air has been displaced, which should be the case under normal conditions. If now after two minutes only a trace of foam has collected, the end of the delivery tube is dipped under the water in a dish as shown in Fig. 48, the pear raised to the highest point of the delivery tube, and the glass cock opened in order that the potash may drive out the air in the delivery tube: when this has been done, the cock is closed again and the pear lowered to the bottom. All the flames but one under the magnesite are now extinguished, and those under the long copper spiral as well as those under four-fifths of the adjacent layer of coarse copper oxide are lighted at the same time; the flames, small at first, are increased in size, after the tube has become somewhat heated, until the copper oxide is heated to dull redness. Concerning the steps taken in heating the tube, refer to Fig. 47 — the numbers on the left indicate the portions of the tube to be heated successively. At this point care is taken that the flames are not so large as to meet above the tube. As before in heating the magnesite, after the first warming the heated portions of the tube are covered on both sides with the tiles. As soon as the forward layer of coarse copper oxide becomes dark red, the burners under the rear layer of coarse oxide adjacent to the magnesite are lighted — small flames at first, which are increased after a time, the tube being covered simultaneously with the tiles. Care must be taken that the flames nearest the layer of substance plus fine copper oxide are not too large, in order that the substance may not yet be burned. Upon the operation which now follows — the gradual heating of the fine oxide containing the substance — virtually depends the success of the analysis. For the proper manipulation of this operation especial care must be taken. It is a rule that the heating had better be somewhat too slow than too rapid. A small flame is now lighted at the point adjacent to the short layer of coarse oxide : an observation of the absorption apparatus will show whether after some time any unabsorbed gas collects. If this is the case, no other burners are lighted until the evolution of gas ceases. When the gas no longer collects, another burner on the opposite side of the fine oxide is lighted. In this way the burners are gradually lighted from both sides, toward the middle of the fine oxide, and after, in each case, the cessation of the evolution of the gas, the flames are gradually made larger until finally the tube covered with tiles is heated with full flames ; thus the substance is regularly and quietly burned. The combustion must be so conducted that the bubbles of gas ascend in the absorption apparatus with a slow regularity. If the single bubbles cannot be counted, or if they are so large as to occupy almost the entire cross-section of the absorption tube, the heating is too strong, and the last burners lighted must be extinguished or lowered, the tiles being also laid back at the same time if necessary, until the generation of the gas is lessened. When this is ended, small flames are again lighted under the entire layer of magnesite, and increased in size after some time. As soon as the evolution of carbon dioxide has become active, the flames under the rear half of the magnesite lighted at the beginning of the analysis are extinguished. After a rapid current of carbon dioxide has passed through the tube for ten minutes, all the nitrogen is carried over to the absorption apparatus. This is shown by the complete absorption of the gas bubbles by the potash, as at the beginning of the analysis, except for a minimum foam-like residue. The absorption apparatus is then closed by the pinch-cock and the rubber stopper bearing the connecting tube is withdrawn from the combustion tube. The gas is not immediately transferred to the eudiometer, but the pear is raised until the surfaces of the liquid in the pear and that in the tube are at the same level: the apparatus is then allowed to stand for at least half an hour. The flames under the combustion tube are not turned out simultaneously, but first one is extinguished, and then after a short time another, and so on. During the cooling of the tube the weighing-flask is weighed again.

**Transferring the Nitrogen.** — After the nitrogen has stood in contact for at least half an hour with the caustic potash, in order that the last portions of carbon dioxide may be absorbed, the end of the delivery tube is dipped under the surface of the water contained in a wide-mouth cylinder, as represented in Fig. 49, care being taken that in the lower bent portion of the delivery tube no air bubbles are present; if there are, they must be removed with a capillary pipette. The eudiometer is now filled with water, the end closed with the thumb, inverted and dipped below the surface of the water, the thumb removed and the tube clamped to the cylinder, at an oblique angle, so that the end of the delivery tube may be passed under it. The pear supported by the clamped ring is raised as high as possible above the delivery tube, and the glass cock gradually opened. The nitrogen is thus transferred to the eudiometer, the cock being left open until the delivery tube is completely filled with the caustic potash. The absorption apparatus is then removed, the eudiometer wholly immersed in the water. To obtain the temperature, a thermometer, held in the clamp which supported the

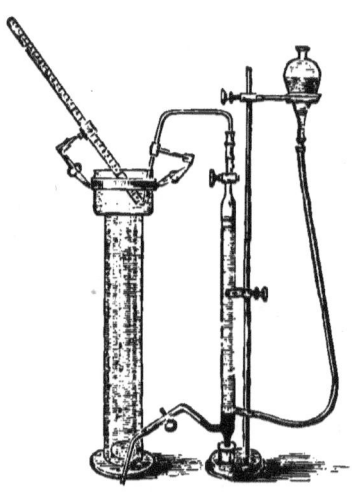

FIG. 49.

delivery tube, is immersed in the water as far as possible. After about ten minutes, the nitrogen has come to the same temperature as the water, the eudiometer is then seized with a clamp especially adapted to this purpose, or crucible tongs, — never with the hands, — and is raised so far out of the water that the level of water inside and outside the tube is the same. The volume of gas thus read off, is under the same pressure as that indicated by a barometer.

**Calculations of the Analysis.** — If $s$ is the amount of substance in grammes, $v$ the temperature at $t°$, and $b$ the weight of the barometer in millimetres, when the volume of nitrogen was read, $w$ the tension of the water vapour in millimetres at $t°$, then the percentage of nitrogen is $p$:

$$p = \frac{v \cdot (b - w) \cdot 0.01256}{760 \cdot (1 + 0.00367 \cdot t) \cdot s}.$$

The calculation of the analysis is rendered easier by referring to the table in which the weight of one cubic centimetre of moist nitrogen is given in milligrammes at different temperatures and pressures. If this, under the observed conditions, is $g$, then the percentage of nitrogen is:

$$p = \frac{100 \times v \times g}{s}.$$

In this formula, $s$ is the weight of the substance in milligrammes.

**Length of Time for an Analysis.** — The following abstract will give an approximate idea of the length of time that the single operations of a well-conducted combustion ought to occupy. From the beginning of the heating of the magnesite to the appearance of a rapid current of carbon dioxide requires about 10 minutes, the first test as to whether air is still present in the tube follows after a further 15 minutes; length of time for various tests, 5 minutes. From the warming of the forward layer of the copper oxide with the spiral to the heating of the rear layer of oxide to a dark red heat, 15 minutes. The actual combustion of the substance requires 30 minutes. The displacement of the last portions of nitrogen by heating the magnesite requires 10 minutes. Total, 1 hour and 25 minutes.

These time figures are, of course, only to be considered as approximate, since they depend upon the efficiency of the furnace, upon the nature of the substance burned, upon the skill of the experimenter, and upon other factors.

WEIGHT OF ONE CUBIC CENTIMETRE OF MOIST NITROGEN IN MILLIGRAMMES.

| t | b. 726 | 728 | 730 | 732 | 734 | 736 | 738 | 740 |
|---|---|---|---|---|---|---|---|---|
| 5° | 1.168 | 1.171 | 1.175 | 1.178 | 1.181 | 1.184 | 1.188 | 1.191 |
| 6° | 1.163 | 1.167 | 1.170 | 1.173 | 1.176 | 1.179 | 1.183 | 1.186 |
| 7° | 1.158 | 1.162 | 1.165 | 1.168 | 1.171 | 1.174 | 1.178 | 1.181 |
| 8° | 1.153 | 1.157 | 1.160 | 1.163 | 1.166 | 1.169 | 1.173 | 1.176 |
| 9° | 1.148 | 1.152 | 1.155 | 1.158 | 1.161 | 1.164 | 1.168 | 1.171 |
| 10° | 1.143 | 1.147 | 1.150 | 1.153 | 1.156 | 1.159 | 1.163 | 1.166 |
| 11° | 1.138 | 1.142 | 1.145 | 1.148 | 1.151 | 1.154 | 1.158 | 1.161 |
| 12° | 1.133 | 1.136 | 1.140 | 1.143 | 1.146 | 1.149 | 1.152 | 1.155 |
| 13° | 1.128 | 1.131 | 1.135 | 1.138 | 1.141 | 1.144 | 1.147 | 1.150 |
| 14° | 1.123 | 1.126 | 1.129 | 1.133 | 1.136 | 1.139 | 1.142 | 1.145 |
| 15° | 1.118 | 1.121 | 1.124 | 1.127 | 1.131 | 1.134 | 1.137 | 1.140 |
| 16° | 1.113 | 1.116 | 1.119 | 1.122 | 1.125 | 1.129 | 1.132 | 1.135 |
| 17° | 1.108 | 1.111 | 1.114 | 1.117 | 1.120 | 1.123 | 1.126 | 1.130 |
| 18° | 1.102 | 1.105 | 1.109 | 1.112 | 1.115 | 1.118 | 1.121 | 1.124 |
| 19° | 1.097 | 1.100 | 1.103 | 1.106 | 1.110 | 1.113 | 1.116 | 1.119 |
| 20° | 1.092 | 1.095 | 1.098 | 1.101 | 1.104 | 1.107 | 1.110 | 1.113 |
| 21° | 1.086 | 1.089 | 1.092 | 1.096 | 1.099 | 1.102 | 1.105 | 1.108 |
| 22° | 1.081 | 1.084 | 1.087 | 1.090 | 1.093 | 1.096 | 1.099 | 1.102 |
| 23° | 1.075 | 1.078 | 1.081 | 1.084 | 1.088 | 1.091 | 1.094 | 1.097 |
| 24° | 1.070 | 1.073 | 1.076 | 1.079 | 1.032 | 1.085 | 1.088 | 1.091 |
| 25° | 1.064 | 1.067 | 1.070 | 1.073 | 1.076 | 1.079 | 1.082 | 1.085 |
| 26° | 1.058 | 1.061 | 1.064 | 1.067 | 1.070 | 1.073 | 1.076 | 1.079 |
| 27° | 1.053 | 1.056 | 1.059 | 1.062 | 1.065 | 1.068 | 1.071 | 1.074 |
| 28° | 1.047 | 1.050 | 1.053 | 1.056 | 1.059 | 1.062 | 1.065 | 1.068 |
| 29° | 1.041 | 1.044 | 1.047 | 1.050 | 1.053 | 1.056 | 1.059 | 1.062 |
| 30° | 1.035 | 1.038 | 1.041 | 1.044 | 1.047 | 1.050 | 1.053 | 1.056 |

## ORGANIC ANALYTICAL METHODS 91

| t | 742 | 744 | 746 | 748 | 750 | b. 752 | 754 | 756 | 758 | 760 | 762 | 764 | 766 | 768 | 770 |
|---|---|---|---|---|---|---|---|---|---|---|---|---|---|---|---|
| 5° | 1.194 | 1.197 | 1.201 | 1.204 | 1.207 | 1.210 | 1.214 | 1.217 | 1.220 | 1.223 | 1.227 | 1.230 | 1.233 | 1.236 | 1.240 |
| 6° | 1.189 | 1.192 | 1.196 | 1.199 | 1.202 | 1.205 | 1.209 | 1.212 | 1.215 | 1.218 | 1.222 | 1.225 | 1.228 | 1.231 | 1.234 |
| 7° | 1.184 | 1.187 | 1.191 | 1.194 | 1.197 | 1.200 | 1.203 | 1.207 | 1.210 | 1.213 | 1.216 | 1.220 | 1.223 | 1.226 | 1.229 |
| 8° | 1.179 | 1.182 | 1.186 | 1.189 | 1.192 | 1.195 | 1.198 | 1.202 | 1.205 | 1.208 | 1.211 | 1.214 | 1.218 | 1.221 | 1.224 |
| 9° | 1.174 | 1.177 | 1.180 | 1.184 | 1.187 | 1.190 | 1.193 | 1.196 | 1.200 | 1.203 | 1.206 | 1.209 | 1.212 | 1.216 | 1.219 |
| 10° | 1.169 | 1.172 | 1.175 | 1.178 | 1.182 | 1.185 | 1.188 | 1.191 | 1.194 | 1.198 | 1.201 | 1.204 | 1.207 | 1.210 | 1.214 |
| 11° | 1.164 | 1.167 | 1.170 | 1.173 | 1.177 | 1.180 | 1.183 | 1.186 | 1.189 | 1.192 | 1.196 | 1.199 | 1.202 | 1.205 | 1.208 |
| 12° | 1.159 | 1.162 | 1.165 | 1.168 | 1.171 | 1.174 | 1.178 | 1.181 | 1.184 | 1.187 | 1.190 | 1.193 | 1.197 | 1.200 | 1.203 |
| 13° | 1.153 | 1.157 | 1.160 | 1.163 | 1.166 | 1.169 | 1.172 | 1.176 | 1.179 | 1.182 | 1.185 | 1.188 | 1.191 | 1.195 | 1.198 |
| 14° | 1.148 | 1.152 | 1.155 | 1.158 | 1.161 | 1.164 | 1.167 | 1.170 | 1.174 | 1.177 | 1.180 | 1.183 | 1.186 | 1.189 | 1.192 |
| 15° | 1.143 | 1.146 | 1.149 | 1.153 | 1.156 | 1.159 | 1.162 | 1.165 | 1.168 | 1.171 | 1.174 | 1.178 | 1.181 | 1.184 | 1.187 |
| 16° | 1.138 | 1.141 | 1.144 | 1.147 | 1.150 | 1.154 | 1.157 | 1.160 | 1.163 | 1.166 | 1.169 | 1.172 | 1.175 | 1.178 | 1.182 |
| 17° | 1.133 | 1.136 | 1.139 | 1.142 | 1.145 | 1.148 | 1.151 | 1.154 | 1.158 | 1.161 | 1.164 | 1.167 | 1.170 | 1.173 | 1.176 |
| 18° | 1.127 | 1.130 | 1.133 | 1.137 | 1.140 | 1.143 | 1.146 | 1.149 | 1.152 | 1.155 | 1.158 | 1.161 | 1.164 | 1.168 | 1.171 |
| 19° | 1.122 | 1.125 | 1.128 | 1.131 | 1.134 | 1.137 | 1.140 | 1.144 | 1.147 | 1.150 | 1.153 | 1.156 | 1.159 | 1.162 | 1.165 |
| 20° | 1.116 | 1.120 | 1.123 | 1.126 | 1.129 | 1.132 | 1.135 | 1.138 | 1.141 | 1.144 | 1.147 | 1.150 | 1.153 | 1.156 | 1.160 |
| 21° | 1.111 | 1.114 | 1.117 | 1.120 | 1.123 | 1.126 | 1.129 | 1.132 | 1.135 | 1.139 | 1.142 | 1.145 | 1.148 | 1.151 | 1.154 |
| 22° | 1.105 | 1.108 | 1.111 | 1.115 | 1.118 | 1.121 | 1.124 | 1.127 | 1.130 | 1.133 | 1.136 | 1.139 | 1.142 | 1.145 | 1.148 |
| 23° | 1.100 | 1.103 | 1.106 | 1.109 | 1.112 | 1.115 | 1.118 | 1.121 | 1.124 | 1.127 | 1.130 | 1.133 | 1.136 | 1.139 | 1.142 |
| 24° | 1.094 | 1.097 | 1.100 | 1.103 | 1.106 | 1.109 | 1.112 | 1.115 | 1.118 | 1.121 | 1.124 | 1.128 | 1.131 | 1.134 | 1.137 |
| 25° | 1.088 | 1.091 | 1.094 | 1.097 | 1.100 | 1.103 | 1.106 | 1.109 | 1.112 | 1.116 | 1.119 | 1.122 | 1.125 | 1.128 | 1.131 |
| 26° | 1.083 | 1.086 | 1.089 | 1.092 | 1.095 | 1.098 | 1.101 | 1.104 | 1.107 | 1.110 | 1.113 | 1.116 | 1.119 | 1.122 | 1.125 |
| 27° | 1.077 | 1.080 | 1.083 | 1.086 | 1.089 | 1.092 | 1.095 | 1.098 | 1.101 | 1.104 | 1.107 | 1.110 | 1.113 | 1.116 | 1.119 |
| 28° | 1.071 | 1.074 | 1.077 | 1.080 | 1.083 | 1.086 | 1.089 | 1.092 | 1.095 | 1.098 | 1.101 | 1.104 | 1.107 | 1.110 | 1.113 |
| 29° | 1.065 | 1.068 | 1.071 | 1.074 | 1.077 | 1.080 | 1.083 | 1.086 | 1.089 | 1.092 | 1.095 | 1.097 | 1.100 | 1.103 | 1.106 |
| 30° | 1.059 | 1.062 | 1.065 | 1.068 | 1.070 | 1.073 | 1.076 | 1.079 | 1.082 | 1.085 | 1.088 | 1.091 | 1.094 | 1.097 | 1.100 |

**Subsequent Operations.** — After the tube has cooled and the copper spiral taken out, all the copper oxide is sifted to separate the coarse from the fine, and may be used again for further analyses as often as desired, provided that it is reheated each time in the nickel crucible to oxidise it. The tube may also be used again if it has not been distorted by high heating. The magnesite is useless for further analyses.

The caustic potash in the absorption apparatus, which can be used a second time, is poured into a bottle, which is then well closed. The absorption apparatus, including the rubber tubing, is washed out repeatedly with water, so that the latter may not be corroded by the caustic potash.

**General Remarks.** — The above-described method of Dumas for nitrogen is used in variously modified forms, but the principle is the same in all. It is preferred in many places to generate the carbon dioxide from acid sodium carbonate or manganese carbonate. A combustion tube open at both ends may be used. In this case one of the ends is narrowed and attached to a Kipp carbon dioxide generator. Further, the mixing of the substance with the fine copper oxide may be done in the tube. Instead of the absorption apparatus of Schiff, described above, a graduated tube from which the volume of the gas may be read directly, thus obviating the necessity of transferring it to a eudiometer, may be used. This modification carries with it, however, the disadvantage that the tension of caustic potash is not exactly known, and therefore a somewhat arbitrary correction must be applied. But as mentioned these modifications do not differ essentially.

## QUANTITATIVE DETERMINATION OF CARBON AND HYDROGEN
### LIEBIG'S METHOD

The essential part of the method consists in completely burning with copper oxide a weighed amount of the substance, and then weighing the combustion products, carbon dioxide and water.

The requisites for analysis are:

1. A hard glass tube open at both ends; outside diameter 12–15 mm. It should be about 10 cm. longer than the furnace.

2. Four hundred grammes of coarse and 50 grammes of fine copper oxide, preserved in bottles closed with tin-foil-covered corks as in the nitrogen determination. But the copper oxide used for the latter purpose and that for the carbon and hydrogen determinations are always kept in separate bottles.
3. A U-shaped- and a straight calcium chloride tube.
4. A caustic potash apparatus. The Geissler form is the most convenient.
5. A drying apparatus for air or oxygen.
6. Two one-hole rubber stoppers fitting the ends cf the combustion tube.
7. A glass tube provided with a cock.
8. Two copper spirals of 10 and 12–15 cm. length, respectively; two short spirals 1–2 cm. long.
9. A piece of good rubber tubing 20 cm. long; six pieces rubber tubing 2 cm. long (thick walled and seamless).
10. A porcelain and a copper boat.
11. A screw pinch-cock.
12. Two asbestos plates for the protection of the rubber stoppers.

**Preparations for the Analysis.** — The sharp edges of the combustion tube are rounded by careful heating in a blast-flame. After cooling the tube is rinsed out with water several times, this is allowed to drain off, and the tube dried by one of the methods given on page 66.

The coarse copper oxide is not previously heated in the nickel crucible as in the determination of nitrogen, but this is done later in the tube itself. If the nature of the substance to be analysed is such that it is necessary to mix it with fine copper oxide, the latter is ignited for a quarter hour in the porcelain crucible and allowed to cool in a desiccator.

The U-tube for the absorption of the water (Fig. 52) is filled with granulated, not fused, calcium chloride, which must be freed from any powder by sifting. In order to prevent the calcium chloride from falling out, both ends of the tube are provided with loose plugs of cotton. The open leg is closed by a rubber stopper or a good cork bearing a glass tube bent at a right angle. The cork stopper is covered with a thin layer of sealing-wax. Calcium chloride tubes, in which the open leg is longer than the other, are very convenient. After the tube is filled the open end may be sealed in a blast-flame. In this case the plug in this end is not cotton, but asbestos or glass-wool. In order that the tube may be suspended from the arm of the balance in weighing, a platinum wire with a loop in the centre is attached to both legs. Calcium chloride often contains basic chlorides, which not only absorb water, but also carbon dioxide, thus causing an error in the results of the analysis; before the filled tube is used a stream of dry carbon dioxide is passed through it for about two hours, dried air is then drawn through for half an hour to displace the carbon dioxide. The two side-tubes of the calcium chloride tube are closed by pieces of rubber tubing 2 cm. long in which is inserted a glass rod rounded at both ends, $1\frac{1}{2}$ cm. long. The tube may be used repeatedly until the calcium chloride begins to liquefy. The straight calcium chloride tube is filled in like manner, but it is unnecessary to pass carbon dioxide through this before using.

The three bulbs of the potash apparatus similar to the one represented in Fig. 50 are three-fourths filled with a solution of caustic potash (2 parts potassium hydroxide, 3 parts water) as follows: The horizontal tube which is to be charged with solid caustic potash is removed, and to the free end of the bulb tube rubber tubing is attached. The inlet tube represented in Fig. 50 at the left is now dipped into the caustic potash solution, contained in a shallow dish, and this is sucked up with the rubber tubing until the three bulbs are three-fourths filled. Care must be taken not to suck too strongly, otherwise some of the caustic potash solution may be drawn into the mouth. This may be prevented by inserting an empty wash-bottle between the potash apparatus and the mouth, or the suction-pump may be used, in which case the water-cock must be opened to a very slight extent. After filling the bulbs that part of the tube immersed in the potash solution is cleaned with pieces of rolled-up filter-paper. The horizontal potash tube, removed before filling the bulbs, is now filled with coarse-grained soda-lime and solid caustic potash in pea-size pieces as follows: In the bulb is placed a plug of glass-wool or asbestos, then follows a layer of the soda-lime, a layer of caustic potash, and finally another plug of glass-wool or asbestos. When this is done, it is closed in the same way as the calcium chloride tube. In handling the Geissler bulbs it is always to be remembered that they are very fragile, and in all cases the lever-arm formed in lifting them should be as short as possible. When the apparatus is to be closed by rubber tubing, *e.g.*, it is not

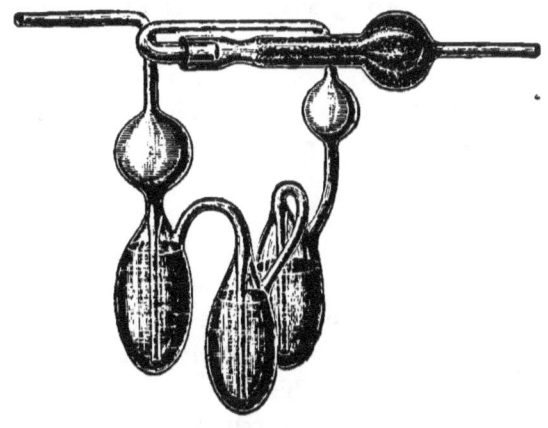

FIG. 50.

grasped by the bulbs, but immediately behind the place over which the tubing is to be drawn or pushed. When the potash apparatus has been used twice, it must be refilled. The longer of the two so-called copper oxide spirals need not be reduced before the combustion; on the contrary, it is oxidised in the combustion tube, as will be pointed out below. In order to be able to remove it from the tube conveniently a loop of copper wire is fastened in the meshes of the gauze near the end, or a not too thin copper wire is passed through the centre of the spiral and bent at one end to a right angle and at the other in the form of a loop. The shorter spiral, which, as in the nitrogen determination, serves to reduce any oxides of nitrogen, is next reduced according to the directions given on page 82. To remove any adhering organic substances like methyl alcohol or its oxidation products the spiral is placed, after cooling, in a glass tube 20 cm. long, one end of which is narrowed; carbon dioxide is passed through it, and as soon as the air has been displaced, it is heated for a few minutes with a Bunsen flame and then allowed to cool in a current of carbon dioxide. To remove the mechanically adhering gas the spiral is placed in a vacuum desiccator. If this is not at hand, an ordinary desiccator containing a small dish of solid caustic potash or unslaked lime is used.

For drying the oxygen or air an apparatus consisting of two wash cylinders and two U-shaped glass tubes mounted on a wooden stand is employed. The gas passes first through a wash cylinder containing a solution of potassium hydroxide (1 : 1), then a tube filled with soda-lime, then one filled with granulated calcium chloride, and finally a wash cylinder containing sulphuric acid.

The legs of the glass tube containing the stop-cock are fused off and slightly narrowed at the ends, so that on either side of the cock the length is 5 cm.

**Filling the Tube.**—The simplest case of combustion with which one can deal is that involving the analysis of a substance containing no nitrogen. In a case of this kind, assuming that the furnace has a flame surface of 75 cm., the tube is filled in the following manner: A short copper gauze roll, of sufficient diameter to fit the tube tightly, and somewhat elastic, is pushed into the tube 5 cm., and then the opposite side of the tube is partially filled with a layer of coarse copper oxide 45 cm. held in position by another small elastic copper spiral at its upper end. Into the tube lying in a horizontal position the copper oxide spiral is pushed so far that its loop is 5 cm. from the mouth of the tube (Fig. 51).

**Igniting the Copper Oxide.**—The charged tube is placed in the furnace, the end nearest the copper oxide spiral is closed by a rubber stopper bearing the glass stop-cock tube, and the latter is connected with the drying apparatus by means of rubber tubing provided with a screw pinch-cock. The other end of the tube is

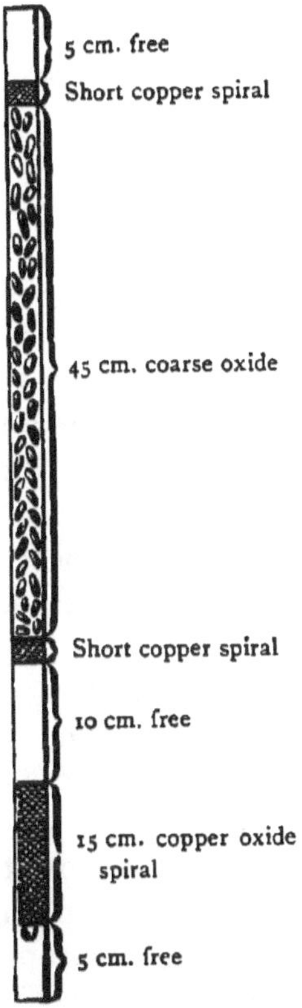

5 cm. free

Short copper spiral

45 cm. coarse oxide

Short copper spiral

10 cm. free

15 cm. copper oxide spiral

5 cm. free

FIG. 51.

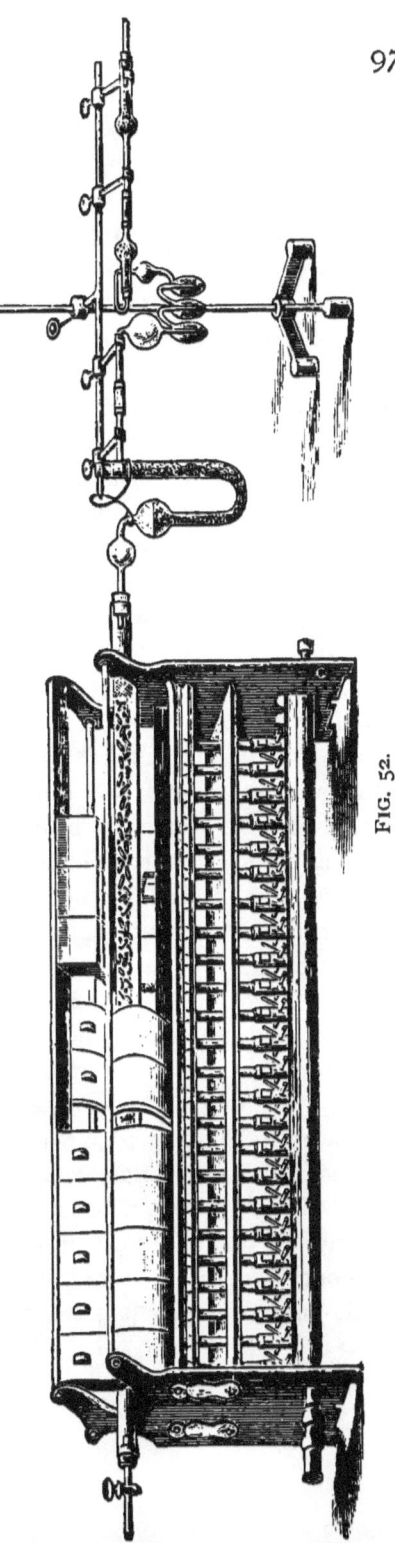

FIG. 52.

allowed to remain open at first; while a current of oxygen is passed through the tube, slow enough to enable one to count the bubbles (the glass stop-cock is opened wide and the current regulated with the pinch-cock), the entire length of the tube is heated, at first with flames as small as possible; these are gradually increased until finally, the tiles being in position, the copper oxide begins to appear dark red. The water deposited at the beginning of the heating, in the forward cool end of the tube, is now removed with filter-paper wrapped around a glass rod. When no more water collects, the front end of the tube is closed by a rubber stopper bearing the straight calcium chloride tube. After about 20 or 30 minutes' heating the burners under the copper oxide spiral, the adjacent empty space, and those under about 5 cm. of the copper oxide layer lying next, are extinguished, and at the same time the current of oxygen is cut off.

**Weighing the Absorption Apparatus and the Substance.** — While the rear part of the tube is cooling, the calcium chloride tube, the potash bulbs, and the substance are weighed. Before the absorption apparatus is weighed, it is wiped off with a clean cloth, free from lint, and the rubber tubing and glass rods removed; after the weighing, these are replaced. The substance, if solid, is weighed in a porcelain boat which has previously been heated strongly, and cooled in a desiccator. The boat is first weighed empty, 0.15 to 0.20 grammes of the substance placed in it, and weighed again; it is then placed on a tin-foil-covered cork, in which a suitable groove has been cut, and transferred to a desiccator.

**The Combustion.** — When the rear end of the tube is cold, the copper oxide spiral is withdrawn with a hooked glass rod or wire, the porcelain boat is inserted as far as the coarse copper oxide, care being taken not to upset the boat, and finally the spiral is replaced. The stop-cock tube, with the cock closed, is then put in position. The straight calcium chloride tube is replaced by the weighed U-tube, with its empty bulb, which will condense the greater portion of the water, nearest the furnace. To the U-tube is connected, by a rubber joint, the potash apparatus, and the soda-lime tube of the latter with the straight calcium chloride tube in the same way (Fig. 52). The connecting of the different parts of the apparatus may be facilitated by blowing air from the lungs through each rubber joint before pushing it on the glass tubes. Especial care is taken to have a good joint between the U-calcium chloride tube and the potash bulbs, since at this point very commonly lies the source of error in analyses not concordant. A thick-walled, seamless rubber tubing is employed; it is drawn over the two ends of the glass tubes until they touch. In order to provide against any possible leak, two ligatures of thin copper wire or "wax ends" are bound around the joints. A test as to whether the apparatus is perfectly tight is not always convincing when the combustion is conducted in an open tube; since, on the one hand, the heating is not constant, and on the other, in consequence of the friction of the solution in the narrow tubes, a leak, at times, may not be detected. The rubber stoppers closing the tube may be protected from the heat by placing on the tube, close to the furnace, an asbestos plate with a circular hole in the centre. After closing the screw pinch-cock, the glass-cock is opened, and a slow current of oxygen (two bubbles per second) is admitted to the tube by carefully opening the pinch-cock. Small flames are now lighted under the copper oxide spiral, which are increased after some time, until, finally, the spiral is brought to a dark red glow. When this is done, the flames under the unheated copper oxide are gradually lighted, care being taken not to allow any flame near the porcelain boat to be too large. Now follows the most difficult operation of the analysis, upon which the success of it virtually depends, viz. the gradual heating of the substance. This is conducted in exactly the same way as that given under the nitrogen determination. The heating is begun, at first, with a single small flame; this is gradually increased in size, or several others may be lighted, then the tube is covered on one side with the tiles, and after a short time, on the other, and finally the full flames are used. With easily volatile substances, the heating at the beginning is not done with the flame, but by covering that portion of the tube containing the boat with hot tiles, taken from the forward highly heated portion of the furnace. Numerous modifications have been applied to this most difficult part of the analysis, concerning which no satisfactory general directions can be given. A valuable rule is to conduct the heating in such a way that the gas bubbles passing through the potash apparatus follow one another with as slow a regularity as possible. If the passage of bubbles becomes too rapid, the heating is moderated. If, during the combustion, water should condense in the glass-cock, or in the rear, cold portion of the tube, as it always does in the front end it is removed by holding a hot tile under it, or by heating with a small flame. When the boat has been heated some time with the full flames, the combustion is considered to be ended. In order to drive the last portions of carbon dioxide and water from the tube into the absorption apparatus, a somewhat more rapid current of oxygen is passed through the tube, until a glowing splinter held before the opening of the straight calcium chloride tube is ignited. During this operation, the water, condensed for the most part in the front end of the tube, is also driven over into the calcium chloride tube, as above described. When this has been done, the rubber stopper is withdrawn from the front end of the combustion tube, care being taken to prevent the water in the calcium chloride tube from running out. To remove the oxygen in the absorption apparatus, a slow current of air which need not be dried is drawn through it for 1–2 minutes, with the mouth or suction. The apparatus is taken apart, closed up as above described, allowed to stand in the weighing-room for half an hour, and is then weighed. From the difference in the weights of the absorption apparatus before and after the combustion, the percentage of carbon and hydrogen is found from the following equations:

$$\text{Percentage of Carbon} = \frac{\text{Wt. } CO_2 \times 300}{\text{Wt. Substance} \times 11}.$$

$$\text{Percentage of Hydrogen} = \frac{\text{Wt. } H_2O \times 100}{\text{Wt. Substance} \times 9}.$$

**Modifications of the Method.** — In many cases instead of using oxygen for the ignition of the copper oxide, the same result may be obtained by using a current of air. The combustion may also be conducted in a current of air; but when the substance is difficult to burn, it is still necessary toward the end of the operation to pass oxygen through the tube for some time. As soon as a glowing splinter held at the end of the straight calcium chloride tube is ignited, the combustion is ended. The combustion may also be conducted without passing a current of air or oxygen into the tube at the beginning, in which case the glass stop-cock is closed. Under these conditions, as soon as the substance has been heated for some time with the full flames, toward the end of the operation the glass stop-cock is opened and a current of air or oxygen passed through the tube. Substances which burn with great difficulty can also be mixed with fine copper oxide in a copper boat (see below), and then burned in the same way in oxygen.

**Combustion of Substances containing Nitrogen.** — Since in the combustion of nitrogenous compounds, the reduced copper spiral serving for the reduction of the oxides of nitrogen, must be used, the combustion tube is charged somewhat differently in this case. The first copper roll is inserted in the tube, not 5 cm., but 15 cm., the space in front of it being reserved for the reduced spiral. Consequently the layer of coarse copper oxide is but 35 cm., and not 45 cm., in length. No change is made in the disposition of the copper oxide spiral. The ignition of the copper oxide is conducted exactly as above, except that a current of air is used. If, however, the ignition should be conducted throughout with oxygen, at the end of the operation this is displaced by air. The further operations are the same as those described above, except that the reduced copper spiral is put in position last — just before connecting the combustion tube with the absorption apparatus. In order to prevent the oxidation of the copper, the combustion proper is performed with the glass-cock closed, and oxygen is not admitted to the tube until at the end. As soon as the oxygen is admitted, the flames under the reduced copper spiral are extinguished. The gas is passed through until it can be detected at the end of the apparatus as above described. In the combustion of substances which leave a charred residue containing difficultly combustible nitrogen, it is necessary to burn this by mixing it with fine copper oxide. Since the porcelain boats are generally too small to contain a sufficient quantity of this, a boat made of sheet copper, 8 cm. long and of a width sufficient to enable it to be just passed into the tube, is used. It is filled as follows: After it has been previously ignited, it is placed upon a sheet of black glazed paper, and half filled with fine copper oxide also previously ignited and afterwards cooled in a desiccator. Upon this is carefully spread the weighed substance from a weighing-tube as in the nitrogen determination, then a layer of fine copper oxide is added until the boat is three-fourths full: the substances are now well mixed by careful stirring with a thick platinum wire. If some of the mixture should fall upon the glazed paper, it is returned to the boat with the aid of a feather or brush. The combustion is made with the glass-cock closed. Oxygen is not admitted until at the end of the operation.

**Combustion of Substances containing Sulphur or a Halogen.** — Sulphur compounds cannot be burned with copper oxide in the manner described, since the copper sulphate formed gives off sulphurous acid at a red heat, which is absorbed by the potash apparatus along with the carbon dioxide, giving a result in which the percentage of carbon is too high. In this case the oxidation is accomplished with granulated lead chromate. The filling of both ends of the tube is done just as described above: copper oxide spiral, empty space for boat, long layer of lead chromate. The ignition in oxygen, etc., is also the same. But two points are here to be observed: (1) the lead chromate is not heated as strongly as the copper oxide, otherwise it fuses in the glass; and (2) the most forward portion of the lead chromate layer, that nearest the calcium chloride tube (that above about three burners), is heated very slightly, since lead sulphate is not completely stable at a red heat. The substance is mixed in the copper boat with powdered, ignited lead chromate.

Halogen compounds can be burned in the usual way with copper oxide; but since the copper halides are partially volatile and give up the halogen on being heated to redness, a silver spiral must be inserted in the tube to retain the halogen. The tube is filled in the same way as for the combustion of a nitrogen compound, only in place of the reduced copper spiral, one of silver is used. But it is better to perform the combustion with lead chromate, in which case it will not be necessary to use a silver spiral. Since the lead halides are also somewhat volatile at a red heat, so, as above, the front part of the tube containing the lead chromate is heated but slightly.

**Combustion of Liquids.** — If the compound to be analysed is a liquid, it can be weighed directly in the porcelain boat provided it is very difficultly volatile. Moderately volatile substances are weighed in a small glass tube which is loosely closed with a glass stopper (see Fig. 46, page 72). In order to introduce this into the tube, it is placed in the porcelain boat in such a position that the mouth of the tube is directed upwards. A preliminary trial will show whether the boat containing the empty tube will pass into the combustion tube. Very easily volatile substances are weighed in small bulb-tubes which are sealed after weighing (Fig. 53). The filling is done as follows: The empty tube is weighed, heated gently, and the open end dipped under the liquid to be analysed. On cooling, the liquid will be drawn up into the bulb. If a sufficient quantity is not obtained the first time, the operation is repeated; before it is sealed care must be taken that the capillary contains none of the liquid; if it does, it must be removed by heating. It is now sealed, and the tube *plus* substance weighed. Care must again be taken to prevent any of the liquid from finding its way into the capillary, due to sudden movements or other causes. To prepare the tube for the combustion, the extreme end is filed and broken off, during which operation the tube is not held by the bulb. It is placed in the boat with its open end elevated and directed toward the front end of the furnace. The precaution to ascertain beforehand whether the boat loaded with the tube will pass into the combustion tube, should always be taken. If necessary, the capillary is shortened.

FIG. 53.

**Calculation of the Atomic Formula.** — In order to calculate the simplest formula of a substance from the figures giving its percentage composition, the method is as follows: If a substance contains, *e.g.*:

$$\text{Carbon} = 49.98\%$$
$$\text{Hydrogen} = 2.72\%$$
$$\text{Chlorine} = 48.30\%$$

the percentage figures are divided by the corresponding atomic weights. There is thus obtained:

$$49.98 \div 12 = 4.08 \text{ C}$$
$$2.72 \div 1 = 2.72 \text{ H}$$
$$48.30 \div 35.5 = 1.36 \text{ Cl}$$

These figures are divided by the smallest — in this case 1.36:

$$4.08 \div 1.36 = 3 \text{ C}$$
$$2.72 \div 1.36 = 2 \text{ H}$$
$$1.36 \div 1.36 = 1 \text{ Cl}$$

The simplest atomic formula, therefore, is $C_3H_2Cl$. If the numbers obtained in the last division are not integers, they are multiplied by the smallest integer which will convert the fractions into whole numbers. If, *e.g.*, the following numbers have been found, 1.25, 1.75, and 0.5, they are multiplied by 4, the results being 5, 7, and 2. The simplest formula thus obtained from the analytical data does not always correspond with the true molecular weight. This must be determined by one of the usual methods, unless it may be inferred from the nature of the reaction by which the substance analysed was produced.

# SPECIAL PART

## I. ALIPHATIC SERIES

### 1. REACTION: THE REPLACEMENT OF AN ALCOHOLIC HYDROXYL GROUP BY A HALOGEN

1. EXAMPLE: **Ethyl Bromide from Ethyl Alcohol**[1]

To 200 grammes of concentrated sulphuric acid contained in a round litre-flask, add gradually, and with constant shaking, 90 grammes of alcohol (about 95%). After cooling the mixture to the room temperature by immersing the flask in water, add 75 grammes of ice; and then 100 grammes of finely pulverised potassium bromide (or sodium bromide obtained as a by-product in the preparation of acetylene, page 165) is added, and the mixture subjected to distillation, which must not be too slow, the flask being heated on a small sand-bath with as large a flame as possible (Fig. 54). Since the boiling-point of ethyl bromide is low (38°), a condenser as long as possible, with a quite rapid current of water passing through it, is used. At the beginning of the operation, the receiver is filled

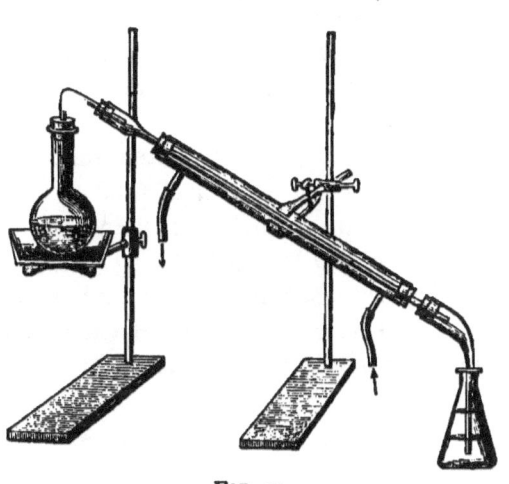

FIG. 54.

[1] J. 1857, 441.

with a sufficient amount of water containing a few pieces of ice, to allow the end of the adapter to dip under the surface. The reaction is ended as soon as the oily drops which sink to the bottom of the receiver cease passing over. If, during the distillation, the contents of the receiver should be drawn up into the condenser, this difficulty may be overcome by placing the receiver in such a position that the end of the adapter reaches just below the surface of the water : the same result may be attained by turning the adapter to one side, so that air may enter it. The lower layer of the distillate consisting of ethyl bromide is washed in the receiver several times with water, and finally with a dilute solution of sodium carbonate, during which the flask must not be closed. The lower layer is then run out of a separating funnel, dried with calcium chloride, and finally distilled, the same precautions as to cooling, mentioned above, being observed. In this case, the free flame is not used for the heating, but the bulb of the fractionating flask is immersed in a vessel filled with water at 60–70° (compare page 22). The ethyl bromide distils between 35–40°, the main portion at 38–39°. In consequence of the low boiling-point of ethyl bromide, it is never allowed to stand in open vessels for any length of time ; during the drying over calcium chloride, the flask must be closed by a tight-fitting cork. The finished preparation, particularly at summer temperature, must not be preserved in thinwalled vessels of any kind, but always in thick-walled, so-called specimen bottles. Yield, 70–80 grammes.

## 2. EXAMPLE : **Ethyl Iodide from Ethyl Alcohol** [1]

To a mixture of 5 grammes of red phosphorus and 40 grammes of absolute alcohol, contained in a small flask of about 200 c.c. capacity, 50 grammes of finely pulverised iodine are added gradually in the course of a quarter hour; the flask is frequently shaken during the addition, and cooled from time to time by immersion in cold water. An air condenser — a straight vertical glass tube is the common form — is connected with the flask, and the reac-

---

[1] A. 126, 250.

tion mixture is allowed to stand at least four hours. (In case the experiment was begun late in the afternoon, it may stand until the next day.) After the reaction is ended, the mixture is heated for two hours on the water-bath, a reflux condenser being attached to the flask. The ethyl iodide is then distilled off conveniently by immersing the flask in a rapidly boiling water-bath. The distillation is facilitated by the use of a frayed thread (see page 32). The distillate, coloured brown by iodine, is washed several times with water to free it from alcohol, and then with water to which a few drops of caustic soda have been added, to remove the iodine; the colourless oil is now separated in a separating funnel, dried with a small quantity of granular calcium chloride, and distilled. If the calcium chloride should float on the oil, it is poured through a funnel containing some asbestos or glass wool into the fractionating flask. The boiling-point of ethyl iodide is $72°$. Yield, about 50 grammes.

Both of these reactions are special cases of a reaction of general application, viz.: the replacement of an alcoholic hydroxyl group with a halogen atom. This may be effected in two ways: (1) by causing the alcohol to react with the halogen hydracid as in the preparation of ethyl bromide, e.g.:

$$C_2H_5 \cdot \boxed{OH + H} Br = H_2O + C_2H_5 \cdot Br$$
$$(HCl, HI)$$

(2) by treating the alcohol with a phosphorus halide, as in the preparation of ethyl iodide, e.g.:

$$3 C_2H_5 \cdot OH + PI_3 = 3 C_2H_5I + P(OH)_3$$
$$(PCl_3, PBr_3)$$

1. The first reaction takes place most easily with hydriodic acid; in many cases saturation with the gaseous acid being sufficient to induce the reaction. Hydrobromic acid reacts with greater difficulty; it is frequently necessary to heat the alcohol saturated with the acid in a sealed tube. The above method for the preparation of ethyl bromide is a very simple case of the general reaction. In place of using hydrobromic acid directly, it can in some cases, like the one given, be generated by warming a mixture of potassium bromide and sulphuric acid:

$$KBr + H_2SO_4 = HBr + KHSO_4.$$

Hydrochloric acid reacts with most difficulty, and it is, *e.g.*, in the preparation of methyl chloride and ethyl chloride, necessary to employ a dehydrating agent — zinc chloride is the best — or with the alcohols of high molecular weights, to heat in a closed vessel under pressure.

This reaction is not only applicable to the aliphatic, but also to the aromatic alcohols; *e.g.*:

$$C_6H_5 \cdot CH_2 \cdot OH + HCl = C_6H_5 \cdot CH_2Cl + H_2O.$$
      Benzyl alcohol                 Benzyl chloride

But phenol hydroxyl groups cannot be replaced by the action of a halogen hydracid.

With di-acid and poly-acid alcohols the reaction takes place, at least with hydrochloric and hydrobromic acids; but, in this case, the number of hydroxyl groups which are replaced by the halogen depends upon the conditions of the experiment, the quantity of the halogen acid, the temperature, etc., *e.g.*:

$$\begin{array}{c} CH_2 \cdot OH \\ | \\ CH_2 \cdot OH \end{array} + HBr = \begin{array}{c} CH_2 \cdot Br \\ | \\ CH_2 \cdot OH \end{array} + H_2O$$
   Ethylene glycol          Ethylene bromhydrine

$$\begin{array}{c} CH_2 \cdot OH \\ | \\ CH \cdot OH \\ | \\ CH_2 \cdot OH \end{array} + 2\,HCl = \begin{array}{c} CH_2 \cdot OH \\ | \\ CHCl \\ | \\ CH_2Cl \end{array} + 2\,H_2O$$
    Glycerol            Dichlorhydrine

$$\begin{array}{c} CH_2 \cdot OH \\ | \\ CH_2 \\ | \\ CH_2 \cdot OH \end{array} + 2\,HBr = \begin{array}{c} CH_2 \cdot Br \\ | \\ CH_2 \\ | \\ CH_2 \cdot Br \end{array} + 2\,H_2O.$$
  Trimethylene glycol     Trimethylene bromide

Hydriodic acid, in consequence of its reducing properties, acts upon the poly-acid alcohols in a different manner. A *single* hydroxyl group, and that particular one in combination with a carbon atom which is in turn in combination with other carbon atoms, is only replaced by iodine, while at times other hydroxyl groups are replaced by the hydrogen of the acid, *e.g.*:

## ALIPHATIC SERIES

$$\underset{\text{Glycerol}}{\overset{\begin{array}{c}CH_2.OH\\|\\CH.OH\\|\\CH_2.OH\end{array}}{}} + 5\,HI = \underset{\text{Isopropyl iodide}}{\overset{\begin{array}{c}CH_3\\|\\CHI\\|\\CH_3\end{array}}{}} + 3\,H_2O + 4\,I$$

$$\underset{\text{Erythrite}}{CH_2(OH).CH(OH).CH(OH).CH_2(OH)} + 7\,HI$$

$$= \underset{\text{Sec. Butyl iodide}}{CH_3.CH_2.CHI.CH_3} + 4\,H_2O + 6\,I$$

With derivatives of alcohols, *e.g.* alcohol acids (hydroxy acids), the first reaction takes place:

$$\underset{\beta\text{-Hydroxyproprionic acid}}{CH_2.OH.CH_2.COOH} + HI \quad = \underset{\beta\text{-Iodoproprionic acid}}{CH_2I.CH_2.COOH} + H_2O$$

$$\underset{\text{Glyceric acid}}{CH_2(OH).CH(OH).COOH} + 2\,HCl = \underset{\alpha\text{-}\beta\text{-Dichlorproprionic acid}}{CH_2Cl.CHCl.COOH} + 2\,H_2O$$

2. The second reaction takes place much more energetically, especially when the phosphorus halide which has been previously made is used. This is not always necessary, at least in introducing bromine and iodine; in many cases it is better to generate the phosphorus halides in the course of the reaction, by adding to the mixture of alcohol and red phosphorus either bromine from a dropping funnel, or as above, finely pulverised iodine. This reaction, as well as the first, is applicable to poly-acid alcohols and substituted alcohols, *e.g.*:

(*a*) $\underset{\text{Ethylene glycol}}{CH_2(OH).CH_2(OH)} + 2\,PCl_5 = \underset{\text{Ethylenechloride}}{CH_2Cl.CH_2Cl} + 2\,POCl_3 + 2\,HCl$

(*b*) $\underset{\text{Dichlorhydrine}}{CH_2(OH).CHCl.CH_2Cl} + PCl_5$

$$= \underset{\text{Trichlorhydrine}}{CH_2Cl.CHCl.CH_2Cl} + POCl_3 + HCl$$

This example illustrates the more energetic action of the phosphorus halide as compared with the corresponding hydrogen halide; it being impossible to replace the third hydroxyl group of glycerol with chlorine by the use of hydrochloric acid.

(*c*) $\underset{\substack{(\text{I})\\\alpha\text{-Hydroxyproprionic acid}}}{CH_3.CH(OH).COOH} + PCl_5 = \underset{\substack{(\text{I})\\\alpha\text{-Chlorproprionic acid}}}{CH_3.CHCl.COOH} + POCl_3 + HCl$

In cases of this kind a complication arises, due to the fact that the phosphorus halide also acts upon the hydroxyl of the carboxyl group, replacing it with the halogen, giving rise to an acid-chloride:

$$CH_3.CHCl.CO.OH + PCl_5 = CH_3.CHCl.CO.Cl + POCl_3 + HCl$$

The acid may be regenerated by treating the acid-chloride with water:

$$CH_3.CHCl.CO.Cl + H_2O = CH_3.CHCl.CO.OH + HCl$$

The action of phosphorus iodide on poly-acid alcohols is similar to the action of hydriodic acid referred to under Reaction 1.

The more energetic action of the phosphorus halides may also be perceived in the fact that phenol-hydroxyl groups can be replaced by a halogen, by the use of the phosphorus compounds, which, as mentioned above, is impossible with the halogen hydracids, *e.g.*:

$$\underset{\text{Phenol}\quad\text{(Br)}}{C_6H_5.OH + PCl_5} = \underset{\text{(Br)}}{C_6H_5.Cl + POCl_3 + HCl}$$

$$C_6H_4{\scriptstyle\diagup NO_2 \atop \diagdown OH} + PCl_5 = C_6H_4{\scriptstyle\diagup NO_2 \atop \diagdown Cl} + POCl_3 + HCl$$

$$\underset{\text{Picric acid}}{C_6H_2{\scriptstyle\diagup OH \atop \diagdown (NO_2)_3}} + PCl_5 = \underset{\text{Picryl chloride}}{C_6H_2{\scriptstyle\diagup Cl \atop \diagdown (NO_2)_3}} + POCl_3 + HCl$$

But the yields are much less satisfactory, the reason being that the phosphorus oxychloride attacks the unacted-upon phenol, forming phosphoric acid esters, *e.g.*:

$$POCl_3 + 3\,C_6H_5.OH = PO.(OC_6H_5)_3 + 3\,HCl$$

In this way a large portion of the phenol is withdrawn from the main reaction.

The monohalogen alkyls $C_nH_{(2n+1)}Cl(Br, I)$ are in most cases colourless liquids, the exceptions being methyl and ethyl chlorides and methyl bromide, which are gaseous at the ordinary temperature, and the members of the series having high molecular weights, like cetyl iodide, $C_{16}H_{33}I$, which are semi-solid, salve-like substances. The iodides are only colourless when freshly prepared; on long standing, especially under the influence of light, a slight decomposition, resulting in the separation of iodine, takes place, imparting to them a faint pink colour at first, which becomes brownish red after a long time. This decomposition can be prevented if some finely divided, so-called molecular silver is added to the liquid. A coloured iodide can be made colourless by shaking it with some caustic soda solution. The halogen alkyls mix readily with organic solvents, as alcohol, ether, carbon disulphide, benzene, etc., but *not* with water. The chlorides are lighter, the bromides and iodides heavier than water; the latter have the highest specific gravities. This property decreases in all three classes with the decrease in halogen percentage from the lower up to the higher members of the series; *i.e.* the higher molecular weight compounds have a smaller specific gravity than the lower members. The chlorides have the lowest boiling-points; the corresponding bromides boil about 25° higher, and the iodides 50° higher, than the chlorides.

The ease with which the monohalogen alkyls react with other compounds gives them great importance; they are used primarily as a means of introducing alkyl groups into other molecules, *i.e.* by replacing an hydrogen atom with an alkyl group. If it is desired, *e.g.*, to replace in an alcohol, mercaptan, phenol, or acid the hydrogen of the (OH)-, (SH)-, or (COOH)-groups by an alkyl radical, *i.e.* to prepare an ether of these substances, the corresponding sodium compound, or in case of acids, better the silver salt, is treated with the halogen alkyl, *e.g.*:

$$C_2H_5 \cdot ONa + IC_2H_5 = C_2H_5 \cdot O \cdot C_2H_5 + NaI$$
Sodium alcoholate                Ethyl ether

$$C_2H_5 \cdot SNa + IC_2H_5 = C_2H_5 \cdot S \cdot C_2H_5 + NaI$$
Sodium ethyl mercaptide       Ethyl sulphide

$$C_6H_5 \cdot ONa + ICH_3 = C_6H_5 \cdot O \cdot CH_3 + NaI$$
Sodium phenolate            Phenyl methyl ether
                                          = Anisol

$$CH_3 \cdot COOAg + IC_2H_5 = CH_3 \cdot COO \cdot C_2H_5 + AgI$$
Silver acetate                 Ethyl acetate

The alkyl groups may be introduced into the ammonia molecule and into organic amine molecules by means of the halogen alkyls; *e.g.*:

$$NH_3 + ICH_3 = NH_2 \cdot CH_3 + HI$$
Methyl amine

Di- and tri-methyl amine are also formed at the same time:

$$C_6H_5 \cdot NH_2 + 2\ CH_3Cl = C_6H_5 \cdot N(CH_3)_2 + 2\ HCl.$$
Aniline                Dimethyl aniline

Hydrogen atoms in combination with carbon may also be replaced by alkyl radicals, by means of the halogen alkyls. Since under these conditions another radical is introduced into the molecule, it presents a method of preparing the higher members of a series from the lower, simpler ones. Various examples of this will be taken up later in laboratory practice. Here it will be sufficient to refer to several equations showing this kind of reaction.

$$CH_3.CO.CHNa.COOC_2H_5 + ICH_3 = CH_3.CO.CH-COOC_2H_5 + NaI$$
Sodium acetacetic ester              | <br>
                                          $CH_3$<br>
Methylacetacetic ether

$$\begin{matrix} COOC_2H_5 \\ | \\ CHNa \\ | \\ COOC_2H_5 \end{matrix} \quad + IC_2H_5 = \begin{matrix} COOC_2H_5 \\ | \\ CH-C_2H_5 \\ | \\ COOC_2H_5 \end{matrix} + NaI$$

Sodium malonic ester      Ethyl malonic ester

$$C_6H_6 + ClC_2H_5 = C_6H_5.C_2H_5 + HCl$$

Benzene                Ethyl benzene<br>
(In presence of $AlCl_3$)

Fittig's Synthesis, to be taken up later, is a case of this kind, by which a halogen atom is replaced by an alkyl group; *e.g.*:

$$C_6H_5Br + BrC_2H_5 + 2\,Na = C_6H_5.C_2H_5 + 2\,NaBr$$

Brom benzene                Ethyl benzene

Further, the halogen alkyls serve for the preparation of the unsaturated hydrocarbons of the ethylene series.

$$CH_3.CHI.CH_3 = \underbrace{CH_3.CH=CH_2}_{} + HI$$

Isopropyl iodide       Propylene

In many cases the alcohols may also be prepared from the halogen alkyls; *e.g.*:

$$CH_3.CHI.CH_3 + HOH = CH_3.CH(OH).CH_3 + HI$$

Isopropyl alcohol

This reaction is obviously only of importance when the halogen alkyl is not obtained from the corresponding alcohol, as is the case in the example given. As above mentioned, the isopropyl iodide is most simply obtained from glycerol and hydriodic acid, so that by this reaction the glycerol can be converted into the isopropyl alcohol. Halogen alkyls also unite directly with other compounds, like sulphides and tertiary amines:

$$C_2H_5.S.C_2H_5 + IC_2H_5 = S\begin{matrix}\diagup C_2H_5 \\ -C_2H_5 \\ \diagdown C_2H_5 \\ \diagdown I \end{matrix}$$

Ethyl sulphide       Triethylsulphine iodide

ALIPHATIC SERIES 113

$$N(CH_3)_3 + CH_3Cl = N(CH_3)_4Cl$$
Trimethyl amine    Tetramethyl ammonium chloride

$$C_5H_5N + ICH_3 = C_5H_5N \cdot ICH_3$$
Pyridine    Methylpyridine iodide

With these examples the list of the many-sided reactions of the halogen alkyls is not exhausted; they are also used for the preparation of the metallic alkyls, *e.g.* zinc alkyls; for the preparation of the phosphines, and for many other compounds. Finally, attention is called to the characteristic difference between the organic and inorganic halides. While, *e.g.*, potassium chloride, bromide, or iodide in solution act *instantly* with a silver nitrate solution to form a *quantitative* precipitate of silver chloride, bromine, or iodide respectively, silver nitrate in a water solution does not act on most organic halides, so that this reagent does not serve in the usual way to show the presence of a halogen.

EXPERIMENT: Treat a solution of silver nitrate with a few drops of ethyl bromide. Not the least trace of silver bromide is formed.[1]

It has been customary to explain this by saying that the affinity between the halogen atom and carbon is greater than that between the halogen and the metallic atom. According to our later views the difference between the two classes of compounds is explained thus: The metallic halides belong to the class of so-called electrolytes, *i.e.* substances which are dissociated in water solution, *e.g.* the molecule KCl is dissociated into its ions, $\overset{+}{K}$ and $\overset{-}{Cl}$. The organic halides are non-electrolytes, *i.e.* the solutions of these contain the undissociated molecules. According to this conception, the potassium chloride reacts with silver nitrate, because no further separation of the potassium from the chlorine (other than that effected by solution) is necessary, while in case of the brom alkyls, the brom-carbon union must first be severed.

---

[1] A noteworthy exception is ethyl iodide, which, when shaken with a water solution of silver nitrate, gives an abundant precipitate of silver iodide. Methyl iodide does not react with silver nitrate.

## 2. REACTION: PREPARATION OF AN ACID-CHLORIDE FROM THE ACID

EXAMPLE: **Acetyl Chloride from Acetic Acid** [1]

To 100 grammes of glacial acetic acid contained in a fractionating flask connected with a condenser, 80 grammes of phosphorus trichloride is added through a dropping-funnel, the flask being cooled by water. The bulb is then immersed in a porcelain dish

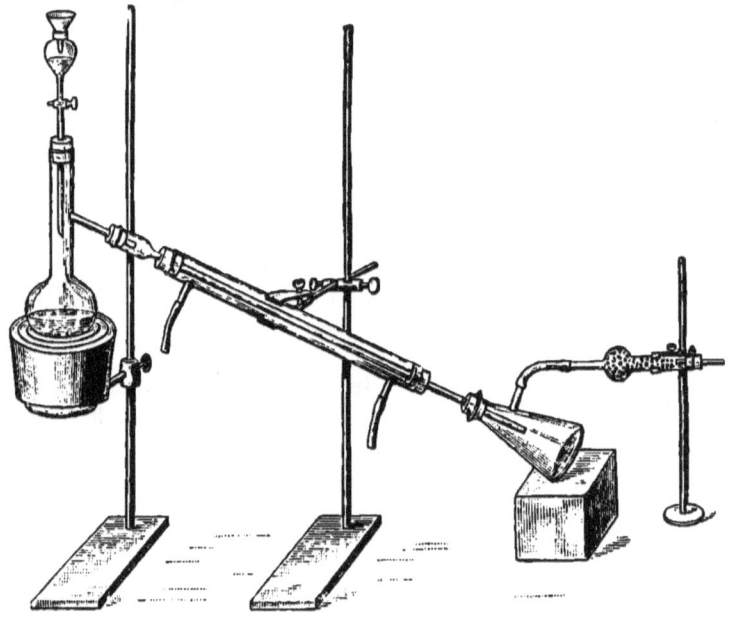

FIG. 55.

filled with water at a temperature of 40–50°, and the heating continued until the active evolution of hydrochloric acid gas slackens, and the liquid which was homogeneous before heating has separated into two layers. To separate the acetyl chloride which forms the upper, lighter layer, from the heavier layer of phosphorous acid, the mixture is heated on a rapidly boiling water-

[1] A. 87, 63.

bath until nothing more passes over. Since acetyl chloride is very easily decomposed by moisture, the distillate must not be collected in an open receiver, but the condenser-tube must be tightly connected with a tubulated flask (suction flask), protected from the air by a calcium chloride tube, as represented in Fig. 55. For complete purification, the distillate is distilled in a similar apparatus, except that the dropping-funnel is replaced by a thermometer (boiling-point, 55°). Yield, 80–90 grammes.

In order to replace the hydroxyl of a carboxyl group (CO.OH)- by chlorine, a reaction, similar to the one employed above for the substitution of an alcoholic hydroxyl group by a halogen, may be used. If, e.g., a mixture of an acid and phosphoric anhydride is treated with gaseous hydrochloric acid (heating if necessary), there is formed an acid-chloride, the reaction being analogous to that by which ethyl bromide was prepared.

$$X.CO.OH + HCl = X.CO.Cl + H_2O$$

This reaction is without practical importance, since the reactions involved in the methods described under (2), page 109, take place much more smoothly and easily, and, therefore, are exclusively used. The above reaction is mentioned here only because it throws some light on one of the methods used in esterification, which will be briefly described below under the characteristics of the acid-chlorides. In practice, the acid-chlorides are almost always prepared by the action of phosphorus tri- or penta-chloride, and, more rarely, the oxychloride on the acid directly, or, in many cases, on the sodium or potassium salt. The selection of the chlorides of phosphorus depends upon (1) the ease with which the acid under examination reacts, and (2) upon the boiling-point of the acid-chloride. If, as in the case of acetic acid and its homologues, the trichloride of phosphorus reacts easily with the formation of the acid-chloride, this is selected in preference to the more energetic pentachloride. The reaction takes place in accordance with the following equation:

$$3\,CH_3.CO.OH + 2\,PCl_3 = 3\,CH_3.CO.Cl + P_2O_3 + 3\,HCl$$

In cases in which the boiling-point of the acid-chloride desired does not lie far from that of phosphorus oxychloride (110°), thus rendering a fractional distillation for the separation of the products difficult, the trichloride is always used. If an acid does not react too energetically, as is the case with the higher members of the acetic acid series with the pentachloride, this is used. With the aromatic acids, the latter is used exclusively, since the trichloride and oxychloride react with great difficulty:

$$C_7H_{15}.CO.OH + PCl_5 = C_7H_{15}.CO.Cl + POCl_3 + HCl$$
Caprylic acid

$$C_6H_5.CO.OH + PCl_5 = C_6H_5.CO.Cl + POCl_3 + HCl$$
Benzoic acid   Benzoyl chloride

Attention is called to the fact that for one molecule of phosphorus pentachloride, but one molecule of the acid-chloride is obtained.

The phosphorus oxychloride is used generally only when dealing with the salts of carbonic acids, upon which it acts as indicated by the equation:

$$2\ CH_3.CO.ONa + POCl_3 = 2\ CH_3.CO.Cl + NaPO_3 + NaCl$$

This reaction may be used with advantage in order to utilise more of chlorine of the phosphorus pentachloride than is the case when the latter acts upon the free acids. If the pentachloride is allowed to act on a sodium salt, as above, there is formed, for an instant, phosphorus oxychloride, and while this no longer acts upon the free acid, it can convert two other molecules of the salt into the chloride:

$$3\ CH_3.CO.ONa + PCl_5 = 3\ CH_3.CO.Cl + NaPO_3 + 2\ NaCl$$

In this way, with the use of one molecule of phosphorus pentachloride, three molecules of the acid-chloride are obtained.

The lower members of the series of acid-chlorides are colourless liquids; the higher, colourless crystalline substances. They boil, generally, at ordinary pressure, without decomposition, but the higher members are more conveniently distilled in a vacuum. The boiling-points of the acid-chlorides are lower than those of the acids; the replacement of hydroxyl by chlorine usually causes a lowering of the boiling-point.

| $CH_3.CO.Cl$ | Boiling-point | $51°$ | $C_6H_5.CO.Cl$ | Boiling-point | $199°$ |
| $CH_3.CO.OH$ | " | $118°$ | $C_6H_5.CO.OH$ | " | $250°$ |

The acid-chlorides possess pungent odours. They fume in the air, since they unite with the moisture and decompose, thus forming the corresponding acid and hydrochloric acid. They are heavier than water, and do not mix with it, but are easily soluble in indifferent organic solvents like ether, carbon disulphide, benzene.

To separate the chlorides from the by-products formed by the phosphorus chloride, one can proceed as in the case of acetyl chloride, by distilling off the volatile chloride from the non-volatile phosphorus acid, either on a water-bath or over a free flame. In order to separate a chloride, in case it distils without decomposition, from the volatile phosphorus oxychloride formed when phosphorus pentachloride is used, a fractional distillation is made. In other cases, the mixture is heated in a vacuum apparatus on an actively boiling water-bath, upon which the phosphorus oxychloride passes over. The non-volatile residue can be used, in many cases, without further purification; it may be obtained perfectly pure by distilling it in a vacuum.

**Chemical Reactions.** — The acid-chlorides are decomposed by water with the formation of the corresponding acid and hydrochloric acid.

$$CH_3.CO.Cl + H_2O = CH_3.CO.OH + HCl.$$

This decomposition takes place often with extreme ease, and the chlorine atom is united to the acid radical much less firmly than it is in the case of an alkyl radical. While it is generally necessary, in order to convert a halogen alkyl into an alcohol, to boil it a long time with water, often with the addition of sodium hydroxide, or potassium hydroxide, a carbonate or acetate, the analogous transformation of an acid-chloride takes place with far greater ease. With the lower members, e.g. acetyl chloride, the reaction begins almost instantly at the ordinary temperature, and continues with violent energy; but it is necessary to heat the higher members, to induce the transformation, e.g. benzoyl chloride, which will be prepared later.

EXPERIMENT: To 5 c.c. of water in a test-tube is gradually added $\frac{1}{2}$ c.c. of acetyl chloride. If the water is very cold, the oily drops, sinking to the bottom, do not mix with it, and may be observed for a short time. On shaking the tube, an energetic reaction sets in with evolution of heat, and the chloride passes into solution, which happens immediately if the water is not very cold.

The acid chlorides react with alcohols and phenols to form acid esters:

$$CH_3.CO.Cl + C_2H_5.OH = CH_3.CO.OC_2H_5 + HCl$$
Ethyl acetate

$$CH_3.CO.Cl + C_6H_5.OH = CH_3.CO.OC_6H_5 + HCl$$
Phenol  Phenyl acetate

EXPERIMENT: To 1 c.c. of alcohol in a test-tube, cooled with water, add an equal volume of acetyl chloride, drop by drop; this mixture is then treated with an equal volume of water, the tube being cooled as before; the liquid is then carefully made weakly alkaline with sodium hydroxide. If the pleasant smelling ethyl acetate does not separate out on the water solution in a mobile layer, finely pulverised salt is added until no more will dissolve. This will cause the ethyl acetate to separate out.

For the preparation of an acid-ester, the previously prepared chloride is rarely used; the method of procedure being to pass gaseous hydrochloric acid into a mixture of the corresponding acid and alcohol to saturation.

There is probably formed, from the carbonic and hydrochloric acids, an intermediate acid-chloride which, as just described, reacts with the alcohol. The acid-chlorides are also used to determine whether a substance under examination contains an alcoholic or a phenol hydroxyl group or not. If the compound reacts readily with an acid-chloride, it contains either alcoholic or phenol hydroxyl, since all compounds containing oxygen in some other form of combination, *e.g.*, as in ethers, do not react with acid-chlorides.

Finally, the action of the acid-chlorides upon alcohols and phenols is made use of to separate the latter from solution, or in order to detect them. However, benzoyl chloride is most generally used for this purpose, concerning the importance of which more will be said later.

Acid-chlorides act upon the salts of carbonic acids, forming anhydrides:

$$CH_3.CO.Cl + CH_3.CO.ONa = CH_3.CO.O.CO.CH_3 + NaCl$$
Acetic anhydride

The next preparation will deal with this reaction. The acid-chlorides react with ammonia as well as with primary and secondary organic bases with great ease:

$$CH_3.CO.Cl + NH_3 \quad = CH_3.CO.NH_2 + HCl,$$
Acetamide

$$CH_3.CO.Cl + C_6H_5.NH_2 = C_6H_5.NH.CO.CH_3 + HCl$$
Aniline              Acetanilide

EXPERIMENT: To 1 c.c. of aniline, acetyl chloride is added in drops; an energetic reaction takes place with a hissing sound, which ceases when about the same volume of the chloride is added. The mixture is cooled with water, and five times its volume of water is added, upon which an abundant precipitate of acetanilide separates out, the quantity of which is increased by rubbing the walls of the test-tube with a glass rod. The precipitate is filtered off, and recrystallised from hot water.

This reaction is also used to characterise organic bases by converting them into their best crystallised acid derivatives, and in order to detect small quantities, especially of liquid bases, by a melting-point determination. Since the tertiary bases do not react with acid-chlorides, because they no longer contain any ammonia hydrogen, this reaction may be employed to decide whether a given base is, on the one hand, primary or secondary, or, on the other, tertiary.

The fact that hydrogen in combination with carbon can be replaced by an acid radical by the use of an acid-chloride is of special importance. In this connection, the Friedel-Crafts ketone synthesis is particularly mentioned. This will be taken up later, in practice. The reaction is indicated by the following equation:

$$C_6H_6 + CH_3 \cdot CO \cdot Cl = C_6H_5 \cdot CO \cdot CH_3 + HCl$$

Benzene (in presence of $AlCl_3$)     Acetophenone

The acid-chlorides are also of service for the synthesis of tertiary alcohols (Butlerow's Synthesis), as well as for that of ketones. The final reactions will only be indicated here; in regard to the details, reference must be made to larger works.

$$CH_3 \cdot CO \cdot Cl + Zn\!\!<\!\!\begin{array}{c}CH_3\\CH_3\end{array} = CH_3 \cdot C\!\!<\!\!\begin{array}{c}CH_3\\CH_3\\Cl\end{array} + ZnO$$

Zinc methyl

$$CH_3 \cdot C\!\!<\!\!\begin{array}{c}CH_3\\CH_3\\Cl\end{array} + H_2O = C\!\!<\!\!\begin{array}{c}CH_3\\CH_3\\CH_3\\OH\end{array} + HCl$$

Trimethyl carbinol

$$2\,CH_3 \cdot CO \cdot Cl + Zn\!\!<\!\!\begin{array}{c}CH_3\\CH_3\end{array} = 2\,CH_3 \cdot CO \cdot CH_3 + ZnCl_2$$

Acetone

## 3. REACTION: PREPARATION OF AN ANHYDRIDE FROM THE ACID-CHLORIDE AND THE SODIUM SALT OF THE ACID

EXAMPLE : Acetic Anhydride from Acetyl Chloride and Sodium Acetate [1]

For the preparation of acetic anhydride an apparatus similar to that used in the preparation of acetyl chloride is employed, except that the fractionating flask is replaced by a tubulated retort (Fig. 56, page 121).

To 70 grammes of finely pulverised, anhydrous sodium acetate (for the preparation of this, see below) contained in the retort, 50 grammes of acetyl chloride is added drop by drop from a separating funnel. As soon as the first half of the chloride is added, the reaction is interrupted for a short time, in order that the pasty mass may be stirred up with a glass rod. The second half is then allowed to run in. If in consequence of a too rapid addition of acetyl chloride some of it should distil over into the receiver undecomposed, this is poured back into the funnel and again allowed to act on the sodium acetate. The separating funnel is then removed, the tubulure closed with a cork, and the anhydride distilled off from the salt residue by means of a *luminous* flame which is kept in constant motion. The distillate is finally purified by distilling in an apparatus similar to the one used in the rectification of the acetyl chloride. Boiling-point of acetic anhydride 138°. Yield, about 50 grammes.

*Preparation of Anhydrous Sodium Acetate :* Crystallised sodium acetate contains three molecules of water of crystallisation. In order to dehydrate it is placed in a shallow iron or nickel dish and heated over a free flame (120 grammes for this experiment). The salt first melts in the water of crystallisation, on further heating steam is copiously evolved, and the salt mass solidifies as soon as the main portion of the water has been driven off, provided the flame is not too large. In order to remove the last portions of the water, the mass is now heated with a large flame, the burner being

---

[1] A. 87, 149.

constantly moved, until the solidified mass melts for the second time; care must be taken not to overheat; in case this should happen, the fact will be recognised by the evolution of combustible gases and the charring of the substance. After cooling, the salt is removed from the dish with a knife. If commercial anhydrous

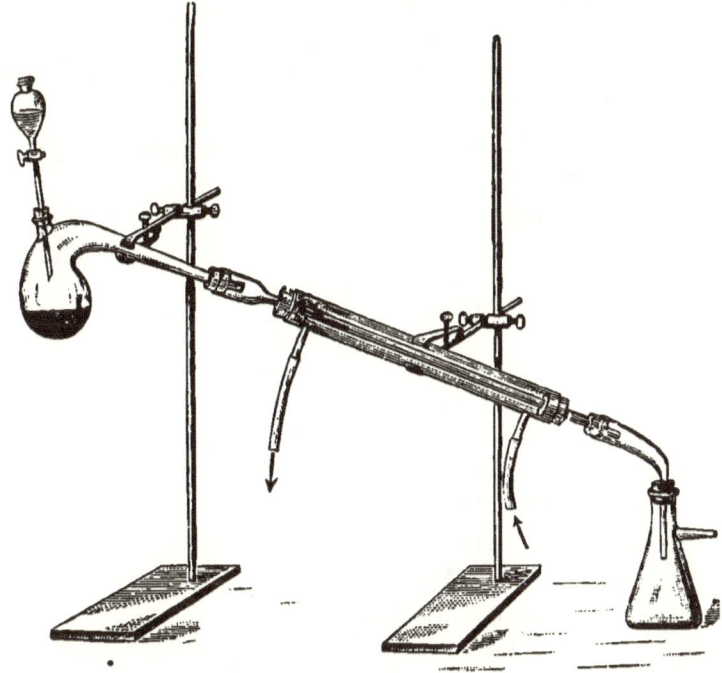

FIG. 56.

sodium acetate is at hand, it is recommended that this also be melted once, since when it is kept for a long time it always takes up water again.

The reaction of acetyl chloride with sodium acetate takes place in accordance with the following equation:

$$CH_3.CO.Cl + CH_3.CO.ONa = \begin{matrix} CH_3CO \\ CH_3CO \end{matrix} \!\!\!\!> O + NaCl$$

This reaction is capable of general application, and the anhydride of the acid may be made by treating its chloride with the corresponding sodium salt. The so-called mixed anhydrides, containing two different acid radicals, can also be prepared by this reaction, by using the chloride and sodium salt of two different acids:

$$CH_3.CO.Cl + CH_3.CH_2.CO.ONa = \begin{matrix}CH_3.CO\\ CH_3.CH_2.CO\end{matrix}\!\!>\!\!O + NaCl$$

Since, as stated, an acid-chloride may be obtained from an alkali salt of the acid and phosphorus oxychloride, it is not necessary for the preparation of an anhydride to first isolate the chloride; it is better to allow the same to act directly on an excess of the salt, so that from the oxychloride and salt an anhydride is directly obtained:

$$2\,CH_3.CO.ONa + POCl_3 = 2\,CH_3.CO.Cl + PO_3Na + NaCl$$

$$2\,CH_3.CO.ONa + 2\,CH_3.CO.Cl = 2\begin{matrix}CH_3.CO\\ CH_3.CO\end{matrix}\!\!>\!\!O + 2\,NaCl$$

---

$$4\,CH_3.CO.ONa + POCl_3 = 2\begin{matrix}CH_3.CO\\ CH_3.CO\end{matrix}\!\!>\!\!O + PO_3Na + 3\,NaCl$$

The lower members of the acid-anhydride series are colourless liquids; the higher members, crystallisable solids. They possess a sharp odour, are insoluble in water, but soluble in indifferent organic solvents. Their specific gravities are greater than those of water. The boiling-points are higher than those of the corresponding acids:

Acetic acid, 118°,
Acetic anhydride, 138°.

The lower members can be distilled without decomposition at ordinary pressure; but the higher members must be distilled in a vacuum.

The chemical conduct of anhydrides toward water, alcohols, and phenols, as well as bases, is wholly analogous to that of the chlorides; but the anhydrides react with more difficulty than the chlorides. Thus with water, the anhydrides yield the corresponding acids:

$$\begin{matrix}CH_3.CO\\ CH_3.CO\end{matrix}\!\!>\!\!O + H_2O = 2\,CH_3.CO.OH$$

EXPERIMENT: 5 c.c. of water are treated with ½ c.c. of acetic anhydride. The latter sinks to the bottom and does not dissolve even on long shaking. It will be recalled that the corresponding chloride reacts instantly with water very energetically. If the mixture be warmed, solution takes place.

In the presence of alkalies, solution takes place much more readily with the formation of the alkali salts:

$$\begin{matrix}CH_3.CO\\CH_3.CO\end{matrix}\!\!>\!\!O + 2\,NaOH = 2\,CH_3.CO.ONa + H_2O$$

EXPERIMENT: Mix 5 c.c. of water with ½ c.c. of acetic anhydride, and add a little caustic soda solution. On shaking, without warming, solution takes place.

Anhydrides of high molecular weight react with water with still greater difficulty, and require a longer heating to convert them into the corresponding acid.

With alcohols and phenols, the anhydrides form acid-esters on heating, while the acid-chlorides react at the ordinary temperature:

$$\begin{matrix}CH_3.CO\\CH_3.CO\end{matrix}\!\!>\!\!O + C_2H_5.OH = CH_3.CO.OC_2H_5 + CH_3.CO.OH$$

$$\begin{matrix}CH_3.CO\\CH_3.CO\end{matrix}\!\!>\!\!O + C_6H_5.OH = CH_3.CO.OC_6H_5 + CH_3.CO.OH$$
Phenyl acetate

It is to be noticed that one of the two acid radicals in the anhydride is not available for the purpose of introducing the acetyl group into other compounds,—acetylating,—since it passes over into the acid.

EXPERIMENT: 2 c.c. of alcohol is added to 1 c.c. of acetic anhydride in a test-tube, and heated gently for several minutes. It is then treated with water and carefully made slightly alkaline. The acetic ester can be recognised by its characteristic pleasant odour. If it does not separate from the liquid, it may be treated with common salt, as in the experiment on page 118.

With ammonia and primary or secondary organic bases, the anhydrides react like the chlorides:

$$NH_3 + \begin{matrix}CH_3.CO\\CH_3.CO\end{matrix}\!\!>\!\!O = CH_3.CO.NH_2 + CH_3.CO.OH$$

$$C_6H_5.NH_2 + \begin{matrix}CH_3.CO\\CH_3.CO\end{matrix}\!\!>\!\!O = C_6H_5.NH.CO.CH_3 + CH_3.CO.OH$$

EXPERIMENT: Add 1 c.c. of aniline to 1 c.c. of acetic anhydride, heat to incipient ebullition, and then, after cooling, add twice the volume of water. The crystals of acetanilide separate out easily if the walls of the vessel be rubbed with a glass rod; these are filtered off, and may be recrystallised from a little hot water.

The acid-anhydrides can, therefore, be used, like the chlorides, for the recognition, separation, characterisation, and detection of alcohols, phenols, and amines.

In order to complete the enumeration of the reactions of the acid-anhydrides, it may be mentioned briefly that they yield alcohols, and the intermediate aldehydes when treated with sodium amalgam:

$$\begin{matrix} CH_3.CO \\ CH_3.CO \end{matrix} \!\!> O + H_2 = CH_3.CHO + CH_3.CO.OH$$
Aldehyde

$$CH_3.CHO + H_2 = CH_3.CH_2.OH$$

It is, therefore, possible to pass from the anhydride of an acid to its aldehyde or alcohol.

### 4. REACTION: PREPARATION OF AN ACID-AMIDE FROM THE AMMONIUM SALT OF THE ACID

EXAMPLE: **Acetamide from Ammonium Acetate**[1]

To 75 grammes of glacial acetic acid heated to 40–50° in a porcelain dish on a water-bath, finely pulverised ammonium carbonate is added (80–90 grammes will be necessary) until a test-portion diluted in a watch-glass with water just shows an alkaline reaction. The viscous mass is then warmed a short time directly over a flame, and as soon as it has become sufficiently liquid, it is poured (without the use of a funnel-tube) directly into two wide bomb-tubes of hard glass, which have been previously warmed in a flame. After the portions of substance adhering to the upper end of the tube have been removed by melting down carefully with a flame, the last traces are removed with filter-paper, the tube sealed and heated for five hours in a bomb-furnace at 220°.

---

[1] B. 15, 979.

The liquid reaction product is fractionated in a distilling-flask provided with a long air condenser (see page 21). There is first obtained a fraction boiling between 100–130°, consisting essentially of acetic acid and water. The temperature then rises rapidly to 180°, at which point the acetamide begins to distil. The fraction passing over between 180–230° is collected in a beaker cooled by ice water; the walls are rubbed with a sharp-edged glass rod; the crystals separating out are pressed on a drying plate to remove the liquid impurities. By another distillation of the pressed-out crystals, the almost pure acetamide boiling at 223° passes over. The product thus obtained possesses an odour very characteristic of mouse excrement; this is not the odour of pure acetamide, but of an impurity accompanying it. In order to remove the impurity, a portion of the distilled amide is again pressed out on a drying plate, and then crystallised from ether. There are thus obtained colourless, odourless crystals, melting at 82°. Yield, about 40 grammes.

The reaction involved in the preparation of an amide from the ammonium salt of the acid is capable of general application. The latter is subjected to dry distillation, or more conveniently, heated in a sealed tube at 220–230° for five hours:

$$CH_3 \cdot CO \cdot ONH_4 = CH_3 \cdot CO \cdot NH_2 + H_2O$$

In order to purify the amide thus obtained, the reaction-mixture may be fractionated, as in the case of acetamide, or if the amide separates out in a solid condition, it may be purified by filtering off the impurities and crystallising. Substituted acid-amides, and especially easily substituted aromatic amides, *e.g.* acetanilide, can also be readily obtained by this method, by heating a mixture of the acid and amine a long time in an open vessel:

$$CH_3 \cdot COOH_3N \cdot C_6H_5 = CH_3 \cdot CO \cdot NH \cdot C_6H_5 + H_2O$$
Aniline acetate      Acetanilide

The ammonium salts of di- and poly-basic acids react in a similar way, *e.g.*:

$$\begin{vmatrix} CO \cdot ONH_4 \\ CO \cdot ONH_4 \end{vmatrix} = \begin{vmatrix} CO \cdot NH_2 \\ CO \cdot NH_2 \end{vmatrix} + 2 H_2O$$
Ammonium oxalate      Oxamide

Concerning further methods of preparation, it may be stated that acid-chlorides or anhydrides when treated with ammonia, primary or secondary bases form acid-amides very easily:

$$CH_3.CO.Cl + NH_3 = CH_3.CO.NH_2 + HCl$$

$$\begin{array}{c}CH_3.CO\\CH_3.CO\end{array}\!\!\!\!>\!\!O + NH_3 = CH_3.CO.NH_2 + CH_3.CO.OH$$

The acid-amides may be furthermore obtained by two methods of general application: (1), by treating an ethereal salt with ammonia, and (2), by treating a nitrile with water:

$$\underset{\text{Ethyl acetate}}{CH_3.CO.OC_2H_5} + NH_3 = CH_3.CO.NH_2 + C_2H_5.OH$$

$$\underset{\text{Acetonitrile}}{CH_3.CN} + H_2O = CH_3.CO.NH_2$$

The acid-amides are, with the exception of the lowest member, formamide, $H.CO.NH_2$, liquid, colourless, crystallisable compounds, the lower members being very easily soluble in water, *e.g.*, acetamide; the solubility decreases with the increase of molecular weight, until finally they become insoluble. The boiling-points of the amides are much higher than those of the acids: ·

| | | | | |
|---|---|---|---|---|
| Acetic acid, Boiling-point, | 118° | Proprionic acid, Boiling-point, | | 141° |
| Acetamide, | „ 223° | Proprionamide, | „ | 213° |

While the entrance of an alkyl residue into the ammonia molecule does not change the basic character of the compound, as will be discussed more fully under methylamine, the entrance of a negative acid radical enfeebles the basic properties of the ammonia residue, so that the acid-amides possess only a very slight basic character. It is true that a salt corresponding to ammonium chloride — $CH_3.CO.NH_2.HCl$ — can be prepared from acetamide by the action of hydrochloric acid; but this shows a strong acid reaction, is unstable, and decomposes easily into its components. If it is desired to assign to the acid-amides a definite character, they must be regarded as acids rather than bases. One of the amido-hydrogen atoms possesses acid properties in that it may be replaced by metals. The mercury salts of the acid-amides may be prepared with especial ease, by boiling the solution of the amide with mercuric oxide:

$$2\,CH_3.CO.NH_2 + HgO = (CH_3.CO.NH)_2Hg + H_2O$$

EXPERIMENT: Some acetanilide is dissolved in water, treated with a little yellow mercuric oxide, and warmed. The latter goes into solution, and the salt of the formula given above is formed.

The amido-hydrogen atoms can also be replaced by the negative chlorine and bromine atoms, as well as by the positive metallic atoms. These substitution compounds are obtained by treating the amide with chlorine or bromine, in the presence of an alkali:

$CH_3.CO.NHCl$    $CH_3.CO.NHBr$    $CH_3.CO.NBr_2$
Acetchloramide    Acetbromamide    Acetdibromamide

The monohalogen substituted amides are of especial interest, since, on being warmed with alkalies, they yield primary alkylamines:

$$CH_3.CO.NHBr + H_2O = CH_3.NH_2 + HBr + CO_2$$

This important reaction will be taken up later, under the preparation of methyl amine from acetamide.

In the acid-amides, the acid radical is not firmly united with the ammonia residue; this is shown by the fact that they are saponified on boiling with water, more rapidly by warming with alkalies, *i.e.* decomposed into the acid and ammonia:

$$CH_3.CO.NH_2 + H_2O = CH_3.CO.OH + NH_3$$

EXPERIMENT: Heat some acetamide in a test-tube with caustic soda solution. An intensely ammoniacal odour is given off, while the solution contains sodium acetate.

If an acid-amide is treated with a dehydrating agent, *e.g.* phosphorus pentoxide, it is converted into a nitrile:

$$CH_3.CO.NH_2 = CH_3.CN + H_2O$$
Acetonitrile

The same result is obtained by treating them with phosphorus pentachloride; but in this case the intermediate products, the amide-chlorides or imide-chlorides are formed:

$$CH_3.CO.NH_2 + PCl_5 = CH_3.CCl_2.NH_2 + POCl_3$$
Amide-chloride

The very unstable amide-chloride then passes over, with the loss of one molecule of hydrochloric acid, into the more stable imide-chloride;

$$CH_3.CCl_2.NH_2 = CH_3.CCl=NH + HCl$$
Imide-chloride

And this finally into the nitrile,

$$CH_3.CCl=NH = CH_3.CN + HCl$$

## 5. REACTION: PREPARATION OF AN ACID-NITRILE FROM AN ACID-AMIDE

EXAMPLE: Acetonitrile from Acetamide[1]

To 15 grammes of phosphoric anhydride, contained in a small, dry flask, 10 grammes of dry acetamide is added. After the two substances are shaken well together, the flask is connected with a condenser, and then heated *carefully*, with a not too large *luminous* flame kept in *constant motion*. The reaction proceeds with much foaming. After the mixture has been heated a few minutes, the acetonitrile is then distilled over with a *large* luminous flame, kept in *constant motion*, into the receiver. The distillate is treated with half its volume of water, and then solid potash is added until it is no longer dissolved by the lower layer of liquid. The upper layer is removed with a capillary pipette, and distilled, a small amount of phosphoric anhydride being placed in the fractionating flask, for the complete dehydration of the nitrile. Boiling-point, 82°. Yield, about 5 grammes.

If an acid-amide is heated with a dehydrating agent (phosphorus pentoxide, pentasulphide, or pentachloride), it loses water, and passes over into the nitrile, *e.g.*:

$$CH_3 \cdot CO\ NH_2 = CH_3 \cdot C \equiv N + H_2O$$
Acetonitrile

Since, as has just been done, the acid-amide may be made by dehydrating the ammonium salt of an acid, thus, in a single operation, the nitrile may be obtained directly from the ammonium salt, if it is treated with a powerful dehydrating agent, *e.g.* ammonium acetate heated with phosphoric anhydride:

$$CH_3 \cdot COONH_4 = CH_3 \cdot CN + 2\,H_2O$$

The acid-nitriles may also be obtained by heating alkyl iodides (or bromides, chlorides) with alcoholic potassium cyanide:

$$CH_3 \boxed{I + K} CN = CH_3 \cdot CN + KI$$

$$\begin{array}{l} CH_2Br \\ | \\ CH_2Br \end{array} + 2\,KCN = \begin{array}{l} CH_2 \cdot CN \\ | \\ CH_2 \cdot CN \end{array} + 2\,KBr$$
Ethylene cyanide

---

[1] A. 64, 332.

$$C_6H_5 \cdot CH_2 \cdot Cl + KCN = C_6H_5 \cdot CH_2 \cdot CN + KCl$$
Benzyl chloride → Benzyl cyanide

or by the dry distillation of alkyl alkali sulphates with potassium cyanide:

$$SO_2\begin{cases}O\boxed{C_2H_5}\\OK\end{cases} + \boxed{CN}K = C_2H_5 \cdot CN + K_2SO_4$$

Ethyl potassium sulphate — Proprionitrile

These two reactions differ from those above in that the addition of a new atom of carbon is brought about. And further, the nitriles appear as cyanides of the alkyls; according to this conception, these compounds may be equally well designated as cyanides, *e.g.*:

$CH_3 \cdot CN$ = Acetonitrile = Methyl cyanide
$C_2H_5 \cdot CN$ = Proprionitrile = Ethyl cyanide
etc.           etc.            etc.

The lower members of the nitrile series are colourless liquids, the higher members, crystallisable solids; the solubility in water decreases with the increase in molecular weights. If they are heated with water up to 180° under pressure, they take up one molecule of water and are converted into the acid-amides:

$$CH_3 \cdot CN + H_2O = CH_3 \cdot CO \cdot NH_2$$

On heating with acids or alkalies, they take up two molecules of water, and pass over into the ammonium salt as an intermediate product:

$$CH_3 \cdot CN + 2 H_2O = CH_3 \cdot COONH_4$$

which immediately reacts with the alkali or acid, in accordance with the following equations:

$$CH_3 \cdot COONH_4 + KOH = CH_3 \cdot COOK + NH_3 + H_2O$$
$$CH_3 \cdot COONH_4 + HCl = CH_3 \cdot COOH + NH_4Cl$$

This process is called "saponification."

If nascent hydrogen (*e.g.* from zinc and sulphuric acid) be allowed to act on nitriles, primary amines are formed (Mendius' reaction):[1]

$$CH_3 \cdot CN + 2 H_2 = CH_3 \cdot CH_2 \cdot NH_2$$
Ethyl amine

---

[1] A. 121, 129.

Further, but of less importance, general reactions may be indicated by the following equations:

$$CH_3.CN + H_2S = CH_3.CS.NH_2$$
Thioacetamide

$$CH_3.CN + CH_3.CO.OH = \begin{matrix} CH_3.CO \\ CH_3.CO \end{matrix}\!\!>\!\!NH = \text{Diacetamide}$$

$$CH_3.CN + \begin{matrix} CH_3.CO \\ CH_3.CO \end{matrix}\!\!>\!\!O = N(CO.CH_3)_3 = \text{Triacetamide}$$

$$CH_3.CN + NH_2.OH = CH_3.C\!\!<^{N.OH}_{NH_2}$$
Hydroxylamine     Acetamide-oxime

$$CH_3.CN + HCl = CH_3.C\!\!<^{NH}_{Cl}$$
Imide-chloride

## 6. REACTION: PREPARATION OF AN ACID-ESTER FROM THE ACID AND ALCOHOL

EXAMPLE: **Acetic Ester from Acetic Acid and Ethyl Alcohol**[1]

A ½-litre flask, containing a mixture of 50 c.c. of alcohol and 50 c.c. of concentrated sulphuric acid, is closed by a two-hole cork; through one hole passes a dropping-funnel, through the other a glass delivery tube connected with a long condenser. The mixture is heated in an oil-bath to 140° (thermometer in oil); when this temperature is reached, a mixture of 400 c.c. of alcohol and 400 c.c. of glacial acetic acid is gradually added through the funnel, at the same rate at which the ethyl acetate (acetic ester), formed in the reaction, distils over. In order to remove the acetic acid carried over, the distillate is treated in an *open* vessel with a dilute solution of sodium carbonate until the upper layer will not redden blue litmus paper. The layers are now separated with a dropping-funnel; the upper layer is filtered through a dry folded filter, and shaken up with a solution of 100 grammes of calcium

---

[1] Bl. 33, 350.

chloride in 100 grammes of water, in order to remove the alcohol. The two layers are again separated with the funnel, the upper one dried with granular calcium chloride and then distilled on the water-bath (see page 18). Boiling-point, 78°. Yield, about 80–90% of the theory.

Acid-esters can be obtained directly by the action of acids on alcohols:

I. $CH_3.CO.OH + C_2H_5.OH = CH_3.CO.OC_2H_5 + H_2O$

A reaction of this kind never takes place to any degree in quantitative proportions, since, as soon as a certain quantity of the ester is formed, and the corresponding quantity of water, the latter saponifies the ester, in accordance with the following equation:

II. $CH_3.CO.OC_2H_5 + H_2O = CH_3.CO.OH + C_2H_5.OH$

In such cases, a condition of equilibrium is reached, in which an equal number of molecules of the ester are formed, according to equation I., and decomposed, according to equation II., at the same time. There are several methods by the use of which this condition may be changed. The saponifying action of the water must be neutralised; accordingly a dehydrating agent is added to the mixture, sulphuric acid being frequently used for this purpose in the preparation of acid-esters. In many cases, where the salts of organic acids are more readily obtained than the free acids, they may be used for the preparation of the esters by heating them directly with alcohol and sulphuric acid. Other methods for the preparation of acid-esters have been referred to, and, in part, carried on practically on the small scale in the foregoing preparations, so that at this point it is only necessary to recall the equations:

(1) $CH_3.CO.OAg + IC_2H_5 = CH_3.CO.OC_2H_5 + AgI$

(2) $CH_3.CO.Cl + C_2H_5.OH = CH_3.CO.OC_2H_5 + HCl$

It will be remembered that, in these reactions, the previously prepared chloride was not used, but that hydrochloric acid was conducted into a mixture of the acid and alcohol (see acetyl chloride).

(3) $\begin{matrix} CH_3.CO \\ CH_3.CO \end{matrix}\!\!>\!\!O + C_2H_5.OH = CH_3.CO.OC_2H_5 + CH_3.CO.OH$

Concerning the purification of the acid-esters, it may be mentioned, that the crude reaction product is shaken with a sodium carbonate solution until the ester no longer shows an acid reaction, in order to remove the free acid. The alcohol may be removed from esters difficultly soluble in water by repeatedly washing with water; in case an ester is moderately soluble in water, as ethyl acetate, it is better to use a solution of calcium chloride to remove the alcohol.

The lower members of the series of acid-esters are colourless liquids with pleasant, fruit-like odours; the higher members, as well as those of the aromatic acids, are crystallisable compounds. The boiling-points of esters containing alkyl residues of small molecular weights ($CH_3$, $C_2H_5$, $C_3H_7$) are lower than those of the corresponding acids; the entrance of more complex alkyl residues raises the boiling-points:

$CH_3.CO.OCH_3$ Boiling-point, 57°
$CH_3.CO.OC_2H_5$ " " 78°
$CH_3.CO.OH$ " " 118°
$CH_3.CO.OC_6H_{13}$ " " 169°
Hexyl acetate

As already mentioned, the esters are saponified by heating with water:

$$CH_3.CO.OC_2H_5 + H_2O = CH_3.CO.OH + C_2H_5.OH$$

The saponification is effected more readily by heating with alkalies:

$$CH_3.CO.OC_2H_5 + KOH = CH_3.CO.OK + C_2H_5.OH$$

Other methods of saponification will be further discussed when a practical example is taken up. The action of ammonia upon acid-esters, forming acid-amides, has already been referred to under acetamide:

$$CH_3.CO.OC_2H_5 + NH_3 = CH_3.CO.NH_2 + C_2H_5.OH.$$

### 7. REACTION: SUBSTITUTION OF HYDROGEN BY CHLORINE

EXAMPLE: **Monochloracetic Acid from Acetic Acid and Chlorine** [1]

A current of dry chlorine is passed into a mixture of 150 grammes of glacial acetic acid and 12 grammes of red phosphorus, contained in a flask provided with a delivery tube and an inverted

---

[1] R. 23, 222; A. 102, 1.

condenser; the flask is heated on a rapidly boiling water-bath, and must be placed in such a position as to receive as much light as possible. The best result is obtained by performing the reaction in the direct sunlight, since the success of the chlorination depends essentially on the action of the sun's rays. The reaction is ended when a small test-portion cooled with ice-water solidifies on rubbing the walls of the vessel (test-tube) with a glass rod. In summer the chlorination may require a single day, while during the cloudy days of winter, two days may be necessary. For the separation of the monochloracetic acid the reaction-product is fractionated from a distilling flask connected with a long air condenser. The fraction passing over from 150–200° is collected in a separate beaker; this is cooled in ice-water and the walls rubbed with a glass rod. The portion solidifying, consisting of pure monochloracetic acid is rapidly filtered with suction, the loose crystals being pressed together with a spatula or mortar-pestle. The suction must not be continued too long, because the monochloracetic acid gradually becomes liquid in warm air. The filtrate is again distilled, and the portion passing over between 170–200° is collected in a separate vessel. This is treated as before (cooling and filtering), and there is obtained a second portion of monochloracetic acid; this is united with the main quantity, which is again distilled. The product thus obtained is perfectly pure. Boiling-point, 186°. Yield varying, 80–125 grammes.

Since monochloracetic acid, especially when warm, attacks the skin with great violence, care must be taken in handling it.

Chlorine or bromine substitution products of aliphatic carbonic acids can be obtained by the direct action of the halogen on the acids:

$$CH_3.CO.OH + Cl_2 = CH_2Cl.CO.OH + HCl$$
$$(Br_2) \quad (Br) \quad\quad (HBr)$$

If the reaction is allowed to continue for a long time, other substitution products can also be obtained. But the action of chlorine or bromine on acids is very sluggish. It may be essentially facilitated by certain conditions. If, *e.g.*, the operation is conducted in direct sunlight, the reaction proceeds much more rapidly than in a dark place. The reaction is assisted more effectively by adding a so-called "carrier." Iodine may be used as such for the introduction of chlorine or bromine. When added in small quantities to the substance to be substituted, it causes the substitution to take place more rapidly and completely. The continuous action of this carrier depends upon the following facts: In the first phase of the reaction, chlorine iodide is formed:

$$(1)\ Cl + I = ICl$$

This acts, then, as a chlorinating agent in the second phase, according to the following reaction:

$$(2)\ CH_3.CO.OH + ICl = CH_2Cl.CO.OH + HI$$

The chlorine acts upon the hydriodic acid as follows:

$$(3)\ HI + Cl_2 = ICl + HCl$$

The molecule of chlorine iodide is thus formed anew (equation 1) and can chlorinate another molecule of acetic acid, and so on. The action of the iodine in the last case depends upon the fact that the molecule of chlorine iodide (ICl) is more easily decomposed into its atoms than the molecule of chlorine ($Cl_2$). The disadvantage necessarily incident to the use of iodine as a carrier is, that the reaction-product is easily contaminated with iodine derivatives, — in small quantities, it is true.

In an entirely different way the chlorination is facilitated by the addition of red phosphorus. In this case, phosphorus pentachloride is first formed from the phosphorus and chlorine; this, acting on the acetic acid, generates acetyl chloride, and this latter, with an excess of the acid, forms the anhydride. Direct experiments have shown that acid-chlorides, as well as anhydrides, are substituted by chlorine with much greater ease than the corresponding acids; in this fact the action of red phosphorus finds its explanation. Since a small amount of phosphorus is sufficient for the chlorination of a large amount of acetic acid, the question as to how this is continuously effected remains to be answered. In accordance with the above statements, the following reactions take place:

$$(1)\ P + Cl_5 = PCl_5$$

$$(2)\ CH_3.CO.OH + PCl_5 = CH_3.CO.Cl + POCl_3 + HCl$$

$$(3)\ CH_3.CO.Cl + CH_3.CO.OH = \begin{matrix} CH_3.CO \\ CH_3.CO \end{matrix}\!\!>\!O + HCl$$

If the chlorine now acts on the anhydride, monochloracetic anhydride is formed:

(4) $\begin{matrix} CH_3.CO \\ CH_3.CO \end{matrix}\Big\rangle O + Cl_2 = \begin{matrix} CH_2Cl.CO \\ CH_3.CO \end{matrix}\Big\rangle O + HCl$

But this reacts directly with the hydrochloric acid, in accordance with this equation:

(5) $\begin{matrix} CH_2Cl.CO \\ CH_3.CO \end{matrix}\Big\rangle O + HCl = CH_2Cl.CO.OH + CH_3.CO.Cl$

There is thus obtained, besides the molecule of chloracetic acid, the molecule of acetyl chloride first formed in reaction 2, which is utilised repeatedly by its regeneration in accordance with reactions 3, 4, and 5.

As a substitute for red phosphorus, sulphur is also recommended for the chlorination of aliphatic acids. This acts in a wholly similar manner, since it first forms sulphur chloride, which, reacting on the acid, like phosphorus chloride, converts it into an acid-chloride. The other phases of the reaction are similar to those given above.

The bromination of aliphatic carbonic acids, which is not only of great importance in preparation work, but also as a means for determining constitution, is also conducted with the addition of red phosphorus (Hell-Volhard-Zelinsky Method).[1]

The halogen atoms always enter the α-position to the carboxyl group. Thus, *e.g.*, when proprionic and butyric acids are brominated, they yield:

$CH_3.CHBr.CO.OH$    $CH_3.CH_2.CHBr.CO.OH$
α-Bromproprionic             α-Brombutyric acid

If no α-hydrogen atom is present, *e.g.*, in trimethyl-acetic acid $(CH_3)_3.\underset{(a)}{C}.CO.OH$, bromination will not occur. The ability of an acid to form a bromine substitution product can, therefore, be used as a test for the presence of an α-hydrogen atom. Iodine cannot be introduced directly into aliphatic acids like chlorine and bromine. To obtain iodine substitution products, it is necessary to treat the corresponding chlorine or bromine compound with potassium iodide:

$$CH_2Cl.CO.OH + KI = CH_2I.CO.OH + KCl$$

The halogen derivatives of the fatty acids are in part liquids, in part

---

[1] B. 14, 891; 21, 1726; A. 242, 141; B. 21, 1904; B. 20, 2026; B. 24, 2216.

solids. In their reactions they resemble the acids, on the one hand, since they form salts, chlorides, anhydrides, esters, etc.; on the other hand, the halogen alkyls. They are of great value in the preparation of oxy- and amido-acids, of unsaturated acids, for the synthesis of polybasic acids, etc. Below are given a few equations capable of general application:

$$CH_2Cl.CO.OH + H_2O = CH_2(OH).CO.OH + HCl$$
Oxyacetic acid = Glycolic acid

$$CH_2Cl.CO.OH + NH_3 = CH_2.NH_2.CO.OH + HCl$$
Amidoacetic acid = Glycocoll

$$\begin{array}{l} CH_2I \\ | \\ CH_2 + KOH \\ | \\ COOH \end{array} \qquad \begin{array}{l} CH_2 \\ \| \\ = CH \;\; + KI + H_2O \\ | \\ CO.OH \end{array}$$
Acrylic acid

$$CH_2Cl.CO.OH + KCN = CH_2.CN.CO.OH + KCl$$
Cyanacetic acid

$$2\,CH_2Br.CO.OH + Ag_2 = \begin{array}{l} CO.OH \\ | \\ CH_2 \\ | \\ CH_2 \\ | \\ CO.OH \end{array} + 2\,AgBr$$
Succinic Acid

### 8. REACTION: OXIDATION OF A PRIMARY ALCOHOL TO AN ALDEHYDE

EXAMPLE: **Acetaldehyde from Ethyl Alcohol**[1]

A 1½-litre flask containing 110 grammes of concentrated sulphuric acid and 200 grammes of water is closed by a two-hole cork; through one hole passes a dropping-funnel, through the other a glass delivery tube connected with a long condenser. To the lower end of the condenser is attached an adapter bent downwards, the narrower portion of which passes through a cork in

---

[1] A. 14, 133; J. 1853, 329.

the neck of a thick-walled suction flask of about ½-litre capacity. (See Fig. 56, page 121.) The latter is placed in a water-bath filled with a freezing mixture of ice and salt. The larger flask is heated over a wire-gauze until the water just begins to boil; a solution of 200 grammes of sodium dichromate in 200 grammes of water which has been treated with 100 grammes of alcohol is then added in a small stream through the dropping-funnel, the lower end of which is about 3 cm. above the surface of the liquid in the flask. During the addition of the mixture, it will be unnecessary to heat the flask, since the heat produced by the reaction is sufficient to cause ebullition. The aldehyde thus formed distils into the receiver, besides some alcohol, water, and acetal. If uncondensed vapours of the aldehyde escape from the receiver, the mixture is admitted to the flask more slowly. On the other hand, if boiling is not caused by the flowing in of the mixture, the reaction is assisted by heating with a small flame. After all of the mixture has been added, the flask is heated for a short time by a flame, until boiling begins.

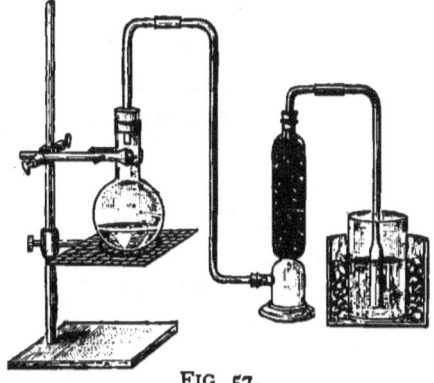

FIG. 57.

Since the aldehyde cannot be obtained easily from the reaction-products by fractional distillation, it is first converted into aldehyde-ammonia, which, on proper treatment, readily yields the pure aldehyde.

The apparatus for this purpose is arranged as follows: A small flask to contain the aldehyde, placed on a wire-gauze, is connected with a moderately large reflux condenser. Into the upper end of the condenser is placed a cork bearing a ⌐-shaped glass tube, which is connected with two wash-bottles, each containing 50 c.c. of dried ether. (See Fig. 59 *a*, page 165.) After the condenser has been filled with water at 30° (the lower side-tube of the condenser is closed with rubber tubing and a pinch-cock), the crude aldehyde is heated for 5–10 minutes, to gentle boiling, and the aldehyde that is not condensed passes over, and is absorbed by the ether. Should the ether begin to ascend in the connecting tube, the flame must be somewhat increased immediately. To obtain aldehyde-ammonia, a current of dry ammonia (see page 319) is conducted, with the aid of a wide adapter or funnel (Fig. 57), into the ethereal solution contained in a beaker surrounded by a freezing mixture of ice and salt, until the liquid smells strongly of it. After an hour, the aldehyde-ammonia which has separated out is scraped from the sides of the vessel and adapter with a spatula or knife, and filtered with suction, washed with a little ether, and then allowed to dry on filter-paper in a desiccator. Yield, about 30 grammes.

In order to obtain pure aldehyde, 10 grammes of aldehyde-ammonia is dissolved in 10 grammes of water, treated with a cooled mixture of 15 grammes of concentrated sulphuric acid and 20 grammes of water, and heated on the water-bath. Since aldehyde has a low boiling-point ($21°$), the receiver is connected with the condenser by a cork, and well cooled with ice and salt.

Aldehydes can be obtained by the use of the general reaction, which in many cases serves as a method of preparation, of extracting two hydrogen atoms from a primary alcohol by oxidation.

$$CH_3 \cdot CH_2 \cdot OH + O = CH_3 \cdot C\!\!\begin{array}{c}\diagup H \\ \diagdown O\end{array} + H_2O$$

The name of the class of compounds is derived from this action: Aldehyde = Al(cohol)dehyd (rogenatum). As an oxidising agent in the above case, chromic acid is the most suitable in the form of potassium, or sodium dichromate in the presence of sulphuric acid:

$$Na_2Cr_2O_7 + 4\,H_2SO_4 = O_3 + (SO_4)_3Cr_2 + Na_2SO_4 + 4\,H_2O.$$

As an oxidising agent, the rather difficultly soluble potassium dichromate (1 part dissolves in 10 parts water) was formerly generally used, but at present the more soluble and cheaper sodium salt is employed wherever it is possible. But in the preparation of the simplest aldehyde (formaldehyde) from an alcohol a different oxidising agent is used, viz. the oxygen of the air. On passing a mixture of the vapour of methyl alcohol and air over a heated copper spiral, formaldehyde is produced.

While by the first reaction one proceeds from substances which in comparison with the aldehydes are oxidation products of a lower order, the aldehydes may also be obtained by a second method involving the use of compounds of the same substitution series, viz. the dihalogen derivatives of the hydrocarbons containing the group $CHCl_2$ or $CHBr_2$. If these are boiled with water, or, better, water containing sodium carbonate, potash, lead oxide, or calcium carbonate, etc., the two halogen atoms are replaced by one oxygen atom:

$$CH_3.CHCl_2 + H_2O = CH_3.CHO + 2\,HCl$$
Ethylidene chloride

$$C_6H_5.CHCl_2 + H_2O = C_6H_5.CHO + 2\,HCl$$
Benzal chloride  Benzaldehyde

This method is used on the large scale for the manufacture of the commercially important benzaldehyde. It will be referred to under benzaldehyde.

Finally, aldehydes can be prepared from their oxidation products, the carbonic acids, by two methods, one of which has been already mentioned under acetic anhydride. If sodium amalgam is allowed to act on acid-anhydrides, an aldehyde is first formed:

$$\begin{matrix} CH_3.CO \\ CH_3.CO \end{matrix}\!\!>\!\!O + H_2 = CH_3.CHO + CH_3.CO.OH$$

But this reaction is of little practical value for the preparation of aldehydes. The second method, which is the real preparation method, consists in the dry distillation of a mixture of the calcium or barium salt of the acid with calcium or barium formate:

$$CH_3.CO.Oca + H.CO.Oca = CH_3.CHO + CaCO_3$$
$$(ca = \tfrac{1}{2}Ca)$$

The lower members of the aldehyde series are colourless liquids, soluble in water, possessing pungent odours. The intermediate members are also liquids, but insoluble in water; the higher members are solid, crystallisable substances. The boiling-points of the aldehydes are lower than those of the corresponding alcohols.

$\begin{cases} CH_3.CHO \\ CH_3.CH_2.OH \\ CH_3.CH_2.CHO \\ CH_3.CH_2.CH_2.OH \end{cases}$ . . . . . Boiling-point, $21°$
" " $78°$
" " $50°$
" " $97°$

Aldehydes are oxidised to acids by free as well as combined oxygen (compare benzaldehyde):

$$CH_3.CHO + O = CH_3.CO.OH$$

Upon this action depends the fact that aldehydes cause metals to separate from certain salts, *e.g.* silver nitrate:

$$2\,CH_3.CHO + 2\,NO_3Ag + H_2O = 2\,CH_3.CO.OH + NO_3H + Ag_2$$

EXPERIMENT: Treat a few cubic centimetres of a diluted silver nitrate solution with a few drops of ammonium hydroxide and 5 drops of aldehyde. Silver will be deposited in the form of a brilliant mirror on the walls of the vessel; the deposition is especially beautiful if the vessel has previously been treated with a solution of caustic soda to remove any fatty matter. The reaction frequently takes place at the ordinary temperature, but in many cases only on gentle warming. The reaction is used for the detection of aldehydes.

Another reaction which can also be employed for the recognition of aldehydes, depends upon the fact that they give a red colour to fuchsine-sulphurous acid. (Caro's reaction.)

EXPERIMENT: Fuchsine-sulphurous acid is prepared by dissolving fuchsine in water; a sufficient quantity of the latter is taken to prevent the solution from being too intense in colour. Sulphur dioxide is conducted into this until a complete decolouration takes place. To a few cubic centimetres of this solution add several drops of aldehyde. On shaking, a violet-red colour will be produced.

Finally aldehydes may also be detected by a method depending upon the fact that when treated with diazobenzene sulphonic acid and sodium amalgam, they give a violet colour.

EXPERIMENT: To as much diazobenzene sulphonic acid as can be held on the point of a knife, add 5 c.c. of water, a few drops of caustic soda, and then a few drops of aldehyde, and finally a piece of solid sodium amalgam as large as a pea. After some time the red-violet colour appears.

That aldehydes upon reduction pass over to primary alcohols has been mentioned under acetic anhydride:

$$CH_3.CHO + H_2 = CH_3.CH_2.OH$$

It is wholly characteristic of the aldehydes that they unite directly (1) with ammonia, (2) sodium hydrogen sulphite, and (3) hydrocyanic acid. The union with ammonia takes place in accordance with the following equation:

$$CH_3.CHO + NH_3 = CH_3.CH\begin{subarray}{l}\diagup OH \\ \diagdown NH_2\end{subarray}$$

Aldehyde-ammonia=
α-amidoethyl alcohol

This reaction is not so common as the second and third. Thus, *e.g.*, formic aldehyde and most of the aromatic aldehydes behave differently toward ammonia. Whenever this reaction does take place it can also be used with advantage for the purification of the aldehyde, as in case of acetaldehyde; by allowing the well-crystallised double compound to separate out, on treating it with dilute sulphuric acid, the free aldehyde is obtained.

The union with sodium hydrogen sulphite takes place in accordance with the following equation:

$$CH_3.CHO + SO_3NaH = CH_3.CH\begin{subarray}{l}\diagup OH \\ \diagdown SO_3Na\end{subarray}$$

Sodium-α-oxyethyl sulphonate

This reaction may also be used for the purification of aldehydes, since when a concentrated solution of the sulphite is employed, the double compound generally separates out in a crystallised condition. The free aldehydes can be obtained from the sulphite compounds by heating with dilute acids or alkali carbonates. (Compare benzaldehyde.)

EXPERIMENT: Treat 5 c.c. of a cooled concentrated solution of sodium hydrogen sulphite with 1 c.c. of aldehyde and shake the mixture. The double compound separates out in a crystallised condition.

As distinguished from the union of aldehyde with ammonia, this reaction is wholly general, and is frequently of great value in dealing with the aldehydes of the aromatic series. It should be noticed in this connection, that the ketones, which are closely related to the aldehydes, show similar reactions:

$$CH_3.CO.CH_3 + SO_3NaH = \underset{\underset{CH_3}{|}}{\overset{\overset{CH_3}{|}}{C}}\!\!\begin{array}{l}OH\\SO_3Na\end{array}$$
Acetone

The addition of hydrocyanic acid to ketones as well as to aldehydes is also general:

$$CH_3.CHO + HCN = CH_3.CH\!\!\begin{array}{l}OH\\CN\end{array}$$
a-oxyproprionitrile

This reaction is of especial interest, since the addition of a new carbon atom is brought about. Concerning the value of this reaction for the synthesis of a-oxyacids, see Mandelic Acid Nitrile, page 261.

Aldehydes possess further a marked tendency to combine with themselves (polymerise).

EXPERIMENT: Treat 1 c.c. of aldehyde with one drop of concentrated sulphuric acid. The aldehyde boils, and polymerisation (condensation) takes place.

The compound thus obtained is called paraldehyde; it boils much higher (124°) than ordinary aldehyde; the determination of its vapour density shows that one molecule is composed of three molecules of ordinary aldehyde. Paraldehyde does not show the characteristic aldehyde reactions; on distillation with dilute sulphuric acid it is converted back into the ordinary variety. For this reason it is believed that no new union of carbon atoms takes place in the molecule, but that three molecules are united by means of the oxygen atoms:

$$CH_3.CH\!\!\begin{array}{l}O-CH-CH_3\\ \phantom{O-CH}\phantom{-}O\\O-CH-CH_3\end{array}$$

If aldehyde is *cooled* and treated with sulphuric acid, or if at the ordinary temperature gaseous hydrochloric acid, sulphur dioxide, or other compounds are passed into it, a solid polymerisation product, metaldehyde, is formed: this can also be converted back into the ordinary variety.

The aldehydes undergo a wholly different kind of polymerisation under certain conditions, concerning which reference must be made to the chemical literature and treatises. For example, two molecules of acetaldehyde can unite with the formation of a compound in which a new carbon union is present.

$$CH_3.CHO + CH_3.CHO = CH_3.CH(OH).CH_2.CHO$$
Aldol

This compound, as distinguished from paraldehyde and metaldehyde, is a true aldehyde, in that it cannot be converted back to acetaldehyde. Aldol loses water easily, and is converted into an unsaturated aldehyde:

$$CH_3.CH(OH).CH_2.CHO = CH_3.CH\!=\!CH.CHO$$
Crotonaldehyde

In connection with these condensations, it may be pointed out that many aldehydes, when heated with alkalies, polymerise to resinous products of high molecular weight (aldehyde resins).

EXPERIMENT: Treat a few cubic centimetres of caustic potash solution with several drops of aldehyde, and warm. A yellow colouration takes place with the separation of a resinous mass.

In order, finally, to represent the great activity of the aldehydes, the following equations are given:

$CH_3.CHO + PCl_5 \qquad = CH_3.CHCl_2 + POCl_3$
Ethylidene chloride

$CH_3.CHO + NH_2.OH \qquad = CH_3.CH\!=\!NOH + H_2O$
Aldoxime

$CH_3.CHO + NH_2.NH.C_6H_5 = CH_3.CH\!=\!N.NH.C_6H_5 + H_2O$
Phenylhydrazine              Hydrazone of aldehyde

$$CH_3.CHO + 2\,C_2H_5.OH = CH_3.CH\!\!\begin{array}{c}OC_2H_5\\OC_2H_5\end{array} + H_2O$$

Acetal, the ether of aldehyde-hydrate $CH_3.CH\!<\!^{OH}_{OH}$ which does not exist in the free condition

$$CH_3.CHO + 2\,C_6H_6 = CH_3.CH\!\!\begin{array}{c}C_6H_5\\C_6H_5\end{array} + H_2O$$
Diphenyl ethane

## 9. REACTION: PREPARATION OF A PRIMARY AMINE FROM AN ACID-AMIDE OF THE NEXT HIGHER SERIES

EXAMPLE: **Methyl Amine from Acetamide**[1]

To a mixture of 25 grammes of acetamide, which has been previously well pressed out on a porous plate, and 70 grammes (23 c.c.) of bromine contained in a $\frac{1}{2}$-litre flask, add a solution of 40 grammes of caustic potash in 350 c.c. of water (the flask is well cooled with water), until the brownish red colour formed at first is changed to a bright yellow, for which the greater portion of the potash solution will be required. This reaction-mixture is then, in the course of a few minutes, allowed to flow from a dropping-funnel in a continuous stream into a litre flask containing a solution of 80 grammes of caustic potash in 150 c.c. of water heated to 70–75°. In case the temperature rises higher than 75°, the flask must be cooled by immersion for a short time, in cold water. The liquid is maintained at this temperature until it becomes colourless, which usually requires a quarter to half hour. The methyl amine is then distilled off with steam, and collected in a receiver containing a mixture of 60 grammes of concentrated hydrochloric acid and 40 grammes of water. In order that the methyl amine may be completely absorbed by the acid, the end of the condenser is connected with an adapter which dips 1 cm. below the surface of the liquid in the receiver. As soon as the liquid condensed in the condenser no longer shows an alkaline reaction, the distillation is discontinued. The methyl amine hydrochloride is partially evaporated over a free flame, then to dryness on the water-bath, and is finally heated for a short time in an air-bath at 100°, to dusty dryness. In order to separate the methyl amine salt from the ammonium chloride mixed with it, the finely pulverised substance is crystallised from absolute alcohol, and the crystals separating out dried in a desiccator. Yield, varying.

---

[1] B. 15, 762; B. 17, 1406 and 1920.

# ALIPHATIC SERIES 145

Under the discussion of acid-amides, it has already been mentioned that the hydrogen of the amido group ($NH_2$) can be substituted by bromine. If a 10 % solution of caustic potash is added to a mixture of one molecule of the amide and one molecule of bromine, until the brownish red colour of the latter has vanished, a monobromamide is formed, e.g., in the above case, acet-monobromamide in accordance with this equation:

$$CH_3.CO.NH_2 + Br_2 + KOH = CH_3.CO.NHBr + KBr + H_2O$$

The monobromamide may be isolated in pure condition in the form of colourless, hydrous crystals. If hydrobromic acid is abstracted, no water being present, the position of the carbonyl group (CO) is changed, and an ester of isocyanic acid is formed:

$$CH_3.CONHBr = CH_3.N{=}CO + HBr$$
Methyl isocyanate

If the attempt is made to eliminate hydrobromic acid in the presence of water by the use of caustic potash solution, the above reaction takes place, but the isocyanate is unstable in the presence of alkalies, and decomposes immediately by taking up the water forming carbon dioxide and a primary amine:

$$CH_3.NCO + H_2O = CO_2 + CH_3.NH_2$$
Methyl amine

This reaction, discovered by A. W. Hofmann, is, therefore, in its last phase, identical with the historical reaction of Wurtz, which led him, in 1848, to the discovery of the primary amines.

The Hofmann reaction is capable of general application. By use of it, the primary amine of the next lower series may be obtained from any acid-amide, since the elimination of carbon dioxide takes place. With the higher members of the series, the reaction in part proceeds still further, since the bromine acts upon the primary amine to form a nitrile:

$$C_7H_{15}.CH_2.NH_2 + Br_4 + 4\,NaOH = C_7H_{15}.C{\equiv}N + 4\,NaBr + 4\,H_2O$$
Octyl amine	Octonitrile

There is thus obtained from the higher members (compounds having five or more carbon atoms) of the amides, first the primary amine, and secondly, the nitrile of the next lower acid. In the aromatic series, the reaction for the preparation of primary amines, which contain the amido group in the benzene ring, is not of general importance, since these

L

may be obtained from the easily accessible nitro-compounds; and since, on the other hand, if the above reaction is employed, bromine substitution products are easily formed. But in those cases in which the nitro-compound corresponding to the amine is not known, or can be prepared only with difficulty, the reaction is also of importance in the aromatic series. Two cases of this kind may be mentioned in this place. If phenyl acetamide is treated in accordance with Hofmann's reaction, there is formed in the usual way benzyl amine:

$$\underset{\text{Phenyl acetamide}}{C_6H_5 \cdot CH_2 \cdot CO \cdot NH_2} + O = \underset{\text{Benzyl amine}}{C_6H_5 \cdot CH_2NH_2} + CO_2$$

Further, the reaction is of practical value in the preparation of o-amidobenzoic acid. If, as above, bromine and caustic potash are allowed to act on phthalimide, there is first formed, by the addition of water, an acid-amide:

$$o\text{-}C_6H_4\!\!\begin{array}{c}CO\\ \diagdown \\ CO\end{array}\!\!NH + H_2O = C_6H_4\!\!\begin{array}{c}CO \cdot NH_2\\ \diagdown \\ CO \cdot OH\end{array}$$

which in accordance with the following reactions gives the amido-acid.

$$C_6H_4\!\!\begin{array}{c}CO \cdot NH_2\\ \diagdown \\ CO \cdot OH\end{array} + Br_2 = C_6H_4\!\!\begin{array}{c}CO \cdot NHBr\\ \diagdown \\ CO \cdot OH\end{array} + HBr$$

$$C_6H_4\!\!\begin{array}{c}CO \cdot NHBr\\ \diagdown \\ CO \cdot OH\end{array} = C_6H_4\!\!\begin{array}{c}N{=}C{=}O\\ \diagdown \\ CO \cdot OH\end{array} + HBr$$

$$C_6H_4\!\!\begin{array}{c}N{=}C{=}O\\ \diagdown \\ CO \cdot OH\end{array} + H_2O = C_6H_4\!\!\begin{array}{c}NH_2\\ \diagdown \\ CO \cdot OH\end{array} + CO_2$$

Since the nitro-acid corresponding to the o-amido benzoic acid is difficult to obtain, and phthalimide is easily prepared, the Hofmann reaction in this case gives a very convenient method of preparation for the amido-acid.

Primary aliphatic amines can also be prepared according to the following equations:

(1) By the action of alcoholic ammonia on halogen alkyls:

$$CH_3I + NH_3 = CH_3 \cdot NH_2 + HI$$

In this case, secondary and tertiary bases, or the corresponding ammonium compounds, are also formed.

(2) From alcohols and zinc chloride-ammonia:

$$C_2H_5.OH + NH_3 = C_2H_5.NH_2 + H_2O$$

(3) By the reduction of nitriles (Mendius' reaction):

$$CH_3.CN + 2H_2 = CH_3.CH_2.NH_2$$

(4) By the reduction of nitro-compounds:

$$CH_3.NO_2 + 3H_2 = CH_3.NH_2 + 2H_2O$$

(5) By the reduction of oximes and hydrazones:

$$CH_3.CH{=}N.OH + 2H_2 = CH_3.CH_2.NH_2 + H_2O$$
Acetaldoxime

$$CH_3.CH{=}N - NH.C_6H_5 + 2H_2 = CH_3.CH_2.NH_2 + C_6H_5.NH_2$$
Ethylidene phenyl hydrazone　　　　　　　　　　　　　　　　Aniline

The lowest members of the amines in the free condition are gaseous compounds soluble in water, possessing odours suggestive of ammonia: they differ from ammonia in being inflammable.

EXPERIMENT : Treat some solid methyl amine hydrochloride in a small test-tube with a concentrated solution of caustic potash, or caustic soda, and warm gently. A gas, smelling like ammonia, is evolved, which burns with a pale flame.

The higher members are liquids or insoluble solids. Since they are derivatives of ammonia, they possess basic properties, and, like ammonia, unite with acids to form salts, the composition of which is analogous to that of the ammonium compounds:

$$NH_3.HCl \longrightarrow CH_3.NH_2.HCl$$
$$(NH_4Cl)_2PtCl_4 \longrightarrow (CH_3.NH_2.HCl)_2PtCl_4$$
$$NH_4Cl, AuCl_3 \longrightarrow CH_3.NH_2.HCl, AuCl_3$$

The hydrochlorides of organic bases are distinguished from ammonium chloride by their solubility in absolute alcohol. Use was made of this property above.

The numerous reactions of the primary amines need not be mentioned here, since, under the aromatic amines, frequent reference will be made to them. At this place, one difference between the aromatic and aliphatic amines will be pointed out. If nitrous acid is allowed to act on an aliphatic primary amine, an alcohol is formed with evolution of nitrogen:

$$CH_3 \cdot NH_2 + NOOH = CH_3 \cdot OH + N_2 + H_2O$$

while, under these conditions, an aromatic amine is converted into a diazo-compound (see Diazo-compounds).

### 10. REACTION: SYNTHESES OF KETONE ACID-ESTERS AND POLYKETONES WITH SODIUM AND SODIUM ALCOHOLATE

EXAMPLE: **Acetacetic Ester from Acetic Ester and Sodium**[1]

For the successful preparation of acetacetic ester, the character of the acetic ester used is of great importance, since, if it is completely free from alcohol, it will be attacked very slowly by sodium, even on heating; if, on the other hand, it contains too much alcohol, the sodium acts easily, but the yield of the product is varying and usually small. According to the experiments of the author, the following method of procedure gives a good yield and is one that does not fail.

*Purification of Acetic Ester:* The acetic ester prepared according to Reaction 6, even after it has been freed from acetic acid and alcohol by shaking with sodium carbonate and calcium chloride respectively, dried over calcium chloride, and finally rectified, is not suitable for this preparation, since it reacts too violently with sodium. But if it is allowed to stand, after distilling, for some hours, over night at least, in a well-closed flask, over about $\frac{1}{8}$ its volume of granulated calcium chloride, and is then filtered, it may be used for the successful preparation of acetacetic ester.

If commercial acetic ester is to be used, it must be shaken with a sodium carbonate solution, as described on page 130, treated with calcium chloride solution, etc.; in short, it is treated as the crude product obtained in the preparation of acetic ester. Obviously, it is also necessary to allow it to stand over night in contact with calcium chloride, after the distillation.

*Preparation of Acetacetic Ester:* 25 grammes of sodium from which the outside layers have been removed is cut with the aid of a sodium knife (Fig. 81) into pieces as thin as possible, and

---

[1] A. 186, 214.

placed in a dry litre-flask. After this is connected with a long reflux condenser, inclined at an oblique angle, 250 grammes of acetic ester is poured into the top of the condenser, by a funnel which must not be attached to the condenser, but is held in the hand, so that the air may escape. If the acetic ester is added properly, no violent ebullition will occur, but at first a gradual, gentle boiling. After 10 minutes, the flask is placed on a previously heated water-bath, the temperature of which is so regulated that the acetic ester boils but gently; the reaction-mixture is heated until all the sodium is dissolved, which will require from 3–4 hours. To the warm liquid is added a mixture of 80 grammes of glacial acetic acid and 80 grammes of water, until it just shows an acid reaction. If a thick, porridge-like mass should separate out, this is again dissolved by vigorous shaking, or carefully breaking up the small lumps with a glass rod. To the liquid is then added an equal volume of a cold saturated solution of sodium chloride, and the lower aqueous layer is separated from the upper one, consisting of acetic ester and acetacetic ester, by allowing it to run off from a dropping-funnel. Should a precipitate settle out on the addition of the salt solution, it is dissolved by adding some water. To separate the acetacetic ester from the main portion of the excess of the acetic ester used, the mixture is distilled from a flask, heated by a free flame over a wire-gauze, or, more conveniently, without the wire-gauze, with a luminous flame. As soon as the thermometer indicates $95°$, the heating is discontinued, and the residue is subjected to vacuum-distillation, as described on page 26. In place of the usual condenser, the outside jacket of a Liebig condenser is pushed over the long side-tube of the distillation flask, and water is allowed to circulate through it. The heating is done in an air-bath. After small quantities of acetic ester, water, and acetic acid have passed over, the temperature becomes constant, and the main portion of the acetacetic ester distils over within one degree. The following table gives the boiling-points at various pressures. A reference to this will show at what approximate points the collection of the preparation should begin :

Boiling-point 71° at 12.5 mm. pressure.
" 74° " 14 " "
" 79° " 18 " "
" 88° " 29 " "
" 94° " 45 " "
" 97° " 59 " "
" 100° " 80 " "

The yield of acetacetic ester amounts to 55–60 grammes.

In the preparation of this substance, it must be borne in mind that the experiment must be completed in one day. The operation should be begun in the morning, the acetic ester heated with sodium at midday, and the experiment completed in the afternoon. If the reaction is discontinued at any point, and the unfinished preparation allowed to stand over night, the yield is essentially diminished.

The formation of acetacetic ester from acetic ester, discovered by Geuther in 1863, takes place in accordance with the following equation:

$$CH_3.CO\boxed{OC_2H_5 + H.}CH_2.CO.OC_2H_5$$
$$= CH_3.CO.CH_2.COOC_2H_5 + C_2H_5.OH$$
Acetacetic ester

But the mechanism of the reaction is much more complicated than here indicated. According to the views of Claisen, the sodium first acts on the alcohol, which, as above mentioned, must be present in small quantities, forming sodium alcoholate, and this unites with the acetic ester as follows:

$$CH_3.CO.OC_2H_5 + C_2H_5.ONa = CH_3.C\begin{subarray}{l}\diagup OC_2H_5\\\leftarrow OC_2H_5\\\diagdown ONa\end{subarray}$$

Reaction then takes place between this addition product and a second molecule of acetic ester, with the elimination of two molecules of alcohol, and the formation of the sodium salt of the acetacetic ester:

$$CH_3.C\begin{subarray}{l}\diagup \boxed{OC_2H_5\ \ H}\\\leftarrow \boxed{OC_2H_5 + H}\\\diagdown ONa\end{subarray}\!\!\!\!\!\!\!\!\!\!\!\!\!\!\!\!\!\! >CH.CO.OC_2H_5$$
$$= CH_3.C{=}CH.CO.OC_2H_5 + 2\,C_2H_5.OH$$
$$\ \ \ \ \ \ \ \ \ \ \ \ \ \ \ \ \ |$$
$$\ \ \ \ \ \ \ \ \ \ \ \ \ \ \ ONa$$

On acidifying with acetic acid, the sodium salt is decomposed with the formation of the free ester, $CH_3.C=CH.CO.OC_2H_5$, which spon-
$\phantom{CH_3.C=CH.CO.}|$
$\phantom{CH_3.C=CH.CO.}OH$
taneously changes into the desmotropic form,

$$CH_3.CO.CH_2.CO.OC_2H_5.$$

In the form indicated above, the reaction is not capable of general application; but a reaction closely related to it, discovered by Claisen and W. Wislicenus, is of general applicability, and is of great value in synthetical operations; for this reason it will be briefly mentioned here. If sodium alcoholate is allowed to act on a mixture of the esters of two monobasic acids, a ketone acid-ester, having a constitution analogous to that of acetacetic ester is formed by the action of the sodium alcoholate on one of the esters with the elimination of alcohol, *e.g.*:

$C_6H_5.CO\,\boxed{OC_2H_5 + H}\,CH_2.COOC_2H_5$
  Benzoic ester          Acetic ester
$\phantom{C_6H_5.CO\,OC_2H_5}= C_6H_5.CO.CH_2.CO.OC_2H_5 + C_2H_5.OH$
$\phantom{C_6H_5.CO\,OC_2H_5 = C_6H_5.CO.}$Benzoyl acetic ester

If one of the compounds is formic ester, esters of aldehyde-acids will be obtained, *e.g.*:

$H.CO.\,\boxed{OC_2H_5 + H}\,CH_2.CO.OC_2H_5 = H.CO.CH_2.CO.OC_2H_5$
  Formic ester        Acetic ester              Formyl acetic ester

If one molecule of a dibasic ester is used, a ketone dicarbonic acid ester will be formed, *e.g.*:

$CO.\,\boxed{OC_2H_5 \quad H}\,.CH_2.CO.OC_2H_5 \quad CO.CH_2.CO.OC_2H_5$
$|\phantom{CO.OC_2H_5}+\phantom{H}\phantom{.CH_2.CO.OC_2H_5}=|\phantom{CO.CH_2.CO.OC_2H_5}+C_2H_5.OH$
$CO.OC_2H_5\phantom{XXXXXXXXXXXXXXXXX}CO.OC_2H_5$
  Oxalic ester        Acetic ester          Oxalacetic ester

In place of the acid ester in the above reaction, which is susceptible of many combinations, a ketone may be used; a ketone acid-ester is not formed, it is true, but polyketones, or ketone-aldehydes:

$CH_3.CO\,\boxed{OC_2H_5 + H}\,CH_2.CO.CH_3$
  Acetic ester        Acetone
$\phantom{CH_3.CO\,OC_2H_5}= CH_3.CO.CH_2.CO.CH_3 + C_2H_5.OH$
$\phantom{CH_3.CO\,OC_2H_5 = CH_3.CO.}$Acetylacetone

$C_6H_5.CO\,\boxed{OC_2H_5 + H.}\,CH_2.CO.CH_3$
  Benzoic ester
$\phantom{C_6H_5.CO\,OC_2H_5}= C_6H_5.CO.CH_2.CO.CH_3 + C_2H_5.OH$
$\phantom{C_6H_5.CO\,OC_2H_5 = C_6H_5.CO.}$Benzoylacetone

$H.CO.\boxed{OC_2H_5 + H.}CH_2.CO.C_6H_5$
　　　Formic ester　　　　Acetophenone

$$= O{=}CH.CH_2.CO.C_6H_5 + C_2H_5.OH$$
　　　　　　　Benzoylaldehyde

These few examples are sufficient to show the many-sided applications of the above reaction.

The most remarkable characteristic of acetacetic ester is that a portion of its hydrogen may be substituted by metals. If sodium is allowed to act on it, the sodium salt is formed with the elimination of hydrogen:

$$CH_3.CO.CHH.CO.OC_2H_5 + Na = CH_3.CO.CHNa.CO.OC_2H_5 + H$$

The same salt is also formed by shaking the ester with a solution of sodium hydroxide. The reason for this phenomenon is to be sought in the acidifying influence of the two neighbouring carbonyl (CO) groups.

The synthetical importance of acetacetic ester depends on the fact that the most various organic halogen substitution products react with sodium acetacetic ester, the halogen uniting with the sodium, with the condensation of the two remaining residues. Thus a large number of compounds may be built up from their constituents. A few typical examples may elucidate these statements:

(1) $CH_3.CO.CHNa.COOC_2H_5 + ICH_3$

$$= CH_3.CO.\underset{\underset{\text{Methylacetacetic ester}}{CH_3}}{\overset{|}{CH}}.CO.OC_2H_5 + NaI$$

(2) $CH_3.CO.CHNa.CO.OC_2H_5 + C_6H_5.CO.Cl$
　　　　　　　　　　　　　Benzoyl chloride

$$= CH_3.CO.\underset{\underset{\underset{\text{Benzoylacetacetic ester}}{C_6H_5}}{\overset{|}{CO}}}{\overset{|}{CH}} - CO.OC_2H_5 + NaCl,$$

(3) $CH_3.CO.CHNa.COOC_2H_5 + Cl.CH_2.CO.OC_2H_5$
　　　　　　　　　　　　　　Chloracetic ester

$$CH_3.CO.\underset{\underset{\underset{\text{Acetsuccinic ester}}{COOC_2H_5}}{\overset{|}{CH_2}}}{\overset{|}{CH}}.CO.OC_2H_5 \quad + NaCl.$$

## ALIPHATIC SERIES

In the compounds thus obtained the second methylene hydrogen atom is also replaceable by sodium, and this salt is likewise capable of entering into similar reactions, by which the number of derivatives is largely increased, *e.g.*:

$$CH_3.CO.\underset{\underset{\text{Sodiummethylacetacetic ester}}{CH_3}}{C}.Na - CO.OC_2H_5 + IC_2H_5 = CH_3.CO.\underset{\underset{\text{Methylethylacetacetic ester}}{CH_3\ \ C_2H_5}}{C}\!\!-\!\!CO.OC_2H_5 + NaI.$$

From all these compounds simpler ones may be obtained on saponification. The acetacetic ester breaks up in one of two ways, depending upon the conditions of the saponification:

$$CH_3.CO.CH_2.\,|\,CO.OC_2H_5 + HOH = \underset{\text{Acetone}}{CH_3.CO.CH_3} + CO_2 + C_2H_5.OH,$$

$$CH_3.CO.\,|\,CH_2.CO.OC_2H_5 + HOH = 2\,\underset{\text{Acetic acid}}{CH_3.CO.OH} + C_2H_5.OH.$$

The first kind of decomposition is called "ketone decomposition," the second, "acid decomposition." Since, as shown above, either one or both of the methylene hydrogen atoms in acetacetic ester can be replaced by different radicals, X or Y, these substances yield either mono- or di- substituted acetones:

$$X.CH_2.CO.CH_3 \quad \text{and} \quad \overset{X}{\underset{Y}{>}}\!\!CH.CO.CH_3,$$

as well as mono- and di- substituted acetic acids,

$$X.CH_2.CO.OH \quad \text{and} \quad \overset{X}{\underset{Y}{>}}\!\!CH.CO.OH.$$

The variety of the acetacetic ester syntheses is still further increased by the fact that *two molecules* of the ester, by reaction with aldehydes or alkylene bromides, may be united with one another by the most various bivalent radicals.

These examples are sufficient to show the value of acetacetic ester and analogous compounds for organic syntheses. Acetacetic ester may not only be employed for the syntheses of carbon compounds, but it also reacts with nitrogen compounds, *e.g.*, aldehyde-ammonia, phenyl hydrazine, aniline, etc. This action will be referred to later in the appropriate places.

## 11. REACTION: SYNTHESES OF THE HOMOLOGUES OF ACETIC ACID BY MEANS OF MALONIC ACID-ESTER

EXAMPLE: **Butyric Acid from Acetic Acid**

(a) *Preparation of Malonic Acid Ester*[1]

Dissolve 50 grammes of chloracetic acid in 100 grammes of water; warm gently and neutralise with solid potassium carbonate, for which 30–40 grammes will be required. The solution is then heated in a large dish on a sand-bath under the hood, and treated with 40 grammes of finely pulverised, pure potassium cyanide; after a short time, decomposition begins with a vigorous ebullition. The solution is evaporated as quickly as possible on the sand-bath until a thermometer placed in the viscous, brownish salt indicates 135°. Since the substance "bumps" and spatters during the evaporation, it is constantly stirred with the thermometer, the hand being protected by a glove or cloth. It is allowed to cool, the stirring being continued during the cooling, otherwise the product bakes into a hard, scarcely pulverisable mass. It is then powdered as finely as possible, and placed in a flask provided with a reflux condenser, with $\frac{2}{3}$ of its weight of alcohol, and saturated with hydrochloric acid while being warmed on the water-bath. This operation requires from one to one and a half days. A *wide* tube must be used for the introduction of hydrochloric acid, since a narrow one generally becomes clogged up by the salt crystallising in it. An adapter, or a drawn-out wide glass tube, connected by a short rubber joint to the end of the delivery tube, may be used. If the introduction of the gas is interrupted at noon or over night, the wash-bottle is disconnected from the flask containing the mixture; otherwise, the alcoholic solution may be drawn back into the wash-bottle. After cooling, pour the contents of the flask into an equal volume of ice water, filter from the undissolved salt with suction, and wash repeatedly with ether; the aqueous filtrate is also shaken with this. The ethereal layer is dried with calcium

---

[1] A. 204, 121.

chloride, the ether distilled off on the water-bath, and the residue rectified. Boiling-point, 195°. Yield, about 40 grammes.

(b) *The Introduction of an Ethyl Group*

Dissolve 2.3 grammes of sodium in 25 grammes of absolute alcohol in a small flask connected with a reflux condenser; treat with 16 grammes malonic ester, and then add, through the condenser, 20 grammes of ethyl iodide, not too rapidly. The mixture is then heated on the water-bath until the liquid no longer shows an alkaline reaction, for which one to two hours may be necessary. The alcohol is then distilled off on an actively boiling water-bath, a thread being placed in the flask to facilitate the boiling; the residue is taken up with water and extracted with ether, the ether evaporated, and the residue distilled. Boiling-point, 206–208°. Yield, about 15 grammes.

(c) *Saponification of Ethyl Malonic Ester*

For the saponification of the ester, a concentrated solution of caustic potash is prepared; for every gramme of the ester, a solution of 1.25 grammes potassium hydroxide in 2 grammes of water is used. The solution is placed in a flask provided with a reflux condenser; through this the ester is gradually added; an emulsion is first formed, which soon solidifies to a white solid mass, probably potassium ethylmalonicester. On shaking the mixture, it becomes so much heated that the alcohol liberated boils vigorously. After about an half-hour's heating on the water-bath, the oily layer disappears, showing that the saponification is complete. It sometimes happens that, even on longer heating, the saponification will not take place. Under these conditions, add about 5 drops of alcohol to the mixture: the reaction will soon begin.

*Detection of the Alcohol:* In order to show that ethyl alcohol has been formed in the reaction, 40 c.c. of water is added to the contents of the flask; this is connected with a condenser and heated on a wire-gauze until the distillate measures exactly 40 c.c.; this contains the alcohol, besides much water. The residue, as described below, is further treated for ethylmalonic acid. In order to concentrate the alcohol, the distillate is again distilled until 10 c.c. of the liquid passes over. This contains all of the alcohol, since it is more volatile than the water, besides a little water. To separate the alcohol, the distillate is placed in a small test-tube, and solid caustic potash added until the lower aqueous layer no longer dissolves it. The upper layer of the lighter alcohol is removed with a pipette and distilled from a small fractionating flask. Boiling-point, 78°.

To obtain the ethylmalonic acid, the residue remaining is diluted with a *little* water, neutralised carefully with concentrated sulphuric acid; and the ethylmalonic acid is precipitated out in the form of its calcium salt, by the addition of a cold solution of calcium chloride, *as concentrated as possible*. This is filtered off, well pressed on a porous plate, and the ethylmalonic acid liberated by treating with concentrated hydrochloric acid. By repeatedly extracting the mixture with ether free from alcohol, the free acid is obtained by evaporating the ethereal solution, at first as an oil, which, on cooling and rubbing the walls of the vessel with a glass rod, soon solidifies. The crude product is purified by recrystallising from water. Melting-point, 111.5°. Yield, 6–7 grammes.

(d) *Elimination of Carbon Dioxide from Ethyl Malonic Acid*

The ethylmalonic acid is placed in a small fractionating flask supported in an oil-bath at an oblique angle, so that its outlet tube is inclined upward. The mouth is closed by a cork bearing a thermometer. The acid is heated at 180°, until carbon dioxide is no longer evolved, which will require about a half-hour. The residue is distilled from the same flask in the usual way; the butyric acid passes over between 162–163°. Yield, about 80–90% of the theory.

(*a*) In the first phase of the reaction which gives ethylmalonic-ester, the potassium cyanide acts on the chloracetic acid, or on its potassium salt, with the formation of cyanacetic acid:

ALIPHATIC SERIES 157

$$CH_2Cl \cdot CO \cdot OH + KCN = CH_2 \cdot CN \cdot CO \cdot OH + KCl$$
Cyanacetic acid

As already mentioned in the preparation of acetonitrile, a halogen united with aliphatic residues may generally be replaced by the cyanogen group, on heating with potassium or silver cyanide. If alcohol and hydrochloric acid are now allowed to act on the cyanacetic acid, three reactions take place. At first, there is an esterification in accordance with the equation:

$$CH_2 \cdot CN \cdot CO \cdot OH + C_2H_5 \cdot OH = CH_2 \cdot CN \cdot CO \cdot OC_2H_5 + H_2O$$
Cyanacetic ester

Under the discussion of acetyl chloride and acetic ester, it has already been brought forward that, in general, acid-esters can be obtained by conducting a current of hydrochloric acid gas into a mixture of the alcohol and acid. In the second place, the hydrochloric acid has a saponifying action on the cyanacetic ester, *i.e.* the cyanogen group is converted into carboxyl (COOH).

$$CH_2 \cdot CN \cdot CO \cdot OC_2H_5 + HCl + 2\,H_2O = \begin{matrix} CO \cdot OH \\ | \\ CH_2 \\ | \\ CO \cdot OC_2H_5 \end{matrix} + NH_4Cl$$
Acid ester of malonic acid

The carboxyl group thus formed is then acted on in the same way as the carboxyl group of cyanacetic acid above, with the formation of an ester:

$$\begin{matrix} CO \cdot OH \\ | \\ CH_2 \\ | \\ CO \cdot OC_2H_5 \end{matrix} + C_2H_5 \cdot OH = \begin{matrix} CO \cdot OC_2H_5 \\ | \\ CH_2 \\ | \\ CO \cdot OC_2H_5 \end{matrix} + H_2O$$
Malonicdiethyl ester

(*b*) The ester of malonic acid, like acetacetic ester, possesses the property in virtue of which one of the two methylene hydrogen atoms can be replaced by sodium, in consequence of the acid properties imparted by the two neighbouring carbonyl (CO) groups. When the sodium compound is treated with organic halides, like alkyl halides, halogen derivatives of acid-esters, acid-chlorides, etc., the sodium is replaced by alkyl residues, acid residues, etc., just as in the case of the closely related acetacetic ester. In the above-mentioned examples, the sodium salt of the malonic ester is first formed from sodium alcoholate and the ester:

$$\underset{\substack{|\\ \text{CO.OC}_2\text{H}_5}}{\overset{\substack{\text{CO.OC}_2\text{H}_5\\|}}{\text{CH}}}\boxed{\text{H} + \text{C}_2\text{H}_5.\text{O}}\text{Na} = \underset{\substack{|\\ \text{CO.OC}_2\text{H}_5}}{\overset{\substack{\text{CO.OC}_2\text{H}_5\\|}}{\text{CHNa}}} + \text{C}_2\text{H}_5.\text{OH}$$

Ethyl iodide reacts on this as follows:

$$\underset{\substack{|\\ \text{CO.OC}_2\text{H}_5}}{\overset{\substack{\text{CO.OC}_2\text{H}_5\\|}}{\text{CH}}}\boxed{\text{Na} + \text{I}}\text{C}_2\text{H}_5 = \underset{\substack{|\\ \text{CO.OC}_2\text{H}_5}}{\overset{\substack{\text{CO.OC}_2\text{H}_5\\|}}{\text{CH.C}_2\text{H}_5}} + \text{NaI}$$

Ethylmalonicdiethyl ester

As in acetacetic ester, the second hydrogen of the malonic ester can also be replaced by sodium; consequently the malonic ester is capable of reacting a second time with organic halides, so that disubstituted malonic esters can also be prepared.

(c) The compounds thus obtained of the general formulæ:

$$\underset{\substack{|\\ \text{CO.OC}_2\text{H}_5}}{\overset{\substack{\text{CO.OC}_2\text{H}_5\\|}}{\text{CH} - \text{X}}} \quad \text{and} \quad \underset{\substack{|\\ \text{CO.OC}_2\text{H}_5}}{\overset{\substack{\text{CO.OC}_2\text{H}_5\\|}}{\text{C}}}\!\!\underset{\text{Y}}{\overset{\text{X}}{\diagup\!\!\!\diagdown}}$$

are distinguished from the corresponding derivatives of acetacetic ester in that on saponification they do not decompose, but yield the free substituted malonic acids. Thus, the ethylmalonicdiethyl ester reacts with caustic potash as follows:

$$\underset{\substack{|\\ \text{CO.OC}_2\text{H}_5}}{\overset{\substack{\text{CO.OC}_2\text{H}_5\\|}}{\text{CH.C}_2\text{H}_5}} + \begin{matrix}\text{KOH}\\\\\text{KOH}\end{matrix} = \underset{\substack{|\\ \text{CO.OK}}}{\overset{\substack{\text{CO.OK}\\|}}{\text{CH.C}_2\text{H}_5}} + 2\,\text{C}_2\text{H}_5.\text{OH}$$

Concerning the detection of alcohol, the following points may be noted: Under the section on "Salting out," page 42, it has already been stated, that substances very easily soluble in water, e.g. alcohol, acetone, etc., cannot be obtained by extraction with ether. If substances of this class are to be separated from the water, the method of salting out is employed; in this case it is most convenient to use potash for the purpose. If the quantity of the substance dissolved in water is considerable, then the solution may be saturated directly with

ALIPHATIC SERIES 159

solid potash, upon which the previously dissolved substances will separate out over the layer of the heavier potash solution. If the quantity of dissolved substance is small, as in the above case, a portion of the solution is first distilled, provided the dissolved substance is easily volatile, like alcohol or acetone. With the first portions of the water, the main quantity of the dissolved substance will pass over. There is thus obtained a more concentrated solution than before. If necessary, the distillate is again distilled, by which the solution of the dissolved substance becomes more concentrated; so that, finally, it it may be obtained by salting out. In the case under discussion, obviously, the detection of the alcohol is of no practical value, since ethyl alcohol was used in the preparation of malonic ester, and would be obtained again, on testing for it. The separation and isolation of the alcohol is only effected to learn the method of salting out with potash.

Since the free ethylmalonic acid is very soluble in water, it cannot be obtained from the above dilute solution by extraction with ether, but must be separated out in the form of its difficultly soluble potassium salt; this is then decomposed with a little hydrochloric acid.

(*d*) From the substituted malonic acid thus obtained, derivatives of acetic acid may be prepared by heating it to a high temperature. It is a general law that one carbon atom cannot hold two carboxyl groups in combination at high temperatures, since carbon dioxide will be eliminated from one. By this means, a dicarbonic acid is converted into a monocarbonic acid, *e.g.* :

$$\begin{array}{c} \boxed{CO.O}H \\ | \\ CH_2 \\ | \\ CO.OH \end{array} = \begin{array}{c} CH_3 \\ | \\ CO.OH \end{array} + CO_2$$

From the mono- or di-substituted malonic acid a substituted acetic acid is obtained of the formula,

$$\begin{array}{c} CH_2.X \\ | \\ CO.OH \end{array} \quad \text{or} \quad \begin{array}{c} CH{<}^X_Y \\ | \\ CO.OH \end{array}$$

Thus from ethylmalonic acid, there is formed ethylacetic acid = butyric acid. If, instead of ethyl iodide, methyl or propyl iodide is used, proprionic acid or valerianic acid respectively is obtained.

If two methyl groups are introduced into malonic ester, then, on decomposition, a dimethylacetic- or isobutyric-acid will be formed.

As shown above, similar acids may be prepared from the acetacetic ester. Since the decomposition of the acetacetic ester derivatives may take place in two different ways (acid- and ketone-decomposition); and since these decompositions frequently take place side by side, while the malonic acid derivatives decompose in only one way, so in most cases it is more advantageous to use the malonic ester for the synthesis of the homologous fatty acids.

### 12. REACTION: PREPARATION OF A HYDROCARBON OF THE ETHYLENE SERIES BY THE ELIMINATION OF WATER FROM AN ALCOHOL. COMBINATION OF THE HYDROCARBON WITH BROMINE

EXAMPLE: **Ethylene from Ethyl Alcohol. Ethylene Bromide**[1]

A mixture of 25 grammes of alcohol and 150 grammes of concentrated sulphuric acid is heated, not too strongly, in a $1\frac{1}{2}$ litre round flask on a wire-gauze covered with thin asbestos paper (Fig. 58). As soon as an active evolution of ethylene takes place, add, through a dropping-funnel, a mixture of 1 part alcohol and 2 parts concentrated sulphuric acid (made by pouring 300 grammes of alcohol into 600 grammes of sulphuric acid, with constant stirring), slowly, so that a regular stream of gas is evolved. If the mixture in the flask foams badly with a separation of carbon, it has been too strongly heated, and it is advisable to empty the flask and begin the operation anew. In order to free the ethylene from alcohol, ether, and sulphur dioxide, it is passed through a wash-bottle containing sulphuric acid, and a second one, provided with three tubulures, the central one supplied with a safety-tube, containing a dilute solution of caustic soda. It then enters two wash-bottles, each containing 50 c.c. of bromine, covered with a layer of water 1 cm. high. Since the combination of ethylene with bromine causes the evolution of heat, the bromine bottles are placed in thick-walled vessels filled with cold water. In order to get rid of the bromine vapours which escape from the last bottle,

---

[1] A. 168, 64; A. 192, 144.

it is connected with the hood or with a flask containing a solution of caustic soda; to prevent the caustic soda from being drawn back into the bromine bottle, the delivery tube must not dip under the surface of the caustic soda. If wash-bottles which have a diameter of at least 6 cm. are not available, a bottle or flask with two tubulures may be used. The ordinary narrow wash-bottles are not suitable, since the height of the bromine layer is so great that the ethylene cannot overcome the pressure. As soon as the bromine is decolourised, the operation is discontinued, care being taken to disconnect all of the vessels immediately;

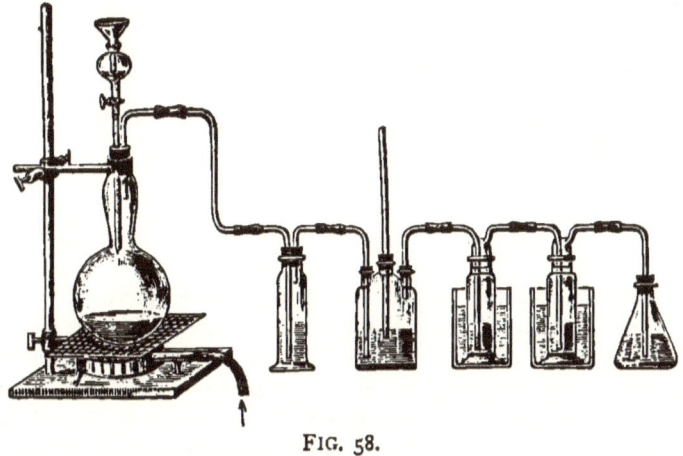

FIG. 58.

otherwise, in consequence of the cooling of the large flask, the contents of the bottles will be drawn back. The ethylene is then washed repeatedly with water in a dropping-funnel, and, finally, with caustic soda solution. It is dried over calcium chloride, and, on distillation, is obtained perfectly pure. Boiling-point, 130°. Yield, 250–300 grammes.

The addition of the alcohol-sulphuric acid mixture is often attended with difficulty, in that as soon as the cock is opened, the gas passes out through the funnel, thus preventing the entrance of the mixture. This difficulty may be obviated by taking the precaution of always keeping the stem of the funnel filled with the mixture. Before the heating is begun, a portion of the mixture is placed in a porcelain dish, the end of the stem of the funnel immersed in it and filled by suction. The cock is then closed, the funnel placed in the cork of the generating flask, and the heating begun.

The hydrocarbons of the ethylene series may be prepared, in general, by abstracting water from the corresponding alcohol, *e.g.*:

$$CH_3 \cdot CH_2 \cdot OH = CH_2 {=} CH_2 + H_2O.$$

If sulphuric acid is used as the dehydrating agent, the reaction does not follow the above equation, but ethylsulphuric acid is first formed, and this, on heating, yields sulphuric acid again:

$$C_2H_5 \cdot OH + SO_2\!\!\begin{array}{c}\diagup OH \\ \diagdown OH\end{array} = SO_2\!\!\begin{array}{c}\diagup OC_2H_5 \\ \diagdown OH\end{array} + H_2O$$
Ethylsulphuric acid

$$SO_2\!\!\begin{array}{c}\diagup OC_2H_5 \\ \diagdown OH\end{array} = C_2H_4 + SO_2\!\!\begin{array}{c}\diagup OH \\ \diagdown OH\end{array}.$$

In many cases the elimination of water takes place so easily that the use of concentrated sulphuric acid is unnecessary, since the diluted acid answers the purpose. With the higher members of the series the reaction is complicated by the fact that the simple alkylenes polymerise under the influence of sulphuric acid. Thus there is formed, besides butylene, $C_4H_8$, hydrocarbons having respectively twice and three times its molecular weight, *e.g.*:

$C_8H_{16}$ Dibutylene<br>$C_{12}H_{24}$ Tributylene

In these cases it is much more convenient to prepare an ester from the alcohol by the action of the chloride of a higher fatty acid, and subjecting this to distillation by which it is decomposed into an hydrocarbon of the ethylene series and the free fatty acid, *e.g.*:

$$\underset{\text{Cetyl palmitate}}{C_{15}H_{31} \cdot CO \cdot OC_{16}H_{33}} = \underset{\text{Palmitic acid}}{C_{15}H_{31} \cdot CO \cdot OH} + \underset{\text{Hexadecylene}}{C_{16}H_{32}}$$

The first four members of the alkylene series are gases at ordinary temperatures, which burn with strongly luminous, smoky flames. The intermediate members are colourless liquids, not miscible with water, which can be distilled at ordinary pressures without decomposition; the higher members are solids, and can only be distilled without decomposition in a vacuum. Chemically these compounds are characterised primarily by the property of uniting with two univalent atoms, or a univalent atom and a univalent radical, upon which the double union is changed to single union.

They take up, especially in the presence of platinum-black, two atoms of hydrogen, thus passing over to the hydrocarbons of the saturated series (paraffins):

$$CH_2{=}CH_2 + H_2 = CH_3 - CH_3.$$

Hydrogen halides may also be added to them; hydriodic acid with the greatest ease, hydrobromic acid with less, and hydrochloric acid only with difficulty:

$$CH_2{=}CH_2 + HI = CH_3.CH_2I.$$
Ethyl iodide

The homologues of ethylene also form addition products; the halogen atom seeks that carbon atom which is combined with the smallest number of hydrogen atoms:

$$\underset{\text{Propylene}}{CH_2{=}CH.CH_3} + HI = \underset{\text{Isopropyl iodide}}{CH_3.CHI.CH_3}.$$

The constituents of water (H and OH) may also be added indirectly to the alkylenes. If concentrated sulphuric acid be allowed to act on one of them, it dissolves, forming a sulphuric acid ester:

$$CH_2{=}CH_2 + SO_2{\genfrac{}{}{0pt}{}{\diagup OH}{\diagdown OH}} = SO_2{\genfrac{}{}{0pt}{}{\diagup OC_2H_5}{\diagdown OH}},$$

If this is boiled with water, the ester is decomposed into alcohol and sulphuric acid:

$$SO_2{\genfrac{}{}{0pt}{}{\diagup OC_2H_5}{\diagdown OH}} + HOH = C_2H_5.OH + SO_2{\genfrac{}{}{0pt}{}{\diagup OH}{\diagdown OH}},$$

so that finally H and OH have been added to ethylene:

$$CH_2{=}CH_2 + H.OH = CH_3.CH_2.OH.$$

Analogous to the halogen atoms, the hydroxyl (OH) group unites with that carbon atom holding in combination the smallest number of hydrogen atoms.

The alkylenes take up two atoms of chlorine or bromine with great ease:

$$CH_2{=}CH_2 + Cl_2 = CH_2Cl - CH_2Cl$$
$$CH_2{=}CH_2 + Br_2 = CH_2Br - CH_2Br.$$

Finally they combine directly with hypochlorous acid to form glycolchlor hydrines.

The reactions taking place in the formation of the alkylenes as well as those in the formation of addition products are not only applicable to the hydrocarbons but also to their substitution products. Thus, e.g., unsaturated acids are commonly obtained from oxyacids by the elimination of water:

$CH_2.OH.CH_2.CO.OH$ $= CH_2{=}CH.CO.OH + H_2O$
β-hydroxyproprionic acid — Acrylic acid

$C_6H_5.CH.OH.CH_2.CO.OH$ $= C_6H_5.CH{=}CH.CO.OH + H_2O.$
Phenyllactic acid — Cinnamic acid

All compounds in which the ethylene condition is present show the addition phenomena, in accordance with the following equations:

$CH_2{=}CH.CH_2.OH + Br_2$ $= CH_2Br - CHBr.CH_2.OH,$
Allyl alcohol — Dibromhydrine

$CH_2{=}CH.CO.OH + Br_2$ $= CH_2Br - CHBr.CO.OH,$
Acrylic acid — Dibromproprionic acid

$C_6H_5.CH{=}CH.CO.OH + Br_2$ $= C_6H_5.CHBr - CHBr.CO.OH,$
Cinnamic acid — Dibromhydrocinnamic acid

$C_6H_5.CH{=}CH.CO.OH + HBr$ $= C_6H_5.CHBr - CH_2.CO.OH,$
Bromhydrocinnamic acid

$C_6H_5.CH{=}CH_2 + Br_2$ $= C_6H_5.CHBr - CH_2Br,$
Styrene — Styrene dibromide

$C_6H_5.CH{=}CH.CO.OH + Cl.OH = C_6H_5.CH.OH.CHCl.CO.OH,$
Phenylchlorlactic acid

$CH_2{=}CH.CO.OH + H_2$ $= CH_3.CH_2.CO.OH.$
Acrylic acid — Proprionic acid

# ALIPHATIC SERIES

## 13. REACTION: PREPARATION OF AN ACETYLENE-HYDROCARBON FROM AN ALKYLENE BROMIDE. ADDITION OF BROMINE[1]

EXAMPLE: **Acetylene from Ethylene Bromide** (Fig. 59, *a* and *b*)

Dissolve 50 grammes of sodium cut in scales the thickness of a five-cent piece in 500 grammes of absolute alcohol in a 1-litre

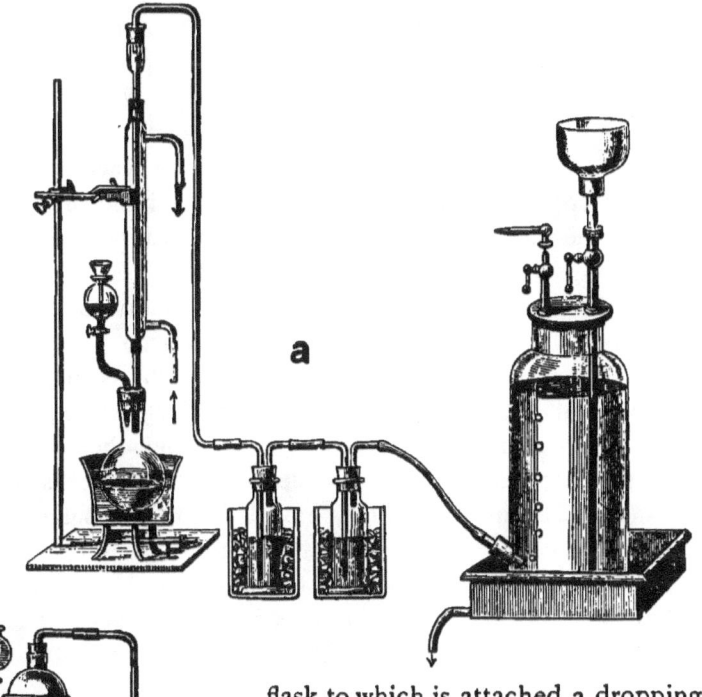

FIG. 59.

flask to which is attached a dropping-funnel and reflux condenser (the Soxhlet bulb condenser, *b*, serves admirably for this purpose). Only 4-5 pieces of sodium are added at one time; a further addition is not made until they are for the most part dissolved. The alcohol begins to boil actively, thus accelerating the solution. Toward the end, the sodium dissolves with evident difficulty; during the addi-

[1] A. 178, 111.

tion of the last fourth of the sodium, the flask is heated on a water-bath. As soon as the sodium has disappeared, add 100 grammes of ethylene bromide through the dropping-funnel, the end of the stem of which has been drawn out to a narrow opening; care is taken to keep the stem filled with the liquid (see the preparation of ethylene bromide, page 160). The flask is then heated *in* a rapidly boiling water-bath, into which several handfuls of common salt have been thrown to increase the boiling temperature. The ethylene bromide is now allowed to flow into the flask slowly, drop by drop, so that about two hours will be required for the operation. The gas escaping from the condenser is conducted through two wash-bottles, each containing 25 c.c. of ordinary alcohol, and cooled by a mixture of ice and salt. At the beginning of the reaction, a test-portion of the evolved gas is collected in a test-tube filled with water; as soon as a test shows that the gas will burn quietly without an explosion, the delivery tube is connected with a gasometer filled with a solution of common salt. The solution is made by shaking a large quantity of salt with water for some time in a large vessel, decanting the solution from the undissolved salt, and then adding a fresh amount of water to the residue. The gasometer is placed in a shallow, pneumatic trough, provided with a tubulure emptying into a vessel placed under it, for the purpose of collecting the salt solution forced out of the gasometer by the entrance of the gas. The solution is again used later to displace the gas.

If the operation is carefully conducted, 8–10 litres of gas will be obtained.

*Reactions of Acetylene.* — (1) Some solid cuprous chloride (see explanation under preparation of tolyl cyanide, page 204) is dissolved in dilute ammonium hydroxide, and acetylene is passed into this solution. A red precipitate of copper acetylide is formed.

(2) Ten drops of a dilute solution of silver nitrate is treated with ammonium hydroxide and acetylene conducted into the solution. A white precipitate of silver acetylide is formed. Filter it off, and wash with water, alcohol, and ether. A *trace* of the *moist* precipitate is placed on a porous plate, and rubbed with a

spatula. A vigorous explosion takes place. A similar phenomenon occurs on heating it over a small flame.

The sodium bromide separating from the alcohol can be re-crystallised, and used for other preparations, *e.g.* for ethyl bromide. It is filtered off with suction, dissolved by heat in as little water as possible, treated with aqueous hydrobromic acid (by-product in the preparation of brombenzene) until the solution shows a distinct acid reaction. It is then heated with animal charcoal for some time, filtered off from this, and allowed to crystallise in a shallow crystallising dish, either in the air, or over sulphuric acid in a desiccator.

## *Vinyl Bromide*[1]

As a by-product in the preparation of acetylene, vinyl bromide is always formed, and remains dissolved in the cooled alcohol in the wash-bottles. In order to obtain this in pure condition, the alcoholic solution is poured into about 500 c.c. of water contained in a dropping-funnel, and cooled by throwing in pieces of ice. The heavy oil separating out is transferred to a tube sealed at one end and drawn out at the other, with the aid of a funnel-tube. The portion of the tube containing the oil is surrounded by a freezing mixture; the tube is then sealed (see page 55).

## *Acetylene Tetrabromide*

To effect a union of bromine with acetylene, the gas is conducted through two wash-bottles containing bromine. For each litre of gas, use 15 grammes (5 c.c.) of bromine. The volume of gas remaining in the gasometer may be obtained by measuring the salt solution forced out, and subtracting from this the amount

---

[1] If it is desired to prepare vinyl bromide rather than acetylene, a similar apparatus is used, but the ethylene bromide is allowed to flow into a solution of 150 grammes of caustic soda in 150 grammes of water, which has been treated with 450 grammes of ordinary alcohol. From 100 grammes of ethylene bromide, 40 grammes of vinyl bromide (70 % of the theory) will be obtained. Under these conditions, only a small quantity of gas will be evolved.

of solution which has been poured back into the gasometer. It is advisable to divide the bromine between the two flasks, each of which contains half as much water as bromine (by volume). Before the gas is allowed to enter the bromine, it is passed through two small wash-bottles made of thick-walled "specimen" tubes. The first contains 10 c.c. of alcohol surrounded by a freezing mixture, in order to condense the last traces of vinyl bromide. The second contains some water, which serves to show the rapidity of the gas-current. In order that the gas may not escape the action of the bromine, it is only allowed to enter one bubble at a time; so long as the first flask still contains bromine, only one bubble should pass over into the second flask from time to time. In order to force the gas from the holder, the basin at the top is filled with the salt solution used in the preparation of acetylene.

To obtain the tetrabromide, pour the contents of both wash-bottles, into a large separating funnel almost filled with water, and then gradually treat with a caustic soda solution until, on shaking, the red colour of the bromine disappears. The alkaline liquid is poured from the heavy oil; this is washed several times with water, dried over calcium chloride, and is obtained completely pure on distillation in a vacuum with the use of the Brühl apparatus:

Boiling-point, $137°$ at 36 mm. pressure
" " $114°$ " 12 " "

In order to test the purity of the acetylene tetrabromide, a specific gravity determination is made with a small pyknometer (Fig. 60, $a$). After this has been dried by carefully heating over a Bunsen burner, the moisture being removed by suction, the weight of the empty glass is found. It is then filled with distilled water, with the aid of a capillary pipette (Fig. 60, $b$), immersed in a beaker filled with water (Fig. 61) at a known temperature up to the mark on the neck of the pyknometer. The latter is then filled just to the mark. The water adhering to the glass walls above the mark is removed by a piece of rolled-up filter-paper; the pyknometer is then removed from the beaker, dried, and weighed. After the water is drawn out by suction, it is dried, filled to the mark with acetylene tetrabromide, again immersed in water at the same temperature as above, dried, and weighed for the third time.

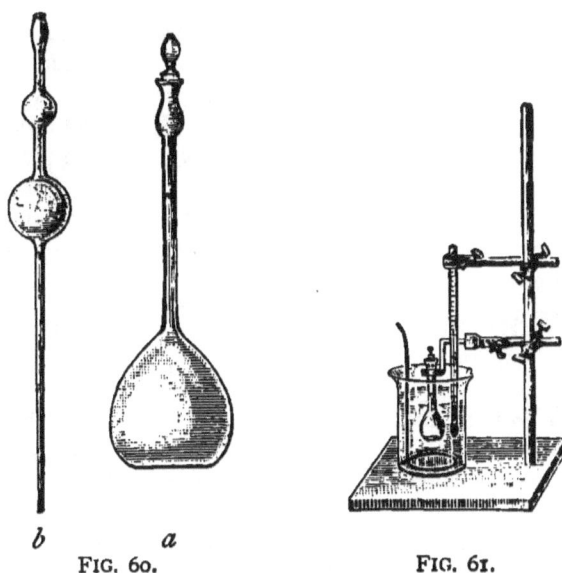

*b*      *a*
FIG. 60.      FIG. 61.

If $a$ is the weight of the pyknometer, $b$ the weight of the same filled with water, and $c$ the weight when filled with acetylene tetrabromide, then $s$ is the specific gravity referred to water at the same temperature:

$$s = \frac{c-a}{b-a}$$

If the temperatures at which the pyknometer is filled with water and the tetrabromide are not the same, or if it is desired to compare the specific gravity with that of water at 4°, reference must be made to text-books on physics for the proper formulæ. The specific gravity of the pure tetrabromide is 2.971 at 17.5°/4° and 2.963 at 21.5°/4°.

If sodium alcoholate dissolved in absolute alcohol is allowed to act on ethylene bromide, a molecule of hydrobromic acid is withdrawn from it:

$$CH_2Br - CH_2Br + C_2H_5ONa = CHBr{=}CH_2 + C_2H_5 \cdot OH + NaBr,$$
Vinyl bromide

while, if this reaction takes place a second time, acetylene is formed:

$$CHBr{=}CH_2 = CH{\equiv}CH + HBr.$$

Since a portion of the easily volatile vinyl bromide is carried away in the last reaction, a mixture of acetylene and vinyl bromide results. In order to obtain the former in pure condition, the mixture is passed into alcohol, well cooled by a freezing mixture. The vinyl bromide, boiling at 16°, is condensed, and remains dissolved in the alcohol, while the acetylene escapes.

Homologues of acetylene may also be prepared by a similar reaction, *i.e.* by abstracting two molecules of hydrobromic acid from one molecule of a homologue of ethylene bromide:

$$CH_3 \cdot CHBr - CH_2Br = CH_3 \cdot C{\equiv}CH + 2\,HBr.$$
Propylene bromide      Methyl acetylene<br>(Allylene)

Ethylene bromide and its homologues may be prepared by the addition of bromine to the ethylene hydrocarbons; this is a general reaction which renders possible a transition from the ethylene series to the acetylene series.

The reaction is also applicable in the aromatic series, as the following equations show:

$$C_6H_5 \cdot CH{=}CH_2 + Br_2 = C_6H_5 \cdot CHBr - CH_2Br,$$
Phenyl ethylene<br>(Styrene)

$$C_6H_5 \cdot CHBr - CH_2Br = C_6H_5 \cdot C{\equiv}CH + 2\,HBr.$$
Phenyl acetylene

Acetylene hydrocarbons may be further prepared, in the aliphatic as well as in the aromatic series, by abstracting two molecules of the halogen halide from dihalogen derivatives in which both the halogen atoms are in combination with the same carbon atom. The action of phosphorus chloride or bromide upon aldehydes yields dichlorides or dibromides, which, upon treatment with alcoholic potash, are converted into acetylene derivatives:

$$C_3H_7 \cdot CH_2 \cdot CHO + PCl_5 = C_3H_7 \cdot CH_2 \cdot CHCl_2 + POCl_3,$$
Valeric aldehyde

$$C_3H_7 \cdot CH_2 \cdot CHCl_2 = C_3H_7 \cdot C{\equiv}CH + 2\,HCl.$$

The chlorides of ketones, containing a carbonyl group in combination with a methyl group, also react in a similar manner:

$$CH_3.CH_2.CCl_2.CH_3 = CH_3.CH_2.C\equiv CH + 2\,HCl,$$
$$C_6H_5.CCl_2.CH_3 = C_6H_5.C\equiv CH + 2\,HCl.$$
Acetophenone chloride

The lowest members of the acetylene series are gases at ordinary temperatures, which burn with strongly smoky flames; the intermediate members are colourless liquids not miscible with water; the higher members are solids.

Acetylene and its homologues can take up either two or four univalent atoms:

$$CH\equiv CH + H_2 = CH_2{=}CH_2,$$
$$CH\equiv CH + 2\,H_2 = CH_3{-}CH_3,$$
$$CH\equiv CH + Br_2 = CHBr{=}CHBr,$$
$$CH\equiv CH + 2\,Br_2 = CHBr_2{-}CHBr_2,$$
$$CH\equiv CH + HBr = CH_2{=}CHBr,$$
$$CH\equiv CH + 2\,HI = CH_3{-}CHI_2.$$

Water may also be added by which aldehyde is formed from acetylene; under the same conditions the homologues of acetylene form ketones:

$$CH\equiv CH + H_2O = CH_3.CHO,$$
$$CH_3.C\equiv CH + H_2O = CH_3.CO.CH_3.$$
Acetone

This addition of water does not follow from mere contact of the two compounds; a crystalline compound of acetylene with a mercuric salt, as mercuric chloride or bromide, is first prepared; this is then boiled with water, and the above reaction takes place.

Acetylene possesses further the remarkable property of forming explosive salts with copper and silver, to which the following formulæ are ascribed:

$$C_2H_2Cu_2O \quad \text{and} \quad C_2H_2Ag_2O\,[1].$$

Cuprous salts especially are well adapted for separating acetylene from a mixture of gases. By warming the cuprous acetylide thus formed, with hydrochloric acid or potassium cyanide, free acetylene is

---

[1] The composition of these compounds deduced from Keiser's investigation is $C_2Cu_2$ and $C_2Ag_2$ respectively. (See American Chemical Journal, Vol. 14, 285.) — *Trans.*

regenerated. The homologues of acetylene also form similar salts provided they contain a true acetylene hydrogen atom, thus, *e.g.*, allylene forms a silver salt of the composition: $CH_3.C\equiv C.Ag$. But if both the acetylene hydrogen atoms have been substituted by alkyl radicals, *e.g.* as in dimethyl acetylene, $CH_3.C\equiv C.CH_3$, the power of forming salts does not exist.

Acetylene and its mono-substitution products also form salts with sodium, *e.g.*:

$$CH\equiv CNa \text{ and } CNa\equiv CNa,$$
$$CH_3.C\equiv CH + Na = CH_3.C\equiv CNa + H.$$

Finally, the polymerisation of the acetylenes into benzene derivatives should be mentioned. If acetylene is heated up to a dark-red heat, it is in part converted into its polymer, benzene:

$$3\,C_2H_2 = C_6H_6.$$

Methylacetylene and dimethylacetylene are condensed by treatment with sulphuric acid without heat to tri- and hexa-methyl benzene, respectively:

$$3\,CH_3.C\equiv CH = C_6H_3.(CH_3)_3,$$
s-Trimethylbenzene<br>Mesitylene

$$3\,CH_3.C\equiv C.CH_3 = C_6.(CH_3)_6.$$
Hexamethylbenzene

---

## II. AROMATIC SERIES

### 1. REACTION: NITRATION OF A HYDROCARBON

EXAMPLES: Nitrobenzene and Dinitrobenzene[1]

*Nitrobenzene*

To 150 grammes of concentrated sulphuric acid contained in a ½-litre flask, add gradually, and with frequent shaking, 100 grammes of concentrated nitric acid (sp. gr. 1.4). After cooling the mixture to the room temperature, by immersion in water,

---

[1] A. 9, 47; 12, 305.

gradually add 50 grammes of benzene, with frequent shaking. If the temperature should rise above 50–60°, the operation is interrupted, and the flask immersed in water for a short time. When all of the benzene has been added, a vertical air condenser is attached to the flask; it is then heated on a water-bath, boiling moderately, for an hour: during the heating the flask is frequently shaken. After cooling, the lower layer, consisting of sulphuric and nitric acids, is separated from the upper layer of nitrobenzene in a separating funnel. The nitrobenzene is then agitated in the funnel several times, with water: it must be borne in mind that the nitrobenzene now forms the *lower* layer. After being washed, it is placed in a dry flask, and warmed on a water-bath with calcium chloride, until the liquid, milky at first, becomes clear. It is finally purified by distillation from a fractionating flask provided with a long air condenser. Boiling-point, 206–207°. Yield, 60–70 grammes.

## Dinitrobenzene

To a mixture of 25 grammes of concentrated sulphuric acid and 15 grammes of fuming nitric acid, 10 grammes of nitrobenzene is gradually added (Hood); the reaction-mixture is then heated for half an hour on a water-bath, with frequent shaking; after cooling somewhat, it is poured, with stirring, into cold water. The dinitrobenzene which solidifies is filtered off, washed with water, pressed out on a porous plate, and recrystallised from alcohol. Melting-point, 90°. Yield, 10–12 grammes.

The property of yielding nitro-derivatives, when treated with nitric acid, is a characteristic of aromatic compounds not possessed by the aliphatic compounds. According to the conditions under which the nitration is carried out, one or more nitro-groups can be introduced at the same time. The above reactions take place in accordance with the following equations:

$$C_6H_6 + NO_2 \cdot OH = C_6H_5 \cdot NO_2 + H_2O,$$
$$C_6H_5 \cdot NO_2 + NO_2 \cdot OH = C_6H_4 \cdot (NO_2)_2 + H_2O.$$

If a saturated aliphatic residue is present in an aromatic compound, the nitration under the above conditions always affects the benzene ring, and not the side-chain. Since the benzene carbon atoms are in combination with only one hydrogen atom, the nitro-compounds obtained on nitration are tertiary; they therefore do not have the power to form salts, nitrolic acids, or pseudo-nitroles, like the primary and secondary nitro-compounds.

Not only can the mother substances, the aromatic hydrocarbons, but all their derivatives, as phenols, amines, aldehydes, acids, etc., undergo similar reactions. But the nitration does not take place in every case with the same ease. In each case, therefore, the most favourable conditions for the experiment must be determined. If a compound is very easily nitrated, the nitration may be effected, according to the conditions, by nitric acid diluted with water, or the substance may be dissolved in a solvent which is not attacked by nitric acid; glacial acetic acid is frequently used for this purpose, and then treated with nitric acid. The reverse process may also be employed, *i.e.* the substance is added to a mixture of nitric acid and water, or nitric acid and glacial acetic acid. If a substance is moderately difficult to nitrate, it is added to concentrated or fuming nitric acid. If the nitration is difficult, the elimination of water is facilitated by the addition of concentrated sulphuric acid to ordinary or fuming nitric acid. In the nitration, the substance may either be added to the mixture of nitric acid and sulphuric acid, or the nitric acid is added to the substance dissolved in concentrated sulphuric acid. In working with sulphuric acid solutions, at times either potassium nitrate or sodium nitrate may be used instead of nitric acid. The three nitration methods just described may be still further modified in two ways: (1) the temperature may be varied; (2) the quantity of nitric acid may be varied. The nitration can be effected in a freezing mixture, in ice, or in water, by gentle heating, or finally, at the boiling temperature. Further, the theoretical amount of nitric acid, or an excess may be used. In order to determine which of these numerous modifications will give the best results, preliminary experiments on a small scale must be made. Since the nitro-compounds are generally insoluble in water, or difficultly soluble, they can be separated from the nitrating mixture by diluting it with water, or in many cases better, with a solution of common salt.

The chemical character of a substance is not changed in kind, but in degree, by the introduction of a nitro-group. Thus, the nitro-derivatives of the hydrocarbons are indifferent compounds like the hydrocarbons. If a nitro-group is introduced into a compound of an acid nature like phenol, it becomes more strongly acid, *e.g.* the nitrophenols are more strongly acid than phenol. When a nitro-group is introduced in a basic compound, the resulting substance is less basic; *e.g.* nitro-aniline is less basic than aniline.

The great importance of the nitro-compounds is due to their behaviour on reduction; this will be considered under the next preparations.

Concerning the introduction of the nitro-group, the following laws are of general application.

The introduction of one nitro-group in the benzene molecule can, obviously, only result in the formation of one mononitrobenzene. If an alkyl radical is present in the benzene molecule, the nitro-groups enter the ortho- and para- but not the meta-position to the radical. On nitrating toluene, *e.g.*, there are formed:

$$\underset{\text{CH}_3}{\bigcirc}-NO_2 \quad \text{and} \quad \underset{\text{CH}_3}{\underset{NO_2}{\bigcirc}}$$

The same position is sought by the nitro-group when a benzene-hydrogen atom has been substituted by hydroxyl. Thus, *e.g.*, phenol gives on nitration a mixture of o- and p-nitrophenol. On the other hand, if a compound contains an aldehyde-, carboxyl-, or cyanogen-group on nitration the nitro-group goes in the meta-position to this. Benzaldehyde, benzoic acid, and benzonitrile give on nitration respectively:

$$\underset{NO_2}{\overset{CHO}{\bigcirc}} \qquad \underset{NO_2}{\overset{CO.OH}{\bigcirc}} \qquad \underset{NO_2}{\overset{CN}{\bigcirc}}$$

If a compound already contains a nitro-group, a second one will take the meta-position to this. Thus on nitrating nitrobenzene, m-dinitrobenzene is formed. O-nitrotoluene or o-nitrophenol yield on nitrating:

$$\underset{NO_2}{\overset{CH_3}{\bigcirc}}NO_2 \qquad \underset{NO_2}{\overset{OH}{\bigcirc}}NO_2$$

respectively.

From m-nitrobenzoic acid the following dinitrobenzoic acid is formed:

$$\underset{NO_2}{\phantom{XX}}\underset{}{\overset{CO.OH}{\bigcirc}}\underset{NO_2}{\phantom{XX}}.$$

The nitro-compounds are in part, liquids, in part solids; in case these latter distil without decomposition, they possess a higher boiling-point than the mother substance.

### 2. REACTION: REDUCTION OF A NITRO-COMPOUND TO AN AMINE

EXAMPLES: (1) Aniline from Nitrobenzene[1]
(2) Nitroaniline from Dinitrobenzene

A mixture of 90 grammes of granulated tin and 50 grammes of nitrobenzene is placed in a 1½-litre round flask. To this is gradually added 200 grammes of concentrated hydrochloric acid in the following manner: At first only about one-tenth of the acid is added; an air condenser, not too narrow, is then attached to the flask and the mixture well shaken. After a short time it becomes warm, and finally an active ebullition takes place. As soon as this happens, the flask is immersed in cold water until the reaction has moderated. The second tenth of the acid is then added, and the above operation repeated. After one half of the acid has been used, the reaction becomes less violent, and the second half may be added in larger portions. In order to effect the reduction of the nitrobenzene completely, the mixture is finally heated one-half hour on the water-bath. To separate the free aniline, the warm solution is treated with 100 c.c. of water, then a solution of 150 grammes of caustic soda in 200 grammes of water is gradually added. If the action of the caustic soda causes the liquid to boil, the flask is cooled by water for a short time, before a further addition of caustic soda. When all of the solution has been added, a long condenser is attached to the flask, and steam is passed into the hot liquid, upon which aniline, as a colourless

---

[1] A. 44, 283.

oil, and water pass over, the aniline collecting under the water. As soon as the distillate no longer appears milky, and becomes clear, the receiver is changed and about 300 c.c. more of the liquid distilled over. The distillates are mixed, treated with 20 grammes of finely powdered sodium chloride for every 100 c.c. of the liquid, shaken until all the salt is dissolved, and the aniline extracted with ether. After the ethereal solution has been dried by treating it with a few pieces of solid potassium hydroxide, the ether is evaporated and the aniline subjected to distillation. Boiling-point, 182°. Yield, 90–100 % of the theory. If the circumstances are such as not to permit the experiment to be completed without interruption, it is so arranged that the neutralisation with sodium hydroxide, and the distillation with steam immediately following, may take place within a short time, so that the heat of neutralisation may be utilised.

To the nitro-compounds of the aromatic series, as well as those of the aliphatic series, belongs the property of being converted into a primary amine on energetic reduction. For the reduction of every nitro-group, six atoms of hydrogen are necessary, and the following equation is the general expression of the reaction:

$$X \cdot NO_2 + 3 H_2 = X \cdot NH_2 + 2 H_2O.$$

For the reduction of nitro-compounds on the small scale in the laboratory, it is most convenient to use, as the reducing agent, granulated tin and hydrochloric acid, or stannous chloride and hydrochloric acid:

(1) $2 C_6H_5 \cdot NO_2 + 3 Sn + 12 HCl = 2 C_6H_5 \cdot NH_2 + 3 SnCl_4 + 4 H_2O$,

(2) $C_6H_5 \cdot NO_2 + 3 SnCl_2 + 6 HCl = C_6H_5 \cdot NH_2 + 3 SnCl_4 + 2 H_2O.$

To 1 molecule of a mononitro-compound, $1\frac{1}{2}$ atoms of tin, or 3 molecules of stannous chloride, are therefore used. In calculating the amount of the latter necessary for a reaction, it is to be remembered that the salt crystallises with 2 molecules of water ($SnCl_2 + 2 H_2O$). If the reduction is to be effected by metallic tin, double the above quantity is frequently used, *i.e.* to 1 nitro-group, 3 atoms of tin. In this case, the tin is not converted into stannic chloride, but into stannous chloride:

$$C_6H_5 \cdot NO_2 + 3 Sn + 6 HCl = C_6H_5 \cdot NH_2 + 3 SnCl_2 + 2 H_2O.$$

Since, in the cases mentioned, hydrochloric acid is always present in excess, and the amines unite with it to form soluble salts, the end of the operation occurs when no more of the insoluble nitro-compound is present, and the reaction-mixture dissolves clear in water. In order to get the free amine from the acid mixture, various methods may be employed. If, as in the above example, the amine is volatile with steam, and insoluble in alkali, then the acid solution is treated with caustic potash, or caustic soda, until the oxide of tin which separates out at first is redissolved in the excess of alkali; the liberated amine is driven over with steam. Further, volatile or non-volatile amines can be extracted from an alkaline solution by a proper solvent, like ether. But this process is often troublesome, since the alkaline tin solution forms an emulsion with ether, which subsides with great difficulty. If the free amine is solid, it may be obtained by filtering off the alkaline liquid. In many cases, where a non-volatile amine is under examination, it is advisable to precipitate the tin before liberating the amine. This is done by diluting the acid solution with much water, heating on the water-bath, and as soon as the liquid has reached the temperature of the bath, hydrogen sulphide is passed into it. The tin is precipitated as stannous or stannic sulphide; this is separated from the amine hydrochloride by filtering. Since tin, in the presence of a large excess of hydrochloric acid, is precipitated only with difficulty by hydrogen sulphide, it is frequently necessary to drive off the excess of the acid before treating with hydrogen sulphide. This is done by evaporating to dryness on the water-bath.

After the tin sulphide has been filtered off, a portion of the filtrate is tested with hydrogen sulphide for tin; if it should be present, the whole filtrate is evaporated on the water-bath, as completely as possible, to remove the hydrochloric acid, then diluted with water, and hydrogen sulphide is again passed into it. At times, the amine forms, with hydrochloric acid, a difficultly soluble salt, or the amine hydrochloride combines with the tin chloride to form a difficultly soluble double salt. In this case, the isolation of the amine may be facilitated by filtering it off, washing with hydrochloric acid, and pressing out on a porous plate, if necessary. If one is dealing with amines, which, like amido-acids, possess an acid character, obviously, these cannot be separated by the use of an alkali, as in the above example. In a case of this kind, the tin is always removed first, the acid solution evaporated to dryness, and the amido-compound is now liberated by the addition of sodium acetate. With amido-phenols, sodium hydrogen carbonate, sodium carbonate, or sodium sulphite, may be used to decompose the hydrochloric acid salt.

In the laboratory, other metals, like iron, zinc, etc., in connection with an acid, are only rarely used in the place of tin or stannous chloride, for the reduction of nitro-compounds. On the large scale, iron, owing to its cheapness, is used in the preparation of bases like aniline, toluidine, α-naphthyl amine, etc., from the corresponding nitro-compounds. By the use of iron and hydrochloric acid, the reduction should theoretically take place in accordance with the following equation:

$$C_6H_5 \cdot NO_2 + 3\,Fe + 6\,HCl = 3\,FeCl_2 + 2\,H_2O + C_6H_5 \cdot NH_2.$$

As a matter of fact, on the large scale, much less hydrochloric acid (only $\frac{1}{40}$) is used than that required by the above equation. In the presence of ferrous chloride, the nitro-compound is reduced by the iron without the action of hydrochloric acid, according to the equation:

$$C_6H_5 \cdot NO_2 + 2\,Fe + 4\,H_2O = C_6H_5 \cdot NH_2 + 2\,Fe(OH)_2.$$

For the neutralisation of the hydrochloric acid, a small quantity of which is always used, on the large scale, slaked lime is employed in preference to the more costly alkalies.

The complete reduction of nitro-compounds containing several nitro-groups is conducted in the same way as for mononitro-compounds. If it is desired to reduce but one or two of several nitro-groups, it cannot be done by adding just the calculated amount of the reducing agent; for cases of this kind, special methods are necessary. For this purpose, hydrogen sulphide in the presence of ammonia is often used for the reduction, or ammonium sulphide:

$$H_2S = H_2 + S.$$

The compound to be reduced is dissolved in water or alcohol, according to circumstances, treated with ammonia, and hydrogen sulphide passed into it on warming; or it is heated in a water or alcohol solution with a previously prepared water or alcohol solution of ammonium sulphide. In this way, *e.g.*, dinitrohydrocarbons may be converted into nitro-amines. A second method, which may be generally used for the reduction, step by step, of compounds containing several nitro-groups, is this: An alcoholic solution of the theoretical amount of stannous chloride saturated with hydrochloric acid is gradually allowed to flow into an alcoholic solution of the substance to be reduced, which is well cooled, and constantly shaken. (B. 19, 2161.)

EXPERIMENT:[1] The recrystallised dinitrobenzene is dissolved in alcohol (4 grammes alcohol to 1 gramme dinitrobenzene), in a flask, and treated with 0.8 gramme of concentrated ammonia for 1 gramme dinitrobenzene (the ordinary dilute solution of ammonia employed as a reagent must not be used). After the flask and its contents have been tared, the mixture is saturated with hydrogen sulphide at the ordinary temperature; the current of hydrogen sulphide is then shut off, and the flask, provided with a reflux condenser, is heated for about half an hour on a water-bath. It is then allowed to cool to the ordinary temperature, and hydrogen sulphide again passed into it to saturation, etc. This operation is repeated until there is an increase of 0.6 gramme in weight for every gramme of dinitrobenze used. The mixture is then diluted with water, the precipitate filtered off, washed with water, and extracted several times by warming with dilute hydrochloric acid. From the acid filtrate, the nitro-aniline is set free by neutralising with ammonium hydroxide; it is recrystallised from water. Melting-point, 114°. Yield, 70–80% of the theory.

$$C_6H_4{<}^{NO_2}_{NO_2} + 3\,H_2S = C_6H_4{<}^{NO_2}_{NH_2} + 2\,H_2O + 3\,S.$$

Special methods are necessary for the reduction of nitro-compounds containing groups capable of being acted upon by hydrogen, *e.g.* an aldehyde-group, an unsaturated side-chain, etc. In cases of this kind, ferrous hydroxide is frequently used:

$$2\,Fe(OH)_2 + 2\,H_2O = 2\,Fe(OH)_3 + H_2.$$

The reduction is effected by adding to the substance to be reduced, in the presence of an alkali (potassium-, sodium-, or barium-hydroxide), a weighed quantity of ferrous sulphate. By this reaction, o-nitrobenzaldehyde is reduced to o-amidobenzaldehyde, o-nitrocinnamic acid to o-amidocinnamic acid.

Besides the reducing agents mentioned, there is still a large number of others which find only an occasional application in reducing nitro-compounds to amines. They will be referred to under the different preparations.

---

[1] A. 176, 44.

The primary mon-amines are in part colourless liquids, e.g. aniline, o-toluidine, xylidine; or colourless solids like p-toluidine, pseudocuminidine, the naphthyl amines, etc. They can be distilled without decomposition, are volatile with steam, and difficultly soluble in water. The di- and poly-amines are for the most part solids, non-volatile with steam, and much more readily soluble in water than the mon-amines. The amines possess a basic character, but the basicity is weaker than that of the aliphatic amines, in consequence of the negative nature of the phenyl group.

Salts: $C_6H_5 \cdot NH_2 \cdot HCl$ . . . . Aniline hydrochloride
$C_6H_5 \cdot NH_2 \cdot HNO_3$ . . . Aniline nitrate
$(C_6H_5 \cdot NH_2)_2 \cdot H_2SO_4$ . . Aniline sulphate

Like ammonia, the amines unite with calcium chloride to form double compounds; for this reason they must not be dried with this substance (see page 45).

The primary mon-amines find numerous applications in the laboratory, as well as on the large scale, in consequence of their great activity. Frequent reference will be made to the subject in the following pages.

With the aniline prepared above, the following experiments are made:

(1) Add 3 drops of aniline to 10 c.c. of water in a test-tube, and shake the mixture. The aniline dissolves. At moderate temperatures, 1 part of aniline dissolves in about 30 parts of water.

(2) Dilute 1 c.c. of this aniline solution with 10 c.c. of water, and add a small quantity of a filtered water solution of bleaching powder. A violet colouration takes place; by this reaction (Runge's), the most minute quantity of free aniline may be detected. If in this experiment the solution should not remain clear, but a dirty violet precipitate separate out, a too concentrated solution has been used; the aniline water is diluted further, and the experiment repeated. If a salt of aniline is to be tested, it is dissolved in water, treated with alkali, the free aniline extracted with ether, this latter evaporated, and the residue dissolved in water. Then proceed exactly as just directed.

This reaction may also be used to detect small quantities of benzene or nitrobenzene. In a test-tube mix 5 drops of concentrated sulphuric acid with 5 drops of concentrated nitric acid, then add 1 drop of benzene, shake, and warm gently by passing the tube through a flame several times. Then add 5 c.c. of water, and extract the nitrobenzene with a little ether; the ether layer is removed with a capillary pipette, and the ether evaporated. The residue is treated with 1 c.c. of concentrated hydrochloric acid, and to this is added a piece of zinc the size of a lentil, to effect the reduction. When the zinc is dissolved, the mixture is diluted with water, and made strongly alkaline, until the hydroxide of zinc precipitated at first is redissolved; the aniline is then extracted with a little ether. Then proceed as just described.

If it is desired to determine whether a given compound is nitrobenzene, it is at once reduced with zinc and hydrochloric acid.

(3) In a small porcelain dish place 5 drops of concentrated sulphuric acid, and with a glass rod add 1 drop of aniline. The aniline sulphate thus formed solidifies for the most part on the rod; remove it by rubbing it against the walls of the dish. Then add 4 drops of an aqueous solution of potassium dichromate, and mix the liquid by revolving the dish. After a short time the liquid assumes a beautiful blue colour. If the reaction does not take place, add 2 more drops of the dichromate, or heat a moment over a small flame.

(4) *Isonitrile Reaction:* Heat a piece of caustic potash the size of a bean with 5 c.c. of alcohol, pour off the solution from the undissolved residue into another test-tube; the warm solution is treated with 1 drop of aniline and 4 drops of chloroform. A reaction takes place immediately, or on gentle warming; this is recognised not only by the separation of potassium chloride, but by a most highly characteristic, disagreeable odour. The odour becomes more pronounced on pouring off the liquid and adding some cold water to the tube. If the vapours of the isonitrile are inhaled through the mouth, a peculiar sweet taste is noticed in the throat.

The reaction must be carried out under a hood with a good draught.

While the two colour reactions with bleaching powder and chromic acid are used especially for the recognition of aniline, the isonitrile reaction will show the presence of any primary amine of the aliphatic or aromatic series. The reaction takes place in accordance with the following equation:

$$C_6H_5 \cdot NH_2 + CHCl_3 = C_6H_5 \cdot NC + 3\,HCl.$$

For the elimination of hydrochloric acid, caustic potash is added. Since all isonitriles or carbylamines possess a characteristic odour, on the one hand the smallest quantity of a primary base may be detected by this reaction, and on the other a base may be shown to be primary. Secondary and tertiary bases do not give the reaction.

In the isonitriles it is very probable that the carbon atom combined with the nitrogen atom is only bivalent: $C_6H_5 \cdot N{=}C{::::}$. The isonitriles are isomeric with the acid-nitriles, e.g. $C_6H_5 \cdot C{\equiv}N$, benzonitrile. While the nitriles on saponification give acids, the isonitriles decompose into a primary amine and formic acid:

$$C_6H_5 \cdot CN + 2\,H_2O = C_6H_5 \cdot CO \cdot OH + NH_3,$$
$$C_6H_5 \cdot NC + 2\,H_2O = C_6H_5 \cdot NH_2 + H \cdot CO \cdot OH.$$

### 3. REACTION: REDUCTION OF A NITRO-COMPOUND TO AN AZOXY-, AZO-, OR HYDRAZO-COMPOUND

EXAMPLES: **Azoxybenzene, Azobenzene, Hydrazobenzene**

(1) *Azoxybenzene:*[1] To 200 grammes of methyl alcohol contained in a ½-litre flask provided with a wide reflux condenser, 20 grammes of sodium in pieces the size of a bean is gradually added; the flask is not cooled (heat being generated by the reaction). Since methyl alcohol frequently contains much water, the first portions of the sodium must not be added too rapidly. When the metal is dissolved, 30 grammes of nitrobenzene is added, and the mixture heated for 3 hours on an actively boiling water-bath (reflux condenser). The greater portion of the methyl alcohol is then distilled off (the flask being placed *in* the water-bath; silk thread). The residue is treated with water, and

---

[1] J. pr. 36, 93; B. 15, 865.

the reaction-mixture poured into a beaker. After long standing, especially in a cool place, the oil at the bottom solidifies to a bright yellow crystalline mass, which is separated from the liquid by decanting the latter; it is washed several times with water and finally pressed out on a porous plate. From ligroïn the azoxybenzene crystallises in bright yellow needles, melting at 36°. Yield, 20–22 grammes.

(2) *Azobenzene:*[1] Five grammes of crystallised azoxybenzene, dried completely by heating on a water-bath for an hour, is finely pulverised and intimately mixed in a mortar with 15 grammes of *coarse* iron filings, which must also be completely dry; the mixture is distilled from a small retort, not tubulated. It is first warmed with a small luminous flame kept in constant motion; the size of the flame is increased after some time; finally the last portions are distilled over with a non-luminous flame. If, on heating, a sudden but harmless explosion should occur, it is due to the fact that the substances were not dry: the experiment must be repeated. The reddish distillate is collected in a small beaker, and, after it has solidified, is washed with hydrochloric acid to remove the aniline, then with water, and pressed out in a porous plate. The experiment is repeated a second time with a fresh quantity of azoxybenzene; by working carefully the same retort can be used again. The two pressed-out crude products are united. Azoxybenzene crystallises from ligroïn, after a partial evaporation of the solvent, in the form of coarse red crystals melting at 68°.

(3) *Hydrazobenzene:*[2] Dissolve 5 grammes of azobenzene in 50 grammes of alcohol (about 95 %) in a flask provided with a reflux condenser, and treat with a solution of 2 grammes of caustic soda in 4 grammes of water. To the boiling solution gradually add zinc dust in small portions (best by occasionally removing the cork) until the orange-coloured solution becomes colourless: about 8 grammes of zinc dust will be necessary. The hot solution is then filtered with suction (Büchner funnel) from the excess of zinc; 5 c.c. of a water solution of sulphur dioxide is previously

---

[1] A. 12, 311; 207, 329.  [2] Z. 1868, 437.

placed in the filter-flask to protect the hydrazobenzene from oxidation. The filtrate is heated to boiling in a beaker and a dilute solution of sulphur dioxide added until the turbidity which rises at first no longer vanishes. The colourless crystals separating out on cooling are filtered off, with suction, washed with alcohol containing sulphur dioxide, and dried best in a vacuum desiccator. Melting-point, 131°. Yield, 80–90 % of the theory.

While on the reduction of nitro-compounds in acid solution the final product is a primary amine, intermediate products between the nitro- and amido-compounds may be obtained by using milder reducing agents. Thus there is formed, first the azoxy-, then the azo-, and finally the hydrazo-compounds:

$$C_6H_5.NO_2 \atop C_6H_5.NO_2 \rightarrow {C_6H_5.N \atop C_6H_5.N}\!\!>\!\!O \rightarrow {C_6H_5-N \atop \underset{\|}{C_6H_5-N}} \rightarrow {C_6H_5.NH \atop C_6H_5.NH} \rightarrow {C_6H_5.NH_2 \atop C_6H_5.NH_2}$$

| 2 Mol. Nitrobenzene | → | 1 Mol. Azoxybenzene | → | 1 Mol. Azobenzene | → | 1 Mol. Hydrazobenzene | → | 2 Mol. Aniline |

In order to reduce a nitro-compound to an azoxy-compound, either sodium amalgam or alcoholic-caustic potash or caustic soda is used. With nitrobenzene, particularly, the reaction takes place most surely by dissolving sodium in methyl alcohol as above. The reducing action of sodium methylate depends upon the fact that it is oxidised to sodium formate, two hydrogen atoms of the methyl group being replaced by one atom of oxygen:

$$CH_3.ONa + O_2 = H_2O + H.CO.ONa.$$

In the operation carried out above the reaction is expressed by the equation:

$$4\,C_6H_5.NO_2 + 3\,CH_3.ONa = 2\ {C_6H_5.N \atop C_6H_5.N}\!\!>\!\!O + 3\,H\,CO.ONa + 3\,H_2O.$$

A few words may be said here concerning the relatively weak reducing power of *previously prepared* alcoholates, in comparison with the extremely energetic action of a mixture of undissolved sodium and alcohol. While the previously prepared alcoholates can generally only abstract oxygen, the mixture just referred to belongs to the class of very strong reducing agents. With the aid of this, it is possible to break up the double or centric union of the benzene ring, and thus prepare hydrogen derivatives of benzene. In this case the alcoholate does not act as a reducing agent as above, but the hydrogen effects the reduction:

$$CH_3.OH + Na = CH_3.ONa + H.$$

The azoxy-compounds are yellow- to orange-red crystallisable substances, which, like the nitro-compounds, are of an indifferent character; but they are not volatile with steam, and cannot be distilled without undergoing decomposition. On reduction they yield first the azo-compounds, then the hydrazo-compounds, and finally two molecules of a primary amine. By heating with sulphuric acid, azoxybenzene is converted into its isomer oxyazobenzene:

$$C_6H_5.\underset{\underset{O}{\vee}}{N-N}.C_6H_5 = C_6H_5.N=N.C_6H_4.OH.$$

If an azoxy-compound is distilled carefully over iron filings, its oxygen atom is removed, and an azo-compound is formed:

$$C_6H_5.\underset{\underset{O}{\vee}}{N-N}.C_6H_5 + Fe = C_6H_5.N=N.C_6H_5 + FeO.$$

Azo-compounds may also be obtained directly from nitro-compounds, since they are reduced by sodium amalgam or an alkaline solution of stannous chloride (sodium stannous oxide). The latter reducing agent acts in accordance with this equation:

$$2\,C_6H_5.NO_2 + 4\,Sn\begin{smallmatrix}ONa\\ \\ONa\end{smallmatrix} = \begin{smallmatrix}C_6H_5.N\\ \|\\C_6H_5.N\end{smallmatrix} + 4\,SnO\begin{smallmatrix}ONa\\ \\ONa\end{smallmatrix}$$
Sodium stannate

Azo-compounds may also be obtained by the oxidation of hydrazo-compounds:

$$C_6H_5.NH.NH.C_6H_5 + O = C_6H_5.N=N.C_6H_5 + H_2O.$$

The azo-hydrocarbons are orange-red to red crystalline substances which can be distilled without decomposition, differing in this respect from the azoxy-compounds.

EXPERIMENT : A few crystals of azobenzene are heated in a test-tube to boiling, over a free flame. A red vapour is evolved, which again condenses to crystals on cooling.

The azo-compounds thus differ in their stability from the very easily decomposable diazo-compounds, which also contain the group $N = N$, but it is in combination with only one hydrocarbon residue and an acid residue ; *e.g.* $C_6H_5.N = N.Cl$.

By the reduction of an azo-compound, a hydrazo-compound is first formed and finally an amine.

---

The hydrazo-compounds are formed by the reduction of azo-compounds with ammonium sulphide or zinc dust and an alkali. Zinc dust with caustic soda acts as follows :

$$Zn + 2\,NaOH = Zn\langle^{ONa}_{ONa} + H_2.$$

The hydrazo-compounds, in contrast with the azoxy-, and especially with the intensely coloured azo-compounds, are colourless. They are derived from hydrazine, $NH_2 - NH_2$, in which one hydrogen atom of the two amido-groups has been replaced by a hydrocarbon radical. The basic character of hydrazine is so weakened by the presence of the negative hydrocarbon residues, that the hydrazo-compounds no longer possess a basic character. On oxidation hydrazo-compounds pass over to azo-compounds, a reaction which takes place slowly but completely, under the influence of the oxygen of the air. The hydrazo-compounds decompose, on heating, into azo-compounds and primary amines.

$2\,C_6H_5.NH.NH.C_6H_5 = C_6H_5.N{=}N.C_6H_5 + 2\,C_6H_5.NH_2.$

EXPERIMENT : A few crystals of hydrazobenzene are heated in a small test-tube to boiling ; the colourless compound becomes red, azobenzene being formed. In order to show the presence of aniline, after cooling, the substance is shaken with water and the bleaching-powder test applied.

If the hydrazo-compounds are treated with concentrated acids like hydrochloric or sulphuric acids, they are converted into derivatives of diphenyl :[1]

---

[1] J. pr. 36, 93 ; J. 1863, 424.

$$C_6H_5 \cdot NH \cdot NH \cdot C_6H_5 = NH_2 \cdot C_6H_4 \cdot C_6H_4 \cdot NH_2$$
p-Diamidodiphenyl = Benzidine

The molecular transformation takes place essentially in para-position to the imide (NH) groups.

EXPERIMENT: Hydrazobenzene is covered with concentrated hydrochloric acid, and allowed to stand for about 5 minutes. It is then treated with water, and half the solution is made alkaline with caustic soda: the free benzidine is extracted several times with ether, the ether evaporated, and the substance crystallised from hot water. Leaflets of a silvery lustre are obtained. Melting-point, 122°. The other half of the solution is treated with dilute sulphuric acid, upon which the difficultly soluble benzidine sulphate separates out.

Benzidine differs from hydrazobenzene, in that it is a strong, di-acid primary base. It is prepared technically, since the azo dyes derived from it possess the important property of colouring unmordanted cotton fibre directly; for most azo dyes the cotton must first be mordanted. The first representative of these dyes made was Congo Red, prepared from the bisdiazo-compound of benzidine and naphthionic acid. In consequence, the entire class of these dyes is called the "Congo Dyes."

$$C_6H_4 \cdot N{=}N \cdot C_{10}H_5{<}{\genfrac{}{}{0pt}{}{NH_2}{SO_3H}}$$
$$C_6H_4 \cdot N{=}N \cdot C_{10}H_5{<}{\genfrac{}{}{0pt}{}{NH_2}{SO_3H}}$$

In a wholly analogous manner, from o-nitrotoluene, or o-nitroanisol, is prepared o-tolidine or dianisidine, respectively.

Tolidine      Dianisidine

If, in hydrazo-compounds, the para-position to the imido (NH) group is occupied as, *e.g.* in p-hydrazotoluene, then the benzidine transformation cannot occur.

In such cases, derivatives of o- and p-amidodiphenyl amine are formed through the so-called "Semidine transformation."[1]

$$CH_3-C_6H_4-NH-NH-C_6H_4-CH_3 \longrightarrow$$

$$\begin{array}{c} CH_3-C_6H_3(NH_2)- \\ \text{o-Semidine} \end{array} NH-C_6H_4-CH_3$$

$$CH_3 \cdot CO \cdot NH-C_6H_4-NH-NH-C_6H_5 \longrightarrow$$

$$\begin{array}{c} CH_3 \cdot CO \cdot NH-C_6H_4-NH-C_6H_4-NH_2. \\ \text{p-Semidine} \end{array}$$

Finally, it is to be mentioned, that recently a further intermediate product between nitro-compounds and amines has been prepared. If nitrobenzene is heated with zinc dust and water, in the presence of certain metallic salts like calcium chloride, the reduction thus taking place in a neutral solution, phenyl hydroxylamine is formed:

$$C_6H_5 \cdot NO_2 + 2 H_2 = C_6H_5 \cdot N {\Large <}^{H}_{OH} + H_2O.$$

## 4. REACTION: PREPARATION OF A THIOUREA AND A MUSTARD OIL FROM CARBON DISULPHIDE AND A PRIMARY AMINE

EXAMPLE: **Thiocarbanilide and Phenyl Mustard Oil from Carbon Disulphide and Aniline.**

*Thiocarbanilide:* A mixture of 40 grammes of aniline, 50 grammes of carbon disulphide, 50 grammes of alcohol, and 10 grammes of finely pulverised potassium hydroxide is gently boiled for 3 hours on a water-bath in a flask provided with a long reflux condenser. The excess of carbon disulphide and alcohol is then distilled off, the residue treated with water, the crystals separating out are filtered off, and washed first with water, then with dilute hydrochloric acid, and finally with water. For

---

[1] B. 26, 681, 688, 699.

the preparation of phenyl mustard oil, the crude product is used directly, after it has been dried on the water-bath. In order to obtain pure thiocarbanilide, 2 grammes of the dried crude product is recrystallised from alcohol. Large, colourless tablets are thus obtained, which melt at 154°. Yield, 30–35 grammes. If a mixture of equal parts, by weight, of aniline, carbon disulphide, and alcohol (40 grammes of each) is heated with the addition of caustic potash, to gentle boiling on the water-bath for 10–12 hours, a better, almost quantitative yield of thiocarbanilide is obtained, although a longer time is required. After the heating, proceed as above.

*Phenyl Mustard Oil:*[1] In a flask of about 400 c.c. capacity, place 30 grammes of the crude thiocarbanilide, and treat with 120 grammes of concentrated hydrochloric acid; the mixture is distilled by heating to the boiling-point of the acid, on a sand-bath, with a large flame under a hood. As soon as about 20 c.c. of the liquid remain in the flask, the distillation is discontinued. The distillate is treated with an equal volume of water, the mustard oil separated in a dropping-funnel, dried with a little calcium chloride, and distilled. Boiling-point, 222°. Yield, almost quantitative.

*Triphenyl Guanidine:* The residue remaining in the flask after the distillation with hydrochloric acid is treated with 100 c.c. of water, and then allowed to stand for several hours, when colourless crystals of triphenylguanidine hydrochloride separate out. These are filtered off, and warmed with some dilute caustic soda solution. The salt is decomposed, and the free base obtained, which on recrystallising from alcohol forms colourless crystals. Melting-point, 143°.

Carbon disulphide acts upon primary amines to form symmetrical disubstituted thioureas, *e.g.*:

$$CS_2 + 2C_6H_5 \cdot NH_2 = C\begin{matrix}\diagup NH \cdot C_6H_5 \\ =S \\ \diagdown NH \cdot C_6H_5\end{matrix} + H_2S.$$

Diphenyl thiourea = Thiocarbanilide

---

[1] B. 15, 986; Z. 1869, 589.

By the addition of caustic potash the elimination of hydrogen sulphide is facilitated, so that the reaction takes place in a shorter time than without the use of the alkali.

From the thioureas thus obtained the mustard oils may be prepared by heating with acids, as hydrochloric acid, sulphuric acid, phosphoric acid. The reaction takes place in accordance with the following equation:

$$\text{CS}\begin{array}{c}\diagup\overline{\text{NH}.\text{C}_6\text{H}_5}\\ \diagdown\underline{\text{N}\,\text{H}}\,\text{C}_6\text{H}_5\end{array} = \text{C}_6\text{H}_5.\text{N}{=}\text{C}{=}\text{S} + \text{C}_6\text{H}_5.\text{NH}_2.$$

Phenyl mustard oil

The primary amine formed in addition to the mustard oil combines with the acid. Besides this reaction a second one takes place, viz.: the amine formed acts upon some still undecomposed thiourea, resulting in the formation of a guanidine derivative:

$$\text{CS}\begin{array}{c}\diagup\text{NH}.\text{C}_6\text{H}_5\\ \diagdown\text{NH}.\text{C}_6\text{H}_5\end{array} + \text{C}_6\text{H}_5.\text{NH}_2 = \text{C}\begin{array}{c}\diagup\text{NH}.\text{C}_6\text{H}_5\\ {=}\text{N}.\text{C}_6\text{H}_5\\ \diagdown\text{NH}.\text{C}_6\text{H}_5\end{array} + \text{H}_2\text{S}.$$

Triphenyl guanidine

Since guanidine $\text{C}\begin{array}{c}\diagup\text{NH}_2\\ {=}\text{NH}\\ \diagdown\text{NH}_2\end{array}$ is an extremely strong base, which, like caustic potash and caustic soda, absorbs carbon dioxide from the air, the introduction of the three negative phenyl groups in the above compound has not neutralised the basic properties entirely, and it still has the power to form salts.

The aromatic mustard oils are in part colourless liquids, in part crystallisable solids, the lower members are easily volatile with steam, and possess a characteristic odour. In chemical behaviour they are very active. If they are warmed for a long time with an alcohol, they combine with the alcohol, addition taking place, and a thiourethane is formed:

$$\text{C}_6\text{H}_5.\text{NCS} + \text{C}_2\text{H}_5.\text{OH} = \text{C}_6\text{H}_5.\text{NH}.\text{CS}.\text{OC}_2\text{H}_5.$$

Phenylthiourethane

In the same way, ammonia and primary bases are added with the formation of thiourea:

$$\text{C}_6\text{H}_5.\text{NCS} + \text{NH}_3 = \text{CS}\begin{array}{c}\diagup\text{NH}_2\\ \diagdown\text{N}\,\text{H}.\text{C}_6\text{H}_5,\end{array}$$

Phenylthiourea

$$C_6H_5 \cdot NCS + C_6H_5 \cdot NH_2 = CS\begin{array}{c}NH \cdot C_6H_5 \\ NH \cdot C_6H_5\end{array}$$
s-Diphenylthiourea

EXPERIMENT : Treat 2 drops of phenyl mustard oil on a watch-glass with 2 drops of aniline, and warm gently over a small flame. On stirring the reaction-product after cooling, with a glass rod, the thiocarbanilide will solidify in crystals, from which in the above reverse reaction the mustard oil itself was prepared.

By heating with yellow mercuric oxide, the sulphur is replaced by oxygen, and an isocyanate is formed, which may be easily recognised by its extremely disagreeable odour:

$$C_6H_5 \cdot NCS + HgO = C_6H_5 \cdot NCO + HgS.$$
Phenyl isocyanate

EXPERIMENT : Heat ½ c.c. of phenyl mustard oil in a test-tube with the same volume of yellow mercuric oxide for some time, until the oil boils. The yellow oxide is changed to the black sulphide, at the same time the extremely disagreeable odour of the phenyl isocyanate arises; the vapour of the compound attacks the eyes, causing tears.

### 5. REACTION: THE SULPHONATION OF AN AMINE

EXAMPLE : **Sulphanilic Acid from Aniline and Sulphuric Acid**[1]

To 100 grammes of concentrated sulphuric acid in a dry flask, 30 grammes of freshly distilled aniline is added gradually, with shaking; the mixture is heated in an oil-bath, up to 180–190°, until, from a test-portion diluted with water and treated with caustic soda, no aniline separates out: about 4–5 hours' heating will be necessary. The cooled reaction-mixture is poured, with stirring, into cold water upon which the sulphanilic acid separates out in crystals. It is filtered off, washed with water, and recrystallised from water, with the addition of animal charcoal, if necessary. Yield, 30–35 grammes.

---

[1] A. 60, 312 ; 100, 163; 120, 132.

When an aromatic compound is treated with sulphuric acid, a portion of the benzene-hydrogen is replaced by a sulphonic acid group, the reaction taking place in accordance with the equation below. The aliphatic compounds do not react in a similar manner. Under the preparation of benzene sulphonic acid, the details of the reaction will be discussed.

$$C_6H_5 \cdot NH_2 + SO_2{\begin{matrix}OH\\OH\end{matrix}} = C_6H_4{\begin{matrix}NH_2\\SO_3H\end{matrix}} + H_2O.$$

p-Amidobenzenesulphonic acid = sulphanilic acid

In the above example, it happens, as in many cases, that the sulphonic acid group enters in the para-position to the amido-($NH_2$) group. The amido sulphonic acids are colourless crystallisable compounds melting with decomposition; they possess acid properties, *i.e.* in dissolving in alkalies. The basic character of the amine is so greatly weakened by the introduction of the negative sulphonic acid group that the amido sulphonic acids cannot form salts with acids. They differ in this from the analogous carbonic acids $\left(e.g., C_6H_4{\begin{matrix}NH_2\\CO.OH\end{matrix}}\right)$, which

Amidobenzoic acid

dissolve in both acids and alkalies.

The amido sulphonic acids, since they are derivatives of a primary amine, may like them be diazotised by the action of nitrous acid; and upon this fact depends their great technical importance. If the diazo-compounds thus obtained are combined with amines or phenols, azo dyes are formed which contain the sulphonic acid group, and in the form of their alkali salts are soluble in water. Sulphanilic acid particularly, and its isomer, metanilic acid, obtained by the reduction of m-nitrobenzenesulphonic acid, as well as the numerous mono- and polysulphonic acids derived from $\alpha$ and $\beta$ naphthyl amines, find extensive technical application in the manufacture of azo dyes.

### 6. REACTION: REPLACEMENT OF THE AMIDO- AND DIAZO-GROUPS BY HYDROGEN

EXAMPLE: **Benzene from Aniline**

Dissolve 5 grammes of aniline in a mixture of 15 grammes of concentrated hydrochloric acid and 30 c.c. of water; cool with ice, and treat with a solution of 5 grammes of sodium nitrite in 15 c.c. of water, until free nitrous acid may be recognised with starch-potassium-iodide paper. The diazobenzenechloride solution thus obtained is allowed to flow carefully into a solution of 10 grammes caustic soda in 30 c.c. of water contained in a 400 c.c. flask which is well cooled with ice. Further, dissolve 20 grammes of stannous chloride in 50 c.c. of water, and treat this solution with a concentrated solution of sodium hydroxide (2 parts to 3 of water), until the precipitate at first formed (stannous oxide) is redissolved in the excess of the alkali. Treat the alkaline diazobenzene solution, cooled with ice-water, gradually with small portions of the sodium-stannous oxide solution, waiting after each addition, until the lively evolution of nitrogen has ceased, before adding more. When all the reducing liquid has been added, the flask is connected with a condenser, and the liquid heated to boiling. The benzene formed passes over first, and is collected in a test-tube. By a careful distillation from a small fractionating flask (without condenser), it is obtained perfectly pure. Boiling-point, 81°. Yield, 3–4 grammes.

As already mentioned, under the preparation of methyl amine, the behaviour of the aliphatic primary amines toward nitrous acid is very different from that of the aromatic compounds. While the former yield alcohols with the elimination of nitrogen, the latter form diazo-compounds discovered by Peter Griess.[1]

$$CH_3 . NH_2 + NOOH = CH_3 . OH + N_2 + H_2O,$$
$$C_6H_5 . NH_2 + NOOH + HCl = C_6H_5 . N{\equiv}N . Cl + 2 HCl.$$
Diazobenzene chloride

The diazo-compounds, like the azo-compounds, contain the bivalent group $N = N$. In the simplest case of the azo-compounds, the double nitrogen group is in combination with *two* hydrocarbon radicals; in the diazo-compounds, it is united with but *one* hydrocarbon radical; for this reason it can, in addition, combine with univalent acid radicals like Cl, Br, $NO_3$, $SO_4H$. The behaviour of the two classes of compounds is entirely different. While the azo-compounds are very stable, and, as pointed out above, the azo-hydrocarbons can be distilled with-

---

[1] A. 137, 39.

out decomposition, the diazo-compounds are extremely unstable, and decompose even in solution; for this reason, care must always be taken in the preparation of them. They are derivatives of the simplest compound, the free diazobenzene:

$$C_6H_5 \cdot N{=}N \cdot OH,$$

which possesses a basic as well as a weakly acid character. In the presence of acids, it forms salts, in accordance with the following equations:

$$C_6H_5 \cdot N{=}N \cdot OH + HCl = C_6H_5 \cdot N{=}N \cdot Cl + H_2O,$$
Diazobenzene chloride

$$C_6H_5 \cdot N{=}N \cdot NO_3 \text{ Diazobenzene nitrate,}$$

$$C_6H_5 \cdot N{=}N \cdot O \cdot SO_2 \cdot OH \text{ Diazobenzene sulphate.}$$

The diazo-compounds can also form double salts, *e.g.*:

$$C_6H_5 \cdot N{=}N \cdot Cl \cdot AuCl_3 \quad \text{and} \quad (C_6H_5 \cdot N{=}N \cdot Cl)_2 \cdot PtCl_4.$$

Diazobromides have the power of taking up two atoms of bromine to form perbromides:

$$C_6H_5 \cdot N{=}N \cdot Br + Br_2 = C_6H_5 \cdot NBr - NBr \cdot Br.$$
Diazobenzene perbromide

EXPERIMENT: Dissolve 1 c.c. of aniline in an excess of hydrochloric acid, diazotise as above, and add 1 c.c. of bromine dissolved in a water solution of hydrobromic acid, or in a concentrated solution of potassium bromide. A dark oil separates out, from which the solution is decanted. It is washed several times with water; on cooling, it solidifies to crystals.

If ammonia is allowed to act on the perbromide, diazobenzeneimide is obtained:

$$C_6H_5 \cdot NBr \cdot NBr_2 + NH_3 = C_6H_5 \cdot N\underset{N}{\overset{N}{\diagup\!\!\!\diagdown}} + 3\,HBr.$$

Diazobenzeneimide

EXPERIMENT: The perbromide just obtained is covered with water, and concentrated ammonium hydroxide added to it. A vigorous reaction takes place with the formation of an oil possessing a strong odour (diazobenzeneimide).

In the presence of strong bases, like caustic potash or caustic soda the diazo-compounds behave like weak acids, forming salts with the alkalies as follows:

$$C_6H_5 \cdot N{=}N \cdot Cl + 2\,NaOH = \underset{\text{Sodium diazobenzene}}{C_6H_5 \cdot N{=}N \cdot ONa} + NaCl + H_2O.$$

Recently the idea has been advanced that the diazo-compounds have the power to undergo a molecular transformation, passing over to the isomeric nitrosoamines.

$$\underset{\text{Diazobenzene}}{C_6H_5 \cdot N{=}N \cdot OH} \longrightarrow \underset{\text{Nitrosoaniline}}{C_6H_5 \cdot N\!\!<\!\!\genfrac{}{}{0pt}{}{H}{NO}}$$

On the other hand, it is maintained that the difference is not due to structural isomerism, but that it is a case of stereo-isomerism. In accordance with these views, the two varieties of the diazo-compound are formulated thus:

$$\underset{\text{Syndiazo-compound}}{\begin{array}{c}C_6H_5 \cdot N\\ \|\\ HO-N\end{array}} \qquad \underset{\text{Antidiazo-compound}}{\begin{array}{c}C_6H_5 \cdot N\\ \|\\ N-OH\end{array}}$$

The salts of the diazo-compounds formed with acids are, in most cases, colourless, crystallisable substances, easily soluble in water, insoluble in ether. In order to prepare them in the solid condition, various methods may be used. Thus, *e.g.*, the very explosive diazobenzene-nitrate may be obtained in colourless needles by conducting gaseous nitrous acid into a well-cooled pasty mass of aniline nitrate and water, and treating the diazo-solution with alcohol and ether. In general, the solid diazo-salts may be prepared by adding to an alcoholic solution of the amine that acid the salt of which is desired, and then treating the well-cooled mixture with amyl nitrite:[1]

$$C_6H_5 \cdot NH_2 + NO_2 \cdot C_5H_{11} + HCl = C_6H_5 \cdot N{=}N \cdot Cl + C_5H_{11} \cdot OH + H_2O.$$

If the solid diazo-compound does not separate out at once, ether is added. On heating, the dry diazo-salts decompose either, as in the case of diazobenzene nitrate, with explosion, or a sudden evolution of gas takes place without detonation. A few diazo-compounds are so stable that they may be recrystallised from water.

---

[1] B. 23, 2994.

In rare cases only, in working with diazo-compounds, is it necessary to isolate them in a pure condition; generally, the very easily prepared water solutions are used. These compounds were formerly obtained by passing gaseous nitrous acid into a salt of the amine until it was diazotised. But at present this method is employed only in rare cases; the free nitrous acid obtained from sodium nitrite is used. In order to diazotise an amine, a solution of it in a dilute acid — most frequently hydrochloric acid or sulphuric acid — is first prepared. Theoretically, two molecules of a monobasic acid are required to diazotise one molecule of a monamine:

$$C_6H_5 \cdot NH_2 + NaNO_2 + 2\,HCl = C_6H_5 \cdot N{=}N \cdot Cl + NaCl + 2\,H_2O,$$

but an excess is always taken, — not less than three molecules of hydrochloric acid or two of sulphuric acid to one molecule of a monamine. In many cases, the hydrochloride or sulphate of the amine is difficultly soluble in water. Under these conditions, it is not necessary to add water until the salt is entirely dissolved, but the solution of the nitrite may be poured into the pasty mass of crystals; when the undissolved salt is diazotised, it passes into solution. For the diazotisation of one molecule of a monamine, one molecule of sodium nitrite is necessary, theoretically; but since the commercial salt is never perfectly pure, it is advisable to weigh off from 5–10 % more than the calculated amount, and to determine by the method given below when a sufficient quantity of this has been added. The nitrite is dissolved in water, generally 10–20 parts of water to 1 part of salt. The nitrite solution must be added gradually to the amine solution, and the liquid must not be allowed to become warm. In many cases, the experimenter is often too careful, in that he cools the amine solution with a freezing mixture, and adds the nitrite solution drop by drop from a separating funnel. Frequently it is sufficient to place the solution in a water-bath filled with cold water, or ice is thrown into the water, or the amine solution is cooled by ice. It is very convenient to cool the solution, not from without, but by throwing into it from time to time small pieces of ice. The nitrite solution may be poured directly from a flask. If the addition causes evolution of gas bubbles or vapours of nitrous acid, the temperature of the solution must be lowered and the nitrite added more slowly. In order to be cognisant of the course of the reaction, as well as to be able to determine when it is completed, starch-potassium-iodide paper, prepared as follows, is used:

A piece of starch the size of a pea is finely pulverised, and added to

100 c.c. of boiling water; it is boiled a short time, with stirring. After cooling, a solution of a crystal of potassium iodide the size of a lentil, in a little water, is added to it. With this mixture, saturate long strips of filter-paper 3 cm. wide; the strips are dried by suspending them from a string in a place free from acids. After drying, the strips are cut up and preserved in a closed vessel.

In order now to diazotise an amine, the cooled solution is first treated with a small portion of the nitrite solution; it is well stirred, and a drop of it transferred with a clean glass rod to the starch-potassium-iodide paper. If the nitrous acid is already used up in the diazotisation, no dark spots appear, and further portions of the sodium nitrite may be added, the test is again repeated, and so on. But if a dark spot is formed at once, the nitrous acid is still present; and in this case, before more of the nitrite is added, one waits until the reaction has been completed, and so on. After the addition of three-fourths of the nitrite solution, larger quantities may be added at one time, but toward the end of the reaction small quantities must again be employed. The diazotisation is ended when, after standing some time, the mixture shows the presence of nitrous acid. Since the diazotisation of the last portions of the amine often requires some time, the addition of the nitrite is not discontinued at once, even if after one minute the test for nitrous acid is obtained, but the solution is allowed to stand 5-10 minutes, and is then tested again. At times, it happens that the weighed-off quantity of sodium nitrite is apparently not sufficient to complete the diazotisation, and that even after the addition of a fresh quantity, the test will not show the presence of nitrous acid. This phenomenon has its cause generally in the fact that the acid (hydrochloric or sulphuric) has been used up, and consequently the nitrite cannot enter into the reaction. Thus, in case the weighed-off amount of sodium nitrite is not sufficient, some acid is first added to a small portion of the liquid, and this is then tested to determine whether the desired reaction has taken place. Further, often the diazo-solution becomes cloudy toward the end of the reaction, or a precipitate separates out. This is the diazoamido-compound; its formation is also caused by the lack of free acid. On the addition of acid and solution of the nitrite, the precipitate disappears. The replacement of the diazo-group by hydrogen in the above reaction takes place in accordance with the following equation:

$$C_6H_5 \cdot N{=}N \cdot OH + H_2 = C_6H_6 + N_2 + H_2O\,[1].$$

[1] B. 22, 587.

# AROMATIC SERIES

In this way it is possible in many cases to replace a primary amido-group by hydrogen. Obviously, such a reaction is superfluous, if, as in the above case, the amine is obtained by the nitration of the hydrocarbon and the reduction of the nitro-compound. But there are cases in which an amine is not obtained in this way, and where it is of importance to prepare the amido-free compound (see below).

The replacement of a diazo-group by hydrogen may be effected by other reducing agents. If, *e.g.*, a diazo-compound is boiled with alcohol, the latter is converted into aldehyde, thus liberating two hydrogen atoms, by which the diazo-compound is reduced:

$$C_6H_5 \cdot N{=}N \cdot OH + CH_3 \cdot CH_2 \cdot OH = C_6H_6 + N_2 + \underset{\text{Aldehyde}}{CH_3 \cdot CHO} + H_2O.$$

The reaction is effected either by conducting gaseous nitrous acid into the boiling alcohol solution of the amine, or by heating the amine with alcohol saturated with ethyl nitrite; or the boiling alcohol solution of the amine, acidified with sulphuric acid, may be treated with sodium nitrite.

At this place, two examples may be mentioned which illustrate the theoretical as well as the practical value of the reaction: by the oxidation of a mixture of aniline and p-toluidine, there is formed a complex dye, para-fuchsine, the constitution of which was unknown for a long time. This was first explained by E. and O. Fischer. They heated the diazo-compound of the leuco-base of the dye, paraleucaniline with alcohol, which gave the mother substance — the hydrocarbon triphenyl methane (A. 194, 270).

As an example of the preparation value of the reaction, the following case is cited:

No method is known by which m-nitrotoluene can be prepared by the nitration of toluene; this results in the formation of the o- and p-compounds only. In order to obtain the m-nitrotoluene, the starting-point is p-toluidine. This is nitrated, upon which a nitrotoluidine of the following constitution is obtained:

$$\underset{NH_2}{\overset{CH_3}{\underset{|}{\bigcirc}}}{-}NO_2.$$

If the amido-group is replaced by hydrogen, using the method last described, the desired m-nitrotolene is obtained.

By boiling a diazo-compound with alcohol the reaction takes place in a different way; at times the diazo-group is not replaced by hydrogen, but by the ethoxy ($-OC_2H_5$) group, thus giving rise to a phenol ether.

$$X.N\equiv N.SO_4H + C_2H_5.OH = X.OC_2H_5 + N_2 + H_2SO_4.$$

In conclusion, special attention is called to the fact that not only aniline and its homologues can be diazotised, but all the derivatives of these, as the nitro-amines, halogen-substituted amines, amido-aldehydes, amido-carbonic acids, etc.

### 7. REACTION: REPLACEMENT OF THE DIAZO-GROUP BY HYDROXYL

EXAMPLE : **Phenol from Aniline**

Pour 20 grammes of concentrated sulphuric acid as rapidly as possible, with stirring, into 50 grammes of water; to the hot solution add 10 grammes of freshly distilled aniline, with stirring, by allowing it to flow down the side of the beaker, then add 100 c.c. of water. After the hot liquid has been cooled by immersion in cold water, it is treated with a solution of 9 grammes of sodium nitrite in 40 c.c. of water, until it shows a blue spot on starch-potassium-iodide paper. The diazobenzene sulphate solution thus obtained is gently heated for half an hour on the water-bath, the phenol is then distilled over with steam, and is extracted several times with ether. The ethereal solution is allowed to stand for some time over fused sodium sulphate. The ether is then evaporated, and the residue of phenol is subjected to distillation. Boiling-point, 183°. Yield, 7–8 grammes.

If a diazo-compound is heated with water, it will pass over to a phenol with the evolution of nitrogen, *e.g.*:

$$C_6H_5.N\equiv N.O.SO_2.OH + HOH = C_6H_5.OH + N_2 + H_2SO_4.$$

For this reaction the diazosulphate is most advantageously used. Under certain circumstances, the diazochloride may also be employed. But the use of the diazonitrate is avoided, since, in this case, the nitric acid liberated, acting upon the phenol, readily forms nitro-compounds. In many cases, it is more convenient not to isolate the diazo-compound, but to add a water solution of the calculated amount of sodium nitrite to a boiling solution of the amine in dilute sulphuric acid. The diazotisation of the substance and the immediate decomposition of the diazo-compound take place in one operation.

The same reactions are also applicable to substituted amines, like amido - carbonic acids, amido - sulphonic acids, halogen substituted amines, etc.

### 8. REACTION: REPLACEMENT OF A DIAZO-GROUP BY IODINE

EXAMPLE : **Phenyl Iodide from Aniline**

*(Phenyliodidechloride, Iodoso-benzene, and Phenyliodite)*

A solution of 10 grammes of aniline in a mixture of 50 grammes of concentrated hydrochloric acid and 150 grammes of water cooled with ice-water is gradually treated with a solution of 9 grammes of sodium nitrite in 30 c.c. of water, until a test will give a blue colour to the starch-potassium-iodide paper. The diazo-solution is then treated in a flask, not too small, with a solution of 25 grammes of potassium iodide in 50 c.c. of water, and heated on the water-bath until the evolution of nitrogen ceases. The dark oil separating out is taken up with ether, washed with water, then with dilute caustic soda, and finally dried over calcium chloride. The residue remaining after the evaporation of the ether is fractionated. Boiling-point, 189–190°. Yield, about 20 grammes.

If a diazoiodide is heated, the diazo-group is replaced by iodine, the reaction taking place smoothly in most cases.

The reaction is effected by diazotising the amine in a hydrochloric acid or sulphuric acid solution, and then treating it with potassium iodide. From diazochloride or diazosulphate there is formed a diazo-iodide, the reaction, in many cases, taking place at the ordinary temperature; in others, on heating, as above. Since the reaction occurs without difficulty, it is used as the method of preparation of many iodides.

The aromatic iodides possess the noteworthy property of combining with two atoms of chlorine, the iodine previously univalent becoming trivalent:

$$C_6H_5 \cdot I + Cl_2 = C_6H_5 \cdot ICl_2.\text{[1]}$$
Phenyliodidechloride

EXPERIMENT: The phenyl iodide obtained is dissolved in five times its volume of chloroform, the solution is cooled by ice-water, and a current of dry chlorine is passed into it, until no more is absorbed. The crystals separating out are filtered off, washed with a fresh quantity of chloroform, spread out in a thin layer on a pad of filter-paper, and allowed to dry in the air.

If caustic soda is allowed to act on an iodochloride, the two chlorine atoms are replaced by one oxygen atom, and an iodoso-compound is obtained:

$$C_6H_5 \cdot ICl_2 + H_2O = C_6H_5 \cdot I{=}O + 2\,HCl.$$
Iodosobenzene

EXPERIMENT: The iodochloride is carefully triturated with dilute caustic soda in a mortar (for 1 gramme of the iodochloride, use a solution of 0.5 gramme sodium hydroxide in 4 grammes of water), and allowed to stand over night. The iodosobenzene is then filtered off, washed with water, and presssd out on a porous plate.

The iodoso-compounds have the power of uniting with acids to form salts, in which they act like a di-acid base, *e.g.*, $C_6H_5 \cdot I\begin{smallmatrix}\diagup OH \\ \diagdown OH\end{smallmatrix}$

EXPERIMENT: Several grammes of iodosobenzene is dissolved with heat in as small a quantity of glacial acetic acid as possible; the solution is evaporated on the water-bath to dryness, in a watch-glass, or shallow dish. The solid residue is pulverised and re-crystallised from a little benzene. Iodosobenzene acetate is thus obtained,

$$C_6H_5 \cdot I\begin{smallmatrix}\diagup OOC \cdot CH_3 \\ \diagdown OOC \cdot CH_3\end{smallmatrix}$$

in the form of colourless prisms, melting at 157°.

---

[1] J. pr. 33, 154. B. 25, 3494; 26, 357; 25, 2632.

The iodoso-compounds, on treatment with hydriodic acid, are reduced to iodides, with a separation of iodine.

$$C_6H_5 \cdot IO + 2\,HI = C_6H_5 \cdot I + I_2 + H_2O.$$

This reaction is used for the quantitative determination of iodoso-oxygen.

EXPERIMENT: Some potassium iodide is dissolved in water, acidified with dilute sulphuric acid, or acetic acid, and a few grains of iodosobenzene is added. The iodine separates out as a brown precipitate.

If an iodoso-compound is heated carefully to 100°, it passes over to an iodite (*Jodoverbindung*):

$$2\,C_6H_5 \cdot IO = \underset{\text{Phenyl iodite}}{C_6H_5 \cdot IO_2} + \underset{\text{Phenyl iodide}}{C_6H_5 \cdot I}.$$

The same compound may also be obtained by treating an iodoso-compound with steam. Of two molecules of the iodoso-compound, one is oxidised to an iodite, and the second is reduced to an iodide:

$$2\,C_6H_5 \cdot IO = C_6H_5 \cdot IO_2 + C_6H_5 \cdot I.$$

EXPERIMENT: Iodosobenzene is treated in a flask with enough water to form a thin paste. Into this steam is conducted (apparatus for distillation with steam), until no more phenyl iodide passes over with the steam and all the iodosobenzene has been dissolved. The solution is then evaporated on the water-bath to a small volume, and allowed to cool, upon which the phenyl iodite separates out in colourless crystals.

The iodites, like the iodoso-compounds, puff up and suddenly decompose on heating. (Try it.) They also abstract iodine from hydriodic acid, and in double the quantity as compared to the similar action of the iodoso-compounds.

$$C_6H_5 \cdot IO_2 + 4\,HI = C_6H_5I + 4\,I + 2\,H_2O.$$

They do not form salts with acids.

#### 9. REACTION: REPLACEMENT OF A DIAZO-GROUP BY CHLORINE, BROMINE, OR CYANOGEN.

EXAMPLE : p-Tolyl Nitrile from p-Toluidine

Dissolve 50 grammes of copper sulphate in 200 grammes of water in a 2-litre flask by heating on the water-bath; then add gradually, with continuous heating, a solution of 55 grammes of potassium cyanide in 100 c.c. of water. Since cyanogen is evolved, the reaction must be conducted under a hood, with a good draught, and the greatest care taken not to breathe the vapours.

While the cuprous cyanide solution is further heated on the water-bath, the diazotoluenechloride solution is prepared in the following way : 20 grammes of p-toluidine is heated with a mixture of 50 grammes of concentrated hydrochloric acid and 150 c.c. of water until solution takes place; the liquid is then quickly immersed in cold water and vigorously stirred with a glass rod, in order that the toluidine hydrochloride may separate out in as small crystals as possible. A solution of 16 grammes of sodium nitrite in 80 c.c. of water is then added to the amine hydrochloride, cooled by ice-water until a permanent reaction of nitrous acid upon the starch-potassium-iodide paper is obtained. The diazotoluene chloride thus formed is poured from a flask into the cuprous cyanide solution, with frequent shaking. After the addition of the diazo-solution, which should require about 10 minutes, the reaction-mixture is heated slightly on the water-bath for about a quarter-hour. The tolyl nitrile is then distilled over with steam. This operation must also be done under a hood with a *good draught*, since hydrocyanic acid passes over. The nitrile distils as a yellow oil, which, after some time, solidifies in the receiver. It is separated by decanting the water, pressed upon a porous plate, and purified by distillation. If the oil will not solidify, the entire distillate may be taken up with ether, the ethereal solution shaken with caustic soda solution to remove the cresol, and then, after evaporating the ether, the residue remaining is distilled directly, or, in case it is solid, it is pressed out on a porous plate, as above, and then distilled. Boiling-point, 218°. Yield, about 15 grammes.

The diazo-group *cannot* be replaced in the same way by iodine as by chlorine, bromine, or cyanogen. If a water solution of a diazo-chloride, -bromide, or -cyanide is heated, a phenol is formed, as is also the case on heating a diazo-sulphate:

$$C_6H_5 \cdot N{=}N \cdot Cl + H_2O = C_6H_5 \cdot OH + N_2 + H_2O.$$

To Sandmeyer[1] we are indebted for the important discovery that, if the heating be done in the presence of cuprous chloride, -bromide, or -cyanide, the reaction taking place is analogous to the one by which phenyl iodide is formed:

$$C_6H_5 \cdot N{=}N \cdot Cl = C_6H_5 \cdot Cl + N_2,$$
$$C_6H_5 \cdot N{=}N \cdot Br = C_6H_5 \cdot Br + N_2,$$
$$C_6H_5 \cdot N{=}N \cdot CN = C_6H_5 \cdot CN + N_2.$$

The manner in which the cuprous salts act is not known; in any case they unite first with a diazo-compound to form a double salt, which plays a part in the reaction.

The reaction in the above preparation of cuprous cyanide takes place in accordance with the following equation:

$$CuSO_4 + 2\,KCN = CuCN_2 + K_2SO_4,$$
$$2\,CuCN_2 = Cu_2CN_2 + 2\,CN.$$

In order to replace the diazo-group by chlorine or bromine, the above method is followed exactly. A diazo-solution is first prepared, and gradually added to a heated solution of cuprous chloride or bromide. With easily volatile chlorine or bromine compounds, it is desirable to use a reflux condenser, and to allow the diazo-solution to flow in from a dropping-funnel. In some cases it is more advantageous not to use a previously prepared diazo-solution, but to proceed as follows: The amine is dissolved in an acid solution of a copper salt; this is heated, and to the hot solution, the solution of nitrite is added from a dropping-funnel. The diazotisation and replacement of the diazo-group then takes place in one reaction. If the reaction-product is not volatile with steam, it may be obtained from the reaction-mixture by filtering, or extracting with ether.

---

[1] B. 17, 1633 and 2650; 18, 1492 and 1496.

The Sandmeyer reaction is capable of general application. Since the yield of the product is generally very good, for many substances it is used as a method of preparation. It should be finally pointed out that by the replacement of the diazo-group by cyanogen a new carbon union takes place.

**10. REACTION: (a) REDUCTION OF A DIAZO-COMPOUND TO A HYDRAZINE. (b) REPLACEMENT OF THE HYDRAZINE-RADICAL BY HYDROGEN**

EXAMPLES: (a) Phenyl Hydrazine from Aniline
(b) Benzene from Phenyl Hydrazine

(a) Add 10 grammes of freshly distilled aniline to 100 c.c. of concentrated hydrochloric acid in a beaker, with stirring; aniline hydrochloride partially separates out in crystals. To the mixture, cooled with ice, add slowly from a dropping-funnel, a solution of 10 grammes of sodium nitrite in 50 c.c. of water, until a test with starch-potassium-iodide paper shows free nitrous acid. In this case the strong acid solution must not be brought directly upon the test-paper, but a test-portion is diluted with water in a watch-glass and then the test applied. To the diazo-solution add, with stirring, a solution of 60 grammes of stannous chloride in 50 c.c. of concentrated hydrochloric acid cooled with ice; a thick paste of crystals of phenyl hydrazine hydrochloride separates out. After standing an hour this is filtered off with suction (Büchner funnel and filter-cloth), the precipitate is pressed firmly together on the filter with a pestle; it is then transferred to a small flask and treated with an excess of caustic soda solution. Free phenyl hydrazine separates out as an oil, it is taken up with ether, the ethereal solution dried with ignited potash, and the ether evaporated. For the later experiments the phenyl hydrazine thus obtained can be used directly. If it is desired to purify the substance, the best method is to distil it in a vacuum, or it can be cooled by a freezing-mixture, and the portions remaining liquid are poured off. Yield, about 10 grammes.

(b) In a 1-litre flask provided with a dropping-funnel and condenser (Fig. 62) 150 grammes of water and 50 grammes of copper sulphate are heated to boiling, then from the funnel add gradually a solution of 10 grammes of free phenyl hydrazine in a mixture of 8 grammes of glacial acetic acid and 75 grammes of water. The oxidation of the phenyl hydrazine proceeds with an energetic evolution of nitrogen; the benzene is immediately distilled over with steam and collected in a test-tube. By another careful rectification from a small fractionating flask (without con-

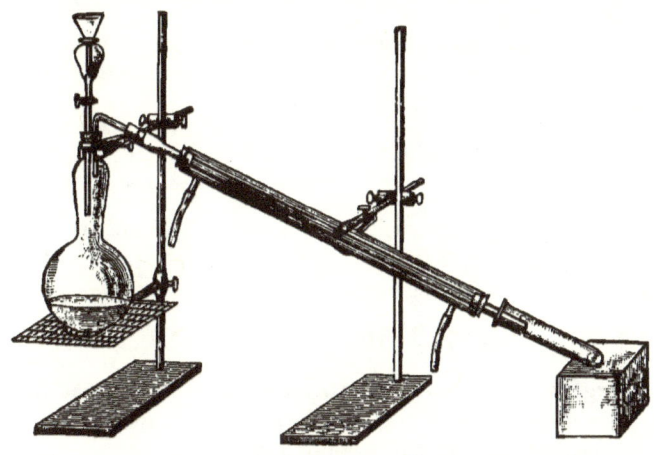

FIG. 62.

denser), benzene, boiling at 81°, is obtained. Yield, about 5 grammes.

Monosubstituted hydrazines of the type of phenyl hydrazine may be obtained according to the method of V. Meyer and Lecco,[1] by reducing the diazo-compounds with stannous chloride and hydrochloric acid:

$$C_6H_5 . N{\equiv}N . Cl + 2 H_2 = C_6H_5 . NH . NH_2, HCl.$$
Phenyl hydrazine hydrochloride

The reaction is always conducted as above: The amine is diazotised in a strong hydrochloric acid solution, and then a solution of stannous chloride in strong hydrochloric acid is added to it. Since the hydrochlorides of the hydrazines are difficultly soluble in concentrated hydrochloric acid, these separate out directly on the addition of the

---

[1] B. 16, 2976.

stannous chloride, and can easily be obtained pure by filtration, as above.

The reduction of the diazo-compounds to hydrazines may be accomplished by the method of Emil Fischer[1] which led to the discovery of this class of compounds, and also by another method. If neutral sodium sulphite is allowed to act on a diazo-salt, the acid radical of the diazo-compound is replaced by a residue of sulphurous acid, *e.g.*:

$$C_6H_5 \cdot N{\equiv}N \cdot Cl + NaSO_3 \cdot Na = C_6H_5 \cdot N{\equiv}N \cdot SO_3Na + NaCl.$$
Sodium diazobenzenesulphonate

If this salt is now reduced with sulphurous acid, or with zinc dust and acetic acid, it takes up two atoms of hydrogen and is converted into a hydrazine sulphonate:

$$C_6H_5 \cdot N{\equiv}N \cdot SO_3Na + H_2 = C_6H_5 \cdot NH \cdot NH \cdot SO_3Na.$$
Sodium phenyl hydrazine sulphonate

If this is heated with hydrochloric acid, the sulphonic acid group is split off, and phenyl hydrazine hydrochloride is formed, which, on evaporation, crystallises out:

$$C_6H_5 \cdot NH \cdot NH \cdot SO_3Na + HCl + HOH = C_6H_5 \cdot NH \cdot NH_2, HCl + NaHSO_4.$$

According to this method, which is slower, but cheaper than the former, phenyl hydrazine is prepared on the large scale.

The monosubstituted hydrazines possess a basic character; in spite of the fact that they contain two ammonia residues, they combine with only one molecule of a monobasic acid, *e.g.*:

$$C_6H_5 \cdot NH \cdot NH_2, HCl.$$
Phenyl hydrazine hydrochloride

Phenyl hydrazine reacts with aldehydes and ketones, the two hydrogen atoms of the amido-groups unite with the oxygen atom of the CHO- or CO-groups, and are eliminated as water:[2]

$$C_6H_5 \cdot CHO + C_6H_5 \cdot NH \cdot NH_2 = C_6H_5 \cdot CH{=}N \cdot NH \cdot C_6H_5 + H_2O,$$
Benzaldehyde ............ Benzylidenephenyl hydrazone

$$C_6H_5 \cdot CO \cdot C_6H_5 + C_6H_5 \cdot NH \cdot NH_2 = \begin{matrix} C_6H_5 \\ C_6H_5 \end{matrix}{>}C{=}N \cdot NH \cdot C_6H_5 + H_2O.$$
Benzophenone

---

[1] A. 190, 67.   [2] B. 16, 2976.

This reaction can be used for the recognition and detection of aldehydes and ketones. In order to prepare a hydrazone, formerly a solution of 1 part of phenyl hydrazine hydrochloride and 1½ parts of crystallised sodium acetate in 10 parts of water was used as a reagent. If this is added to an aldehyde or ketone, there is formed, in many cases at the ordinary temperature, but in others only on heating, the hydrazone. Since, at present, perfectly pure free phenyl hydrazine may be purchased in the market, a mixture of equal volumes of phenyl hydrazine and 50 % acetic acid, diluted with three times its volume of water, is used as the reagent.

EXPERIMENT : To a mixture of 4 drops of phenyl hydrazine and 5 c.c. of water, add 3 drops of glacial acetic acid. To this is added 2 drops of benzaldehyde (from a glass rod), and the mixture shaken. At first there appears a milky turbidity, but very soon a flocculent precipitate of benzylidenephenyl hydrazone separates out. The smallest quantity of benzaldehyde may be recognised in this way.

Phenyl hydrazine is of extreme importance in the chemistry of the sugars for the separation, recognition, and transformation of the different varieties of the sugars. Without this reagent, the fundamental explanations of the last few years in this field could scarcely have been made. If one molecule of phenyl hydrazine is allowed to act on one molecule of a sugar, a normal hydrazone is formed, *e.g.* :

$$CH_2.OH(CH.OH)_4.CHO + C_6H_5.NH.NH_2$$
$$\text{Grape sugar} \quad = CH_2.OH(CH.OH)_4.CH + H_2O$$
$$\qquad\qquad\qquad\qquad\qquad \underset{N - NH.C_6H_5}{\|}$$

But if the phenyl hydrazine is used in excess, it acts as an oxidising agent toward the sugar, since, *e.g.*, in the above case, one of the secondary alcohol, $(CH.OH)$ groups adjoining the aldehyde $(CHO)$ group, is oxidised to a ketone group, which again reacts with the hydrazine. The compound thus obtained is called an "Osazone." In the above example, there is obtained a compound of this composition:

$$CH_2.OH.(CH.OH)_3.\underset{\underset{NH.C_6H_5}{|}}{\underset{N}{\overset{\|}{C}}} - CH = N.NH.C_6H_5.$$

P

If this compound is heated with hydrochloric acid, it acts in the same way as all hydrazones, and phenyl hydrazine is eliminated; the original unchanged sugar is not formed again, but an oxidation product of it is obtained, a so-called "Osone." In the example selected, the osone is:

$$CH_2.OH.(CH.OH)_3.CO.CHO.$$

If this compound is treated with a reducing agent, the aldehyde group and not the ketone group is reduced, and the original sugar is not obtained:

$$CH_2.OH(CH.OH)_3.CO.CH_2.OH.$$

The aldose is converted into a ketose, the grape sugar into fruit sugar. The general importance of the reaction as applied to the sugars may be inferred from these brief statements.

EXPERIMENT: A cold solution of 2 grammes of phenyl hydrazine hydrochloride and 3 grammes of crystallised sodium acetate in 15 c.c. of water is treated with a solution of 1 gramme of pure grape sugar in 5 c.c. of water, and warmed on the water-bath. After about 10 minutes, the fine, yellow needles of the osazone begin to separate out; the quantity is increased by a longer heating. After heating an hour, the crystals are filtered off, washed with water, and allowed to dry in the air. Melting-point, 205°.

Phenyl hydrazine undergoes condensation with $\beta$-diketones and $\beta$-ketone-acid-esters with the formation of ring compounds containing nitrogen — the so-called pyrazoles and pyrazolones. The phenyl methyl pyrazolone formed from acetacetic ester and phenyl hydrazine is of importance:

$$CH_3.CO.CH_2.CO.OC_2H_5 + C_6H_5.NH.NH_2$$
$$= CH_3\!-\!\underset{\substack{\|\\N\!-\!-\!-\!N\!-\!C_6H_5}}{C}\!-\!CH_2\!\diagdown\!CO + H_2O + C_2H_5.OH,$$

from which, by the action of methyl iodide, the important febrifuge — "Antipyrine" — dimethyl phenyl pyrazolone is obtained:

$$\underset{\text{Antipyrine}}{CH_3\!-\!N\!\underset{\phantom{x}}{\overset{CH_3.C\!=\!CH\!-\!CO}{|\qquad\qquad|}}\!N\!-\!C_6H_5.}$$

If the primary hydrazines are boiled with copper sulphate,[1] or ferric chloride,[2] the hydrazine radical is replaced by hydrogen, and there is obtained, *e.g.*, from phenyl hydrazine, benzene:

$$C_6H_5 \cdot NH \cdot NH_2 + 2\,CuO = C_6H_6 + N_2 + H_2O + Cu_2O,$$
or $\quad C_6H_5 \cdot NH \cdot NH_2 + CuO = C_6H_6 + N_2 + H_2O + Cu.$

The statements made above concerning the replacement of a diazo-group by hydrogen are also applicable to this reaction. If it is desired to prepare an amido-compound from an amido-free compound, and if the direct reduction of the diazo-compound by sodium stannous oxide has been shown to be impracticable, then, as above, the hydrochloric acid salt of the corresponding hydrazine is prepared, the free hydrazine is liberated, and oxidised with caustic soda. The amido-free substance is not always easily volatile, as in the example cited. In a case of this kind, the oxidation may be effected in an open vessel; the reaction-product is obtained either by filtering or by extracting with ether. It may be pointed out here that it is more convenient to separate the hydrazine from the hydrochloric acid salt, and subject this to oxidation. If a hydrochloric acid salt of a hydrazine is oxidised, it may happen that the hydrazine radical will be replaced by chlorine:

$$C_6H_5 \cdot NH \cdot NH_2, HCl + O_2 = C_6H_5 \cdot Cl + N_2 + 2\,H_2O,$$

which may give rise to complications.

## 11. REACTION: (*a*) PREPARATION OF AN AZO DYE FROM A DIAZO-COMPOUND AND AN AMINE. (*b*) REDUCTION OF THE AZO-COMPOUND

EXAMPLES: (*a*) **Helianthine from Diazotised Sulphanilic Acid and Dimethyl Aniline**
(*b*) **Reduction of Helianthine**

(*a*) Dissolve 10 grammes of sulphanilic acid, dried on the water-bath, in a solution of 3.5 grammes of dehydrated sodium carbonate in 150 c.c. of water, and treat with a solution of 4.2 grammes of pure sodium nitrite in 20 c.c. of water. To this mixture, after being cooled by water, is added a quantity of hydrochloric acid solution corresponding to 2.5 grammes of an-

---

[1] B. 18, 90.  [2] B. 18, 186.

hydrous hydrochloric acid. For this purpose, concentrated hydrochloric acid is diluted with an equal volume of water, and the specific gravity of the dilute acid is determined by a hydrometer. Consult a table, to find the amount of anhydrous hydrochloric acid corresponding to the reading of the hydrometer. (See Graham-Otto, Vol. II. p. 318.)[1]

Before diazotising the sulphanilic acid, a solution of 7 grammes of dimethyl aniline in the theoretical amount of hydrochloric acid is prepared. Aromatic bases cannot be neutralised with hydrochloric acid in the same way as caustic potash, caustic soda, ammonium hydroxide, by gradually treating with the acid and testing with blue litmus-paper until the liquid is just acid. In consequence of the weak basic character of the amine, their hydrochlorides still give an acid reaction with blue litmus-paper, therefore an acid reaction can be obtained even at the beginning of the neutralisation. Red fuchsine-paper possesses the property of becoming decolourised by free hydrochloric acid, which converts the red monoacid fuchsine into a colourless polyacid salt. The hydrochloric acid salts of bases, on the contrary, do not produce this decolourisation. In order to neutralise the dimethyl aniline (7 grammes), it is treated with 25 c.c. of water, and, with stirring, small quantities of concentrated hydrochloric acid are added; after each addition a test is made to show whether or not the fuchsine-paper is decolourised.

The dimethyl aniline hydrochloride thus obtained is added to the diazo-solution, and the mixture is made distinctly alkaline by the addition of not too much caustic soda solution. The dye separates out directly; the quantity can be increased if 25 grammes of finely pulverised sodium chloride is added to the solution. After filtering off and pressing out on a porous plate, the dye is recrystallised from a little water.

---

[1] If the specific gravity of the hydrochloric acid has been determined, the percentage of free anhydrous acid may be found without a table, by the following calculation: The decimal number is multiplied by 2, and a decimal point placed after the first two figures thus obtained, e.g., sp. gr. = 1.134; 2 × 134 = 268. Percentage contents = 26.8. If the sp. gr. is greater than 1.18, a table must be consulted.

*Preparation of Fuchsine-Paper:* A crystal of fuchsine, the size of a lentil, is pulverised, dissolved by heating in 100 c.c. of water, and the solution filtered.

Into this immerse strips of filter-paper 2 cm. wide; they are dried either by suspending from a string in an acid-free place, or on the water-bath. The paper must not be an intense red, but only a faint rose colour. If the colour is too intense, the fuchsine solution must be correspondingly diluted with water.

Instead of the fuchsine-paper, the commercial Congo-paper will serve, the red colour of which is changed to blue by free acid.

(*b*) Dissolve 2 grammes of the dye in the least possible amount of water by heating; while the solution is still hot, treat with a solution of stannous chloride in 20 grammes of hydrochloric acid until decolourisation takes place. The colourless solution is then well cooled, upon which, especially if the sides of the vessel are rubbed with a glass rod, sulphanilic acid separates out: it is filtered off through asbestos or glass-wool. The filtrate is diluted with water, and caustic soda solution is added until the oxyhydrate of tin separating out at first is again dissolved. It is then extracted with ether several times, the ethereal solution dried with potash, and the ether evaporated, upon which the p-amidodimethyl aniline remains as an oil: on cooling and rubbing with a glass rod, it solidifies.

*Reactions of p-Amidodimethyl Aniline:*[1] The amidodimethyl aniline is treated gradually with small quantities of dilute sulphuric acid until it is just dissolved. Add a few drops of this solution to a dilute solution of hydrogen sulphide in a beaker which has been treated with $\frac{1}{50}$ of its volume of concentrated hydrochloric acid. To this mixture now add several drops of a dilute solution of ferric chloride. An intensely blue colouration, due to the formation of methylene blue, takes place.

·Diazo-compounds react with amines, as well as phenols, to form the Azo dyes:[2]

---

[1] B. 16, 2235.  [2] A. 137, 60; B. 3, 233.

(1) $C_6H_5 \cdot N=N \cdot Cl + C_6H_5 \cdot N(CH_3)_2$
$= C_6H_5 \cdot N=N \cdot C_6H_4 \cdot N(CH_3)_2 + HCl,$
<br>Dimethylamidoazo benzene

(2) $C_6H_5 \cdot N=N \cdot Cl + C_6H_5 \cdot OH = C_6H_5 \cdot N=N \cdot C_6H_4 \cdot OH + HCl.$
(In presence of alkali) — Oxyazo benzene

In accordance with these two typical reactions, the vast number of monoazo dyes are prepared. The great number of possible combinations can be inferred from the following considerations: In equation (1), instead of diazotised aniline, other bases, like o-toluidine, p-toluidine, xylidine, cuminidine, α-naphthyl amine, β-naphthyl amine, etc., may be used. In addition, the most varied derivatives of these bases, especially their sulphonic acids, like sulphanilic acid, metanilic acid, the vast number of α- and β-naphthyl amine, mono- to poly-sulphonic acids, may also be employed. Instead of dimethyl aniline, the diazo-compound can be combined or "coupled" with other tertiary, and in part also with secondary and primary amines, like diphenyl amine, or m-diamines, etc. In the second equation, the diazo-compounds of the just mentioned bases can be employed as the starting-point, and these can be combined with mon-acid phenols, like cresol, naphthols, or di-acid phenols like resorcinol, or the sulphonic acids of these phenols, especially the numerous sulphonic acids of both naphthols. Since a dye must be soluble in water, and the alkali salts of the sulphonic acids of the dyes are more easily soluble than the mother substance containing no sulphonic acid groups, therefore, in the preparation of the azo dyes, the starting-point is usually a sulphonic acid. A few examples will explain these statements:

### I. *Amidoazo Dyes*

$$p\text{-}C_6H_4 \underset{N=N}{\overset{SO_3H}{\diagdown}} \cdot C_6H_4 \cdot N(CH_3)_2 = \text{Helianthine,}$$

Diazotised sulphanilic acid + dimethyl aniline

$$m\text{-}C_6H_4 \underset{N=N}{\overset{SO_3H}{\diagdown}} \cdot C_6H_4 \cdot NH \cdot C_6H_5 = \text{Metanilic Yellow,}$$

Diazot. Metanilic acid + diphenylamine

$$C_6H_5 \cdot N=N \cdot C_6H_3 \underset{NH_2}{\overset{NH_2}{\diagdown}} = \text{Chrysoidine.}$$

Diazot. Aniline + m-phenylenediamine

## II. Oxyazo Dyes

$$p\text{-}C_6H_4\langle{}^{SO_3H}_{N=N}\cdot C_{10}H_6\cdot OH = \text{Orange II.,}$$

Diazot. Sulphanilic acid + β-naphthol

$$C_{10}H_6\langle{}^{SO_3H}_{N=N}\cdot C_{10}H_6\cdot OH = \text{Fast Red (first red azo dye),}$$

Diazot. α-Naphthionic acid + β-naphthol

$$C_6H_5\cdot N=N\cdot C_{10}H_5\langle{}^{OH}_{SO_3H} = \text{Croceïne Orange,}$$

Diazot. Aniline + croceïne acid
(β-Naphthol sulphonic acid)

$$(CH_3)_2\cdot C_6H_3\cdot N=N\cdot C_{10}H_4\langle{}^{OH}_{(SO_3H)_2} = \text{Xylidine Ponceau.}$$

Diazot. Xylidine + β-naphthol disulphonic acid

Concerning the constitution of the azo dyes, provided the components are known, the only question to solve is: which hydrogen atom of the undiazotised component combines with the acid radical of the diazo-compound (the acid thus formed being eliminated). The question may be answered by investigating the reduction products of the azo dyes. By energetic reduction, best in acid solution with stannous chloride, the double $N=N$ union is broken up, thus forming, with the addition of 4 atoms of hydrogen, two molecules of a primary amine, e.g.:

$$C_6H_4\langle{}^{SO_3H}_{N=N}\cdot C_6H_4\cdot N(CH_3)_2 + 2\,H_2 = C_6H_4\langle{}^{SO_3H}_{NH_2} + C_6H_4\langle{}^{NH_2}_{N(CH_3)_2}.$$

From this equation it is evident that by reduction, the amine which was diazotised — in the above case sulphanilic acid — may be obtained again on the one hand, on the other an amido-group is introduced into the second component. If the constitution of this second product can be determined, then the constitution of the azo dyes is also determined. It may be stated as a general rule that, when a diazo-compound combines with an amine or phenol, the hydrogen atom in the para-position to the amido- or hydroxyl-group is always substituted. In accordance with this, in the above case, p-amidodimethyl aniline ought to be obtained on reduction. If the para-position is already occupied, then the o-hydrogen atom unites with the acid radical.

In some cases, the formation and consequent reduction of an azo dye with the introduction of an amido-group into a phenol or amine is of practical value.

Azo dyes which contain two "chromophore groups," $N{=}N$, and which are called dis- or tetr-azo dyes, can be prepared; two methods may be employed: (1) The starting-point is an amido-azo-compound which already contains one azo group; this is diazotised, and then united with an amine or phenol. "Biebrich scarlet" is obtained in this way, by diazotising the disulphonic acid of amidoazobenzene, and combining it with $\beta$-naphthol:

$$C_6H_4{<}{\overset{SO_3H}{\underset{N{=}N}{}}} \cdot C_6H_3{<}{\overset{SO_3H}{\underset{N{=}N}{}}} \cdot C_{10}H_6 \cdot OH.$$

Diazot. Amidoazobenzene disulphonic acid + $\beta$-naphthol

(2) A diamine is the starting-point; this is diazotised, and the bisdiazo-compound is combined with two molecules of an amine or phenol. To this class belong the important dyes of the Congo group, prepared from the benzidine bases (see page 188), *e.g.*:

$$\begin{array}{l}C_6H_4{-}N{=}N{-}C_{10}H_5{<}{\overset{NH_2}{\underset{SO_3H}{}}} \\ | \\ C_6H_4{-}N{=}N{-}C_{10}H_5{<}{\overset{NH_2}{\underset{SO_3H}{}}}\end{array} = \text{Congo,}$$

Diazot. Benzidine + 2 mol. α-naphthionic acid

$$\begin{array}{l}C_6H_4 \cdot N{=}N \cdot C_6H_3{<}{\overset{OH}{\underset{CO \cdot OH}{}}} \\ | \\ C_6H_4 \cdot N{=}N \cdot C_6H_3{<}{\overset{OH}{\underset{CO \cdot OH}{}}}\end{array} = \text{Chrysamine.}$$

Diazot. Benzidine + 2 mol. salicylic acid

These Congo dyes possess the noteworthy property of colouring vegetable fibres (cotton) directly, whereas, with all other azo dyes, the cotton must be mordanted before dyeing.

In conclusion, the above dye-stuff reaction (which may be used for detecting the smallest amount of hydrogen sulphide) is technically carried out on the large scale, for the manufacture of the important methylene blue. The reaction takes place as follows: From two molecules of the diamine there is split off an oxidation with ferric chloride, one molecule of ammonia, while a derivative of diphenyl amine is formed:

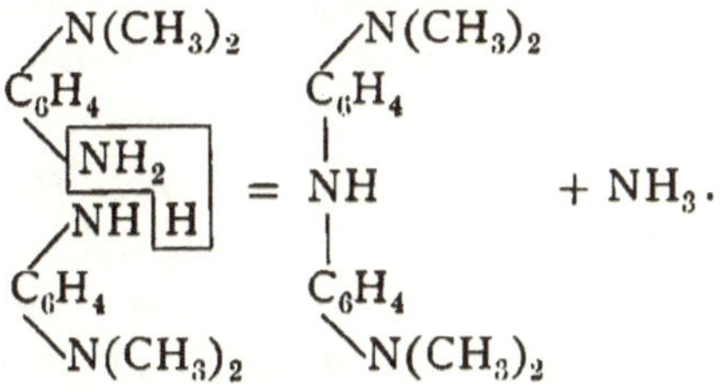

$$\begin{array}{c} N(CH_3)_2 \\ C_6H_4 \\ | \\ \boxed{NH_2} \\ NH\boxed{H} \\ | \\ C_6H_4 \\ N(CH_3)_2 \end{array} = \begin{array}{c} N(CH_3)_2 \\ C_6H_4 \\ | \\ NH \\ | \\ C_6H_4 \\ N(CH_3)_2 \end{array} + NH_3.$$

In the presence of hydrogen sulphide and hydrochloric acid, there is formed from this, by the oxidising action of ferric chloride, a derivative of thiodiphenylamine, as follows:

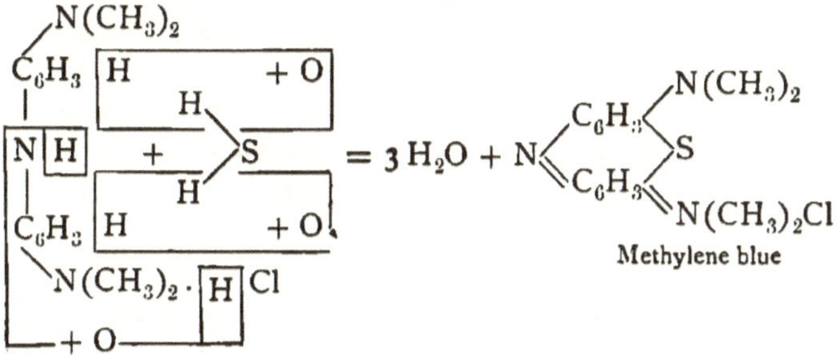

$$\begin{array}{c} N(CH_3)_2 \\ C_6H_3 \ \boxed{H \quad\quad +O} \\ | \quad\quad H \\ N\boxed{H} \ + \ \searrow S \\ | \quad\quad H \nearrow \\ C_6H_3 \ \boxed{H \quad\quad +O} \\ N(CH_3)_2 \cdot \boxed{H}\, Cl \\ \boxed{+O \quad\quad\quad} \end{array} = 3\,H_2O + N\!\!\begin{array}{c} C_6H_3 \\ \diagdown \\ \diagup \\ C_6H_3 \end{array}\!\!\begin{array}{c} N(CH_3)_2 \\ \diagdown \\ S \\ \diagup \\ N(CH_3)_2Cl \end{array}$$

Methylene blue

## 12. REACTION: PREPARATION OF A DIAZOAMIDO-COMPOUND

EXAMPLES: **Diazoamidobenzene from Diazobenzenechloride and Aniline** [1]

Dissolve 10 grammes of freshly distilled aniline in a mixture of 100 c.c. of water, and that quantity of concentrated hydrochloric acid corresponding to 12 grammes anhydrous hydrochloric acid (determine the sp. gr. by a hydrometer). The solution is cooled with ice-water, and diazotised with a solution of 8 grammes of sodium nitrite in 50 c.c. of water, in the manner already described. A solution of 10 grammes of aniline in 50 grammes of water is previously prepared according to the directions already given, and exactly the theoretical amount of hydrochloric acid,

---

[1] A. 121, 257.

after it has been well cooled with ice-water, is added to the diazo-solution, with stirring. Further, 50 grammes of sodium acetate are dissolved in the least possible amount of water, the cooled solution is added, with stirring, to the mixture of the diazo-compound with aniline hydrochloride. After standing half an hour, the diazoamidobenzene separates out, and is filtered off with suction, washed several times with water, well pressed out on a porous plate, and recrystallised from ligroïn, benzene, or alcohol. Melting-point, 98°. Yield, almost theoretical.

If one molecule of a diazo-compound is allowed to act on one molecule of a primary amine, the acid radical of the former unites with the hydrogen atom of the latter, upon which the organic residues combine, as in the formation of the azo dyes. In this case, an amido-hydrogen atom is eliminated, so that a compound containing a chain of three nitrogen atoms is formed; in the formation of an azo dye, one of the benzene-hydrogen atoms of the amine is eliminated:

$$C_6H_5 \cdot N{=}N \cdot Cl + C_6H_5 \cdot NH_2 = \underset{\text{Diazoamidobenzene}}{C_6H_5 \cdot N{=}N \cdot NH \cdot C_6H_5} + HCl.$$

Mixed diazo-compounds may also be prepared by causing the diazo-derivative of an amine to combine with another amine:

$$C_6H_5 \cdot N{=}N \cdot Cl + C_6H_4{\genfrac{}{}{0pt}{}{CH_3}{NH_2}} = \underset{\text{Benzenediazoamidotoluene}}{C_6H_5 \cdot N{=}N \cdot NH \cdot C_6H_4 \cdot CH_3} + HCl.$$

Diazo-compounds combine only with the *free* amines to form diazoamido compounds; the object of the addition of sodium acetate at the end of the reaction (see above) is to set free the base from aniline hydrochloride.

The diazoamido-compounds are yellow substances which do not dissolve in acids. They are far more stable than the diazo-compounds, and may be recrystallised without decomposition. Still, if they are heated rapidly they puff up suddenly and decompose. In their reactions they behave like a mixture of a diazo-compound and an amine. If, *e.g.*, they are boiled with hydrochloric acid, they decompose with evolution of nitrogen, and form a phenol and an amine:

$$C_6H_5 \cdot N{=}N \cdot NH \cdot C_6H_5 + H_2O = C_6H_5 \cdot OH + C_6H_5 \cdot NH_2 + N_2.$$

On heating with cuprous chloride and hydrochloric acid, the Sandmeyer reaction takes place:

$C_6H_5.N{=}N.NH.C_6H_5 + HCl = C_6H_5.Cl + C_6H_5.NH_2 + N_2.$

By reduction with acetic acid and zinc dust, they form a hydrazine:

$C_6H_5.N{=}N.NH.C_6H_5 + 2 H_2 = C_6H_5.NH.NH_2 + C_6H_5.NH_2.$

But, in addition to the reaction-product of the diazo-radical, there is always formed one molecule of an amine.

Under the influence of nitrous acid, they decompose, the amine residue being diazotised, into two molecules of a diazo-compound:

$C_6H_5.N{=}N.NH.C_6H_5 + HNO_2 + HCl = 2 C_6H_5.N{=}N.Cl + 2 H_2O.$

If a diazoamido-compound is warmed with an amine in the presence of some amine hydrochloride, transformation to the isomeric amidoazo-compound takes place:

$C_6H_5.N{=}N.NH.C_6H_5 = C_6H_5.N{=}N.C_6H_4.NH_2.$
Amidoazobenzene

The next preparation deals with this reaction.

The diazo-compounds also have the power of combining with secondary amines to form diazoamido-compounds, — the combinations with an alkaloid base, piperidine $C_5H_{11}N$:

$$\begin{array}{c} CH_2 \\ H_2C \diagup \diagdown CH_2 \\ H_2C \diagdown \diagup CH_2 \\ NH \end{array}$$

are of especial value for preparations. If this is gradually warmed with hydrofluoric acid, it is decomposed with the evolution of nitrogen into piperidine and a fluoride:[1]

$C_6H_5.N{=}N.N.C_5H_{10} + HFl = C_6H_5.Fl + C_5H_{11}N + N_2.$
Benzenediazopiperidine    Fluorbenzene

In this way it has been possible to prepare the aromatic fluorides; the corresponding chlorides, bromides, and iodides can *not* be prepared by an analogous method from the diazo-compounds.

---

[1] A. 243, 239.

### 13. REACTION: THE MOLECULAR TRANSFORMATION OF A DIAZO-AMIDO-COMPOUND INTO AN AMIDOAZO-COMPOUND

EXAMPLE: **Amidoazobenzene from Diazoamidobenzene**

To a mixture of 10 grammes of crystallised diazoamidobenzene, finely pulverised, and 5 grammes of pulverised aniline hydrochloride, contained in a small beaker, add 25 grammes of freshly distilled aniline; the mixture is then heated, with frequent stirring, one hour, on the water-bath, at $45°$. It is then transferred to a larger vessel, and treated with water; dilute acetic acid is added, until all the aniline has passed into solution, and the undissolved precipitate remaining is completely solid. This is filtered off, washed with water, heated in a large dish with a large quantity of water, and gradually treated with hydrochloric acid until the greatest portion of the precipitate is dissolved. From the filtered solution, steel-blue crystals of amidoazobenzene separate out, on long standing; these are filtered off, and washed with dilute hydrochloric acid, *not* with water.

If aniline hydrochloride is not at hand, prepare it by adding aniline to concentrated hydrochloric acid, with stirring. After cooling, the pasty mass of crystals separating out is filtered on glass-wool, pressed firmly together on the filter with a pestle, and then spread in thin layers on a porous plate.

In order to obtain the free amidoazobenzene, the hydrochloride is warmed with dilute ammonia, the free base filtered off, dissolved in alcohol by heating, and hot water is added until the liquid begins to be turbid. Melting-point, $127-128°$. Yield, 6–8 grammes.

If a diazoamido-compound is heated with an amine and some amine hydrochloride, it goes over to an amidoazo-compound. The most probable cause of the reaction is that the amine residue of the diazoamido compound unites with a benzene-hydrogen atom of the amine hydrochloride, upon which the diazo-residue unites with the residue of the amine salt to form amidoazobenzene:

$$C_6H_5.N{=}N.NH.C_6H_5 + H.C_6H_4.NH_2$$
$$= C_6H_5.N{=}N.C_6H_4.NH_2 + C_6H_5.NH_2.$$
Amidoazobenzene

While the amidoazobenzene does not unite with hydrochloric acid, the new molecule of the amine formed in the reaction does, and thus there is a molecule of the amine hydrochloride present, which again causes the transformation, so that a small amount of the hydrochloride may transform an indefinitely large amount of the diazoamido-compound.

If amidoazobenzene is reduced, p-phenylene diamine and aniline are obtained. The transformation accordingly results in the formation of a compound in which the amido-groups are in the para-position, which always happens when the para-position is unoccupied. The amidoazo-compounds possess weakly basic properties; but if their salts are treated with much water, they partially dissociate.

The amidoazobenzene hydrochloride came into the market, formerly, as a yellow dye, under the name of "Aniline Yellow." At present, it is scarcely used, but there is prepared from it, by heating with sulphuric acid, a mono- or di-sulphonic acid, which in the form of its alkali salts finds application as a dye under the name of "Acid Yellow," or "Fast Yellow." As already mentioned under the dis-azo dyes, from the diazo-compound of this dye, "Biebrich Scarlet" may be made by combination with $\beta$-naphthol. Finally, the amidoazobenzene is still used for the preparation of the Induline dyes.

### 14. REACTION: OXIDATION OF AN AMINE TO A QUINONE

EXAMPLE: Quinone from Aniline [1]

To a solution of 25 grammes of aniline in a mixture of 200 grammes of concentrated pure sulphuric acid and 600 c.c. of water contained in a thick-walled beaker (a small battery jar), cooled to 5° by being surrounded with ice, add gradually, with constant stirring (use a small motor), from a dropping-funnel, a solution of 25 grammes of sodium dichromate in 100 c.c. of water (Fig. 63).

---

[1] A. 27, 268; 45, 354; 215, 125; B. 19, 1467; 20, 2283.

Should the temperature rise above 10°, the addition of the dichromate must be discontinued for a short time and a few pieces of ice thrown into the beaker. The reaction-mixture is then allowed to stand over night in a cool place, and the next morning it is again cooled and stirred, while a solution of 50 grammes of sodium dichromate in 200 c.c. of water is added. After the mixture has been allowed to stand until midday, it is divided into two equal parts, one of which is worked up into quinone as follows: In a large separating funnel one half is treated with $\frac{2}{3}$ its volume of ether, and the two layers thus formed are *carefully*

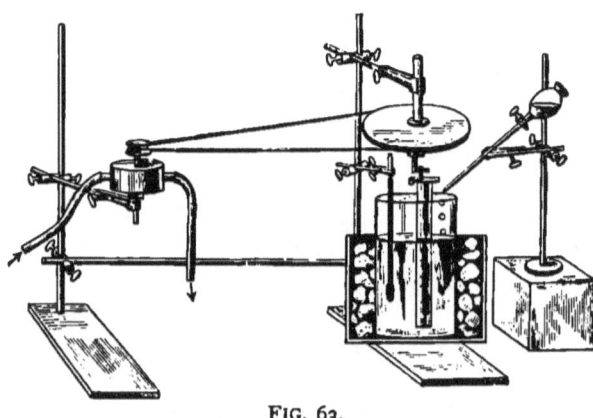

FIG. 63.

shaken together. If the shaking is too vigorous, the layers will not readily separate. After allowing it to stand for half an hour, the lower layer is run off (see page 42, Separation of coloured liquids), the ethereal solution is filtered through a folded filter, and the ether distilled off (water-bath with warm water). The water solution is again extracted with the condensed ether, and the ether again distilled from the same flask as before. In order to obtain perfectly pure quinone, a rapid current of steam is passed over the crude product — it is not treated with water; the pure quinone is carried over with the steam to the condenser and receiver, where it crystallises in the form of golden-yellow needles; they are filtered off and dried in a desiccator. Melting-point, 116°. Yield, 10–12 grammes.

If sodium dichromate is not at hand, the potassium salt may be used for the oxidation. In this case, 25 grammes of aniline are dissolved in a mixture of 200 grammes of sulphuric acid and 800 c.c. of water; then add, as above, with stirring and good cooling, 25 grammes of potassium dichromate, powdered extremely fine. On the next day add 50 grammes of this salt. In other respects, proceed as above.

Many primary aromatic amines yield quinones on oxidation with chromic acid. But the reaction cannot be expressed in a simple equation; still it is always true that the amido-group and the hydrogen atom in the para-position to this are each replaced by an oxygen atom, e.g.:

$$C_6H_5 \cdot NH_2 \longrightarrow C_6H_4O_2,$$
$$\text{Quinone}$$

$$O-C_6H_4\!\!<\!\!\begin{array}{l}CH_3\\NH_2\end{array} \longrightarrow C_6H_3\!\!<\!\!\begin{array}{l}CH_3\\O_2\end{array},$$
$$\text{Toluidine} \qquad \text{Tolyl quinone}$$

$$C_6H_3\!\!<\!\!\begin{array}{l}(CH_3)_2\\NH_2\end{array} \longrightarrow C_6H_2\!\!<\!\!\begin{array}{l}(CH_3)_2\\O_2\end{array}.$$
$$\text{Xylyl quinone}$$

If the para-position to the amido-group is occupied by an alkyl residue, a halogen atom, a carboxyl-group, etc., the quinone formation will not take place; p-toluidine, e.g., in contrast with the o- and m-varieties, forms no quinone. But if the p-position is occupied by an amido-, oxy-, or sulphonic acid-group, these are eliminated and a quinone formed:

$$\text{p-}C_6H_4\!\!<\!\!\begin{array}{l}NH_2\\NH_2\end{array}$$
$$\text{p-}C_6H_4\!\!<\!\!\begin{array}{l}NH_2\\OH\end{array} \longrightarrow C_6H_4O_2.$$
$$\text{p-}C_6H_4\!\!<\!\!\begin{array}{l}NH_2\\SO_3H\end{array}$$

From these methods of formation it follows that the two quinone-oxygen atoms are in the para-position to each other. The quinone reaction can be used in doubtful cases to decide whether a diamine or an amidophenol belongs to the para-series. The quinones can also be obtained very easily from p-dioxy-compounds as well as from the p-sulphonic acids of mon-acid phenols:

$$\text{p-}C_6H_4{<}^{OH}_{OH} + O = C_6H_4O_2 + H_2O$$

$$\text{p-}C_6H_4{<}^{OH}_{SO_3H} \longrightarrow C_6H_4O_2.$$

Two formulæ for the quinones have been proposed — the Peroxide- and Ketone-formula:

Peroxide formulæ        Ketone formulæ

According to the former the quinones still contain the true benzene ring with either three double or six centric bonds. The two oxygen atoms are only singly united with the benzene-carbon atoms, and are united to each other as in the peroxides. According to the second formula, the quinones do not contain the true benzene ring, but they are derived from a dihydrobenzene,

and are regarded as the di-ketone derivative of this. According to this conception, the oxygen atoms are connected by two bonds, as in the ketones, with the carbon atoms of the benzene nucleus. The facts in favour of the first formula are these: In many reactions both of the oxygen atoms are replaced by two univalent atoms or radicals. Thus, *e.g.*, by the action of phosphorus pentachloride on quinone, p-dichlorbenzene is formed, while the second formula would lead one to expect a tetra-chloride. In support of the second formula is the fact that hydroxylamine acts directly on quinones, as on ketones, with the formation of a mono- or di-oxime.

The quinones are yellow compounds, possessing a characteristic odour; they are easily volatile with steam, but with a slight decomposition. They are somewhat volatile even with the vapour of ether, as one observes in the preparation of quinone. On reduction they take up two hydrogen atoms and pass over to hydroquinones. (See the next preparation) *e.g.*:

$$C_6H_4O_2 + H_2 = C_6H_4{\diagup OH \atop \diagdown OH}$$
Hydroquinone

## 15. REACTION: REDUCTION OF A QUINONE TO A HYDROQUINONE

EXAMPLE: **Hydroquinone from Quinone**

Conduct sulphur dioxide into the second half of the quinone solution obtained above, until the liquid smells intensely of the gas, then allow it to stand for 1–2 hours. Should the odour of sulphur dioxide vanish, it is passed in again and the mixture allowed to stand for some time as before. It is then extracted with the ether distilled from the quinone in the preceding experiment, several times; the ether is evaporated or distilled, and the hydroquinone, well pressed out on a porous plate, is crystallised with the use of animal charcoal from a little water. Melting-point, 169°. Yield, 8–10 grammes.

Since the hydroquinone solution may be extracted with ether with much greater ease than the quinone solution, and since the hydroquinone is smoothly oxidised to quinone, the preparation of quinone may be done as follows: The *entire quantity* of the oxidation product is saturated with sulphur dioxide, and as just described the hydroquinone may be obtained by repeated extraction with ether. In order to convert it into quinone it is dissolved in the least possible amount of water, to which is added 2 parts of concentrated sulphuric acid to 1 part of hydroquinone; the well-cooled liquid is treated with a water solution of sodium dichromate until the green crystals of quinhydrone (an intermediate product between quinone and hydroquinone) separating out in the beginning have changed into pure yellow quinone.

The equation for the formation of hydroquinone from quinone has been given above. All homologous quinones react in the same way. The hydroquinones are di-acid phenols, which dissolve in alkalies and show all the properties of phenols. They are not volatile with steam.

### 16. REACTION: BROMINATION OF AN AROMATIC COMPOUND

EXAMPLE: **Mono- and Di-brombenzene from Bromine and Benzene**

A wide-neck 250 c.c. flask is connected with a vertical tube 50 cm. long and $1\frac{1}{2}$ cm. wide, the upper end of which is closed by a cork bearing a glass tube, not too narrow, bent twice at right angles. The other end is connected with a flask containing 250 c.c. of water, by a cork having a *small canal* in the side (Fig. 64). The tube does not touch the liquid, but the end is about 1 cm. above the surface. After 50 grammes of benzene and 1 gramme of coarse iron filings (the bromine carrier) have been placed in the flask, it is cooled in a large vessel (battery jar) filled with ice-water; through the vertical tube there is added 40 c.c. = 120 grammes of bromine: the narrow tube is immediately connected with the vertical tube. After some time an extremely energetic reaction will begin, generally spontaneously, with the evolution of hydrobromic acid, which is completely absorbed by the water. Should the reaction not begin at once, the ice-water is removed for a short time, and if necessary the flask is immersed for a moment in slightly warm water. But as soon as even a weak gas evolution begins, the flask is at once cooled again, since

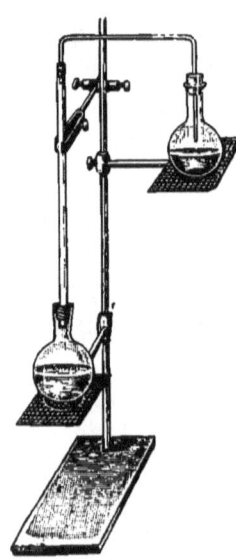

FIG. 64.

otherwise the reaction easily becomes too violent. When the main reaction is over, the ice-water is removed, the flask dried and heated over a small flame until the red bromine vapours are no longer visible above the dark-coloured liquid. The reaction-product is washed several times with water and then distilled with steam. As soon as crystals of dibrombenzene separate out in the condenser, the receiver is changed and the distillation continued until all the dibrombenzene has passed over. The liquid monobrombenzene is separated from the water, dried with calcium chloride, and subjected to a fractional distillation, the portion passing over between 140–170° is collected separately. This is again distilled and the portion going over between 150–160° collected. The boiling-point of the pure monobrombenzene is 155°. Yield, 60–70 grammes. The residue boiling above 170° remaining in the flask after the two distillations, is poured while still warm on a watch-glass, and after cooling is pressed out, together with the separately collected dibrombenzene, on a porous plate. On crystallising from alcohol, coarse colourless crystals are obtained which melt at 89°.

The by-product, hydrobromic acid, is purified as described in the Inorganic Part. (See page 316.)

A portion of the hydrogen of the aromatic hydrocarbons is very easily replaced by bromine, especially in the presence of a carrier, even at low temperatures; while in the aliphatic series the direct substitution of bromine is not used as a preparation method for alkyl bromides, the aromatic bromides are readily prepared in this way. According to the amount of bromine used one or more hydrogen atoms may be substituted; it may happen, *e.g.*, particularly with benzene under the influence of an energetic bromination, that all the hydrogen atoms may be replaced by bromine. A single bromide, even on using only the theoretical amount of bromine, is never formed; but rather a portion of the hydrocarbon is brominated short of the theoretical action, and another portion is always acted upon farther, with the formation of a higher bromine substitution product. Thus in the example above cited, besides the principal product, monobrombenzene, a small quantity of dibrombenzene is formed:

$$C_6H_6 + Br_2 = C_6H_5Br + HBr,$$
$$C_6H_6 + 2\,Br_2 = C_6H_4Br_2 + 2\,HBr.$$

In most cases, however, the principal product may be separated from the by-product without difficulty by distillation or crystallisation. Since the hydrogen atoms substituted by bromine combine with bromine to form hydrobromic acid, therefore, for the introduction of each bromine atom, a molecule (two atoms of bromine) must be used.

The introduction of bromine can be essentially facilitated by the use of a so-called bromine carrier. As such, the bromides of metalloids, or metals, are used; (1) either in the already prepared condition, or (2) they can be generated from their elements in the reaction. To the first class belong ferric bromide and aluminium bromide. The action of ferric bromide depends on the fact that on being reduced to ferrous bromide, it yields bromine in *statu nascendi*:

$$FeBr_3 = FeBr_2 + Br.$$
Ferric bromide    Ferrous bromide

Since the ferrous bromide unites with bromine again, to form ferric bromide, a small quantity of this has the power to transfer an indefinitely large quantity of bromine:

$$FeBr_2 + Br = FeBr_3.$$

Instead of ferric bromide, ferrous bromide or anhydrous ferric chloride may be used. The latter decomposes with hydrobromic acid to ferric bromide and hydrochloric acid:

$$FeCl_3 + 3\,HBr = FeBr_3 + 3\,HCl.$$

The activity of aluminium bromide is explained by the fact that it unites with the hydrocarbon to form a double compound which is more capable of reacting with other substances than the hydrocarbon itself.

To the second class belong iodine, sulphur, phosphorus, iron, aluminium, etc. If these elements are added to the brominating mixture, the corresponding bromides are formed, *e.g.*:

$$I + Br = IBr[1].$$

While these give up all their bromine, or a portion of it, as is the case with ferric bromide, in the atomic condition, the residue again unites with bromine, and as above, a small quantity of the carrier may transfer large quantities of atomic bromine.

Bromine can also act on aromatic hydrocarbons to form addition products, since it may be *added* in one, two, or three molecules, and thus break up the double or centric union. Thus, *e.g.*, the hexabrom-

---

[1] Compare page 134.

addition product, $C_6H_6Br_6$, is obtained from the action of bromine on benzene in the sunlight. Since the addition products render difficult the purification of substitution products, especially on distillation (they decompose when distilled), it is often necessary to remove them before the purification, by long boiling with alcoholic caustic potash, or alcoholic caustic soda. Under these conditions, one-half of the bromine atoms added in common with the same number of hydrogen atoms are abstracted as hydrobromic acid; the residue of the molecule is converted into a substitution derivative, which is not troublesome in the purification:

$$C_6H_6Br_6 = C_6H_3Br_3 + 3\,HBr.$$

On brominating benzene, the same products will be formed, whether the temperature is high or low, but when its homologues are treated with bromine, the nature of the products depends upon the temperature. As will be pointed out more fully, under the chlorination of toluene, the law holds here, that at low temperatures the bromine enters the ring; at high temperatures, the side-chain, e.g.:

$$C_6H_5 \cdot CH_3 + Br_2 = C_6H_4\!\!<^{CH_3}_{Br} + HBr,$$
Ordinary temperature      Bromtoluene

$$C_6H_5 \cdot CH_3 + Br_2 = C_6H_5 \cdot CH_2 \cdot Br + HBr$$
Boiling temperature      Benzylbromide

The aromatic bromides which contain bromine in the benzene nucleus are either colourless liquids or crystals, which in contrast with the side-chain substituted isomers in part possess an aromatic odour, and their vapours do *not* attack the eyes and nostrils. The bromine is held very firmly in them, more firmly than in the aliphatic bromides, and cannot be detected by silver nitrate. While the aliphatic bromides, as mentioned under bromethyl, decompose with ammonia, alcohol, alkalies, etc., to form amines, ethers, alcohols, etc., respectively, these reagents do not act on the aromatic bromides. The bromides containing the bromine in the side-chain, behave like their aliphatic analogues.

By the action of sodium amalgam, the bromine may be replaced by hydrogen, e.g.:

$$C_6H_5 \cdot \underset{\substack{\text{From the}\\ \text{amalgam}}}{Br} + H_2 = C_6H_6 + HBr.$$

The aromatic bromides are of synthetical importance, especially for the building up of homologous hydrocarbons and the preparation of carbonic acids:

$$C_6H_5 \cdot Br + BrC_2H_5 + Na_2 = C_6H_5 \cdot C_2H_5 + 2\,NaBr,$$

$$C_6H_5 \cdot Br + Na_2 + CO_2 = C_6H_5 \cdot CO \cdot ONa + NaBr.$$

The next preparation will take up the first of these reactions in detail.

The hydrocarbons and most of their derivatives, like nitro-, amido-compounds, aldehydes, acids, etc., may be brominated with greater or less ease. At this place, the various modifications by which the bromination may be effected will be mentioned. If a substance is very easily brominated, the bromine may be used in a diluted condition. For this purpose, either bromine water or a mixture of bromine with carbon disulphide or glacial acetic acid may be employed. In many cases a bromination may be very well effected by using gaseous bromine. The method of procedure is as follows: The substance is spread out in thin layers on a watch-glass and placed under a glass bell-jar, under which is also a small dish containing bromine. If it is desired to cause bromine to act gradually, it is allowed to drop from a separating funnel, in concentrated form or in solution, on the compound to be brominated. If an extremely slow and very careful bromination is desired, the bromine may be allowed to flow drop by drop from a siphon-shaped capillary tube. If bromination takes places with difficulty, the brominating mixture is heated either in an open vessel or in a sealed tube. In the first case the condensing apparatus cannot, as usual, be connected to the flask with a cork or rubber stopper, since this is soon attacked and destroyed by the bromine. Instead, the condenser is well wrapped with asbestos twine and then pushed into the conical part of the neck of the flask, the asbestos being pressed in with a knife. A condenser of the kind represented in Fig. 65 can also be used. A long tube $c$, sealed at one end, is closed by a two-hole cork, through one of which passes a long glass tube reaching almost to the bottom $a$; the other bears a short tube just passing through the cork. Water is caused to flow through $a$; it flows

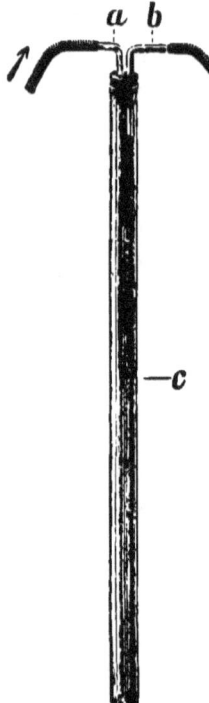

FIG. 65.

out of *b*. This cooling apparatus is suspended in the heating-flask, which is selected with as long a neck as possible.

## 17. REACTION: FITTIG'S SYNTHESIS OF A HYDROCARBON

EXAMPLE: **Ethyl Benzene from Brombenzene and Bromethyl** [1]

In a dry, round, $\frac{1}{2}$-litre flask, provided with a long reflux condenser (the flask is supported on a straw ring in an empty water-bath), place 27 grammes of sodium, cut in scales as thin as possible with a sodium knife, and add 100 c.c. of alcohol-free, dry ether prepared as described below. As soon as this has been completely dried by the sodium, which may be recognised by the fact that the upper surface is no longer disturbed by wave-like motions (after several hours' standing), pour through the condenser a mixture of 60 grammes of brombenzene and 60 grammes of bromethane, and allow to stand until the next day. Should the liquid begin to boil gently, which may easily happen at a summer temperature, cold water is poured into the water-bath. Water is not allowed to run through the condenser over night. During the reaction, the bright sodium will be changed to a blue powder, and an ethereal solution of ethylbenzene is formed. The ether is then distilled off on a water-bath, and the condenser is replaced with an air condenser 40-50 cm. long and 1 cm. wide, containing a short bend. After the flask has been placed in an oblique position, the extreme end of its neck is clamped *loosely*, and the ethylbenzene is distilled from the sodium bromide and sodium *by a large, luminous flame, which is kept in constant motion.* With the use of a Linnemann apparatus, the crude product is finally subjected to two fractional distillations. The boiling-point of pure ethylbenzene is 135°. Yield, about 25 grammes.

The residue of sodium bromide and sodium remaining in the flask must be handled with extreme caution. Water must not be added to it, nor must it be thrown into the sink or waste-jars, nor allowed to stand a long time; it is better to throw the flask, which

---

[1] A. 131, 303.

cannot be used again, and its contents into some open place. The sodium residue may be rendered harmless by throwing water on it from a great distance.

### Preparation of Anhydrous, Alcohol-free Ether

Shake 200 grammes of commercial ether in a separating funnel with half its volume of water; the latter is allowed to run off, and the operation repeated a second time with a fresh quantity of water, by which the alcohol is removed. The ether is dried by standing over calcium chloride, not too little, two hours. It is then filtered through a folded filter, and can now be used for the above reaction.

Fittig's synthesis of the aromatic hydrocarbons is the analogue of Wurtz' synthesis of the aliphatic hydrocarbons, *e.g.*:

$$2\,C_2H_5I + 2\,Na = C_2H_5 \cdot C_2H_5 + 2\,NaI,$$
Ethyl iodide      Butane

$$C_6H_5 \cdot Br + C_2H_5Br + 2\,Na = C_6H_5 \cdot C_2H_5 + 2\,NaBr.$$
Ethylbenzene

The bromides of the homologues of benzene react in a similar way, *e.g.*:

$$C_6H_4\!\!<^{CH_3}_{Br} + ICH_3 + Na_2 = C_6H_4\!\!<^{CH_3}_{CH_3} + NaBr + NaI.$$
Bromtoluene          Xylene

The three isomeric bromtoluenes do not react with the same ease. While the p-bromtoluene gives a good yield of p-xylene, the o-compound does not give good results, and the m-compound generally forms no xylene. Two alkyl residues can also, in many cases, be introduced into a hydrocarbon simultaneously, *e.g.*:

$$p\text{-}C_6H_4\!\!<^{Br}_{Br} + 2\,ICH_3 + 2\,Na_2 = p\text{-}C_6H_4\!\!<^{CH_3}_{CH_3} + 2\,NaBr + 2\,NaI.$$

The great number of hydrocarbons which may be prepared by Fittig's reaction is apparent from the above examples. The value of the reaction is still further increased by the fact that a halogen atom in the side-chain of an aromatic hydrocarbon also reacts in the same way.

Though the halogen cannot be replaced by a methyl or ethyl radical, yet the reaction for the introduction of the higher alkyl residues is of great service, e.g.:

$C_6H_5 \cdot CH_2Cl + CH_3 \cdot CH_2 \cdot CH_2Br + Na_2$
Benzyl chloride    Propyl bromide
$= C_6H_5 \cdot CH_2 \cdot CH_2 \cdot CH_2 \cdot CH_3 + NaCl + NaBr.$
Butylbenzene

Also by means of this reaction, two aromatic residues may be made to combine, and thus form the hydrocarbons of the diphenyl series, e.g.:

$2\, C_6H_5 \cdot Br + Na_2 = C_6H_5 \cdot C_6H_5 + 2\, NaBr.$
Diphenyl

Finally, the hydrocarbons of the dibenzyl series can also be prepared, e g.:

$2\, C_6H_5 \cdot CH_2Cl + Na_2 = C_6H_5 \cdot CH_2 \cdot CH_2 \cdot C_6H_5 + 2\, NaCl.$
Dibenzyl

In conducting operations involving the Fittig reaction, various modifications may be introduced, according to the ease with which the reaction takes place. If the reaction occurs at the ordinary temperature easily, then an indifferent diluent like ether, ligroïn, carbon disulphide, or benzene is employed. These substances are not alike in their activity, since ligroïn and benzene generally prolong the reaction, and on this account find application in a very energetic reaction; ether does not retard the reaction, but causes it to be more regular. At times, the reaction-mixture will not act, even on long standing. In this case, the reaction can frequently be started by a short heating, or the addition of a few drops of ethyl acetate. Since the use of this compound, at times, causes a very stormy action, it is more advantageous to wait for the reaction to begin spontaneously, even if a long time is necessary. In syntheses which are moderately difficult, the reaction-mixture, treated with a diluent, can be heated on the water-bath or in an oil-bath; while, if the reaction takes place with great difficulty, the mixture, generally without dilution, must be heated in an oil-bath. In the latter case, the reaction may be still further facilitated by heating under pressure of a mercury column. By this means, it is possible to heat the reacting substances in an open vessel above their boiling-points. (Fig. 66.)

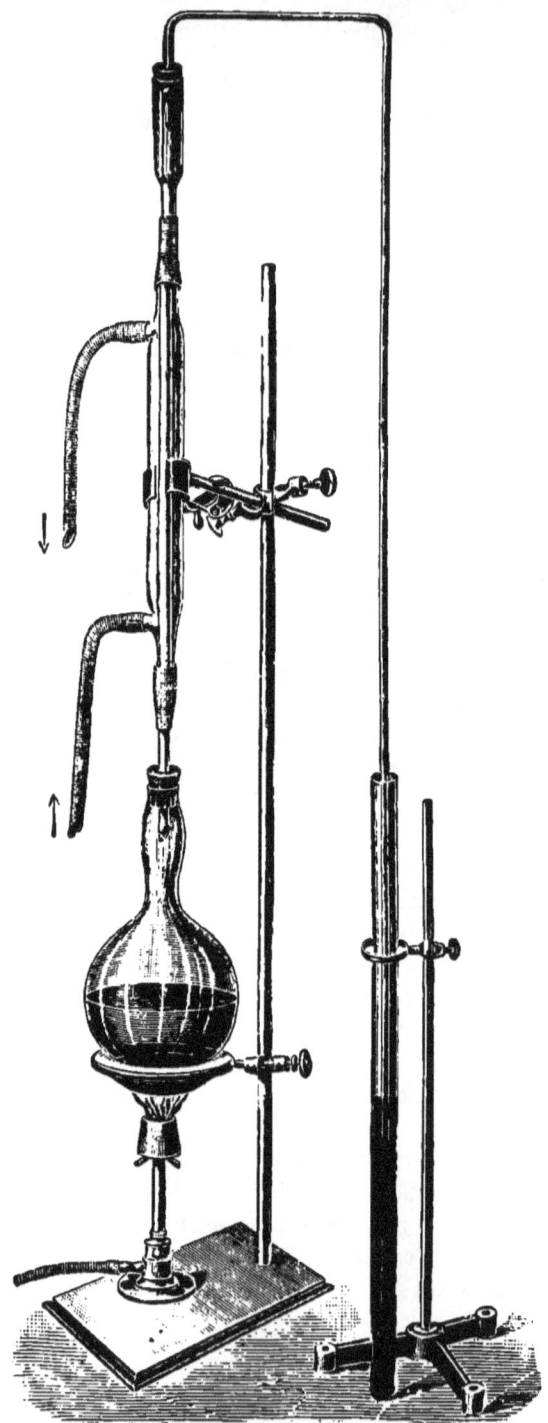

FIG. 66.

## 18. REACTION: SULPHONATION OF AN AROMATIC HYDRO-CARBON (I)

EXAMPLE: (*a*) Benzenemonosulphonic Acid from Benzene and Sulphuric Acid [1]
(*b*) Sulphobenzide. Benzenesulphonchloride. Benzenesulphonamide

(*a*) To 150 grammes of liquid fuming sulphuric acid, containing from 5–8 % of anhydride, placed in a 200 c.c. flask provided with an air condenser, gradually add, with good shaking, 40 grammes of benzene; before the addition of a new portion, always wait until the last portion, which at first floats on the surface of the acid, dissolves on shaking. Should the temperature rise above 50°, the flask is immersed in cold water for a short time. The sulphonation requires about 10–15 minutes. The cold reaction-mixture is then added, with stirring, drop by drop, from a separating-funnel, to three to four times its volume, of a cold, saturated solution of sodium chloride contained in a beaker. In order that the solution may not be heated above the room temperature, the beaker is placed in a large water-bath filled with cold water. After some time, but with especial ease when the walls of the vessel are rubbed with a sharp-edged glass rod, the sodium salt of benzenesulphonic acid separates out in the form of leaflets of a fatty lustre; the quantity is increased, on long standing, to such an extent that the beaker may be inverted without spilling its contents. If the separation of crystals does not begin, 10 c.c. of the liquid is shaken in a corked test-tube, and cooled by immersion in water. The solidified contents of the tube is then added to the main quantity in the beaker. In summer, at times, it may require a several hours' standing before the separation of crystals is ended. The pasty mass of crystals is then filtered off with suction on a Büchner funnel, firmly pressed together with a pestle, washed with a saturated sodium chloride solution, and finally pressed out on a porous plate.

---

[1] P. 31, 283 and 631; A. 140, 284; B. 24, 2121.

In order to obtain pure sodium benzenesulphonate, 5 grammes of the crude product is crystallised from absolute alcohol, upon which it is noticed that the sodium chloride mixed with it is insoluble in alcohol. (*b*) Heat 30 grammes of the well pressed out crude product on a watch-glass in an air-bath up to 110°, until it is completely dry. In order to obtain the by-product, sulphobenzide, the pulverised salt is warmed with 50 c.c. of ether, filtered with suction while hot, and washed with ether. After evaporating the ether, a small amount of a crystalline residue is obtained; this is recrystallised in a test-tube from a little alcohol with the addition of hot water. Melting-point, 129°.

To prepare benzenesulphonchloride from sodium benzenesulphonate, the just extracted salt is warmed for some time on the water-bath to evaporate the ether; it is then treated in a dry flask (under the hood) with 50 grammes of finely powdered phosphorus pentachloride, and the mixture is warmed $\frac{1}{4}$ to $\frac{1}{2}$ hour on an actively boiling water-bath. The cold reaction-product is then poured gradually into 300 c.c. of cold water in a flask; it is shaken up from time to time, and, after standing for two to three hours, the sulphonchloride is taken up with ether and the generally turbid ethereal solution filtered; the ether is then evaporated off.

In a porcelain dish 10 grammes of finely powdered ammonium carbonate is treated with about 1 c.c. of benzenesulphonchloride, and rubbed together intimately; the mixture is heated, with good stirring, over a small flame, until the odour of the sulphonchloride has vanished. After cooling, it is treated with water, filtered with suction, washed several times with water, and the benzenesulphonamide crystallised from alcohol to which hot water is added until turbidity begins. Melting-point, 156°.

Under the sulphonation of aniline it was mentioned that the aromatic compounds differ from the aliphatic compounds, in that they can be sulphonated by the action of sulphuric acid; *i.e.* the benzene-hydrogen atoms are replaced by the sulphonic acid group, $SO_3H$. Thus the above reaction takes place in accordance with the following equation:

$$C_6H_6 + SO_2{\scriptstyle\begin{matrix}OH\\OH\end{matrix}} = C_6H_5 \cdot SO_3H + H_2O.$$

Since, in the sulphonation, an excess of sulphuric acid is always used, after the reaction is complete it is necessary to separate the sulphonic acid from the excess of sulphuric acid. Many sulphonic acids, especially those of the hydrocarbons, are very easily soluble in water, so that the pure acid cannot be separated out on mere dilution with water, as is the case with sulphanilic acid. There are three methods in common use for the isolation of sulphonic acids soluble in water. The sulphonic acids obtained most easily are those difficultly soluble in cold sulphuric acid. In this case it is only necessary to cool the sulphonating mixture, and filter off the sulphonic acid separating out, with suction over asbestos or glass-wool. A second method consists in allowing the sulphuric acid solution to flow into a saturated solution of common salt; in many cases the difficultly soluble (in sodium chloride solution) sodium salt of the sulphonic acid separates out. Frequently it is more advantageous to use sodium acetate, potassium chloride, ammonium chloride, or other salts, instead of sodium chloride. Almost all soluble sulphonic acids, in the form of their alkali salts, can be separated by these two methods in the shortest time. In dealing with a new substance, preliminary experiments with small quantities of the substance are made to determine which salt is best adapted for the separation. The third method, which is the one generally applicable, depends upon the property of sulphonic acids, of forming soluble salts of calcium, barium, and lead in contradistinction to sulphuric acid. If the sulphuric acid solution, diluted with water, is neutralised with the carbonate of one of these metals and then filtered, the filtrate contains only the corresponding salt of the sulphonic acid, while the sulphuric acid in the form of calcium, barium, or lead sulphate remains on the filter. If the alkali salts of the sulphuric acids are desired, the water solution of the above salt is treated with the alkali carbonate until a precipitate is no longer formed. The precipitate is filtered off, and the pure alkali salt of the sulphonic acid is obtained in solution, which, on evaporation to dryness, yields the salt in the solid condition.

In order to obtain the free sulphonic acid, the lead salt is prepared and then decomposed with sulphuretted hydrogen.

The sulphonic acids of the hydrocarbons are generally colourless, crystallisable compounds, very easily soluble in water, behaving like strong acids. By heating with hydrochloric acid, under pressure if necessary, or by the action of steam, the sulphonic acid group may be split off, *e.g.*:

$$C_6H_5 \cdot SO_3H + H_2O = C_6H_6 + H_2SO_4.$$

This reaction is of importance in many cases for the separation of hydrocarbon mixtures. If under certain conditions one hydrocarbon is sulphonated, and another is not, the latter can be separated from the former by removing the sulphuric acid solution of the sulphonic acid of the first, and from this the original hydrocarbon may be regenerated by one of the methods mentioned.

Of particular importance is the behaviour of sulphonic acids when fused with caustic potash or caustic soda, by which the sulphonic acid group is eliminated and a phenol formed:

$$C_6H_5 \cdot SO_3K + KOH = C_6H_5 \cdot OH + K_2SO_3.$$

With benzenesulphonic acid this important reaction does not take place smoothly; for this reason the directions for carrying it out practically will be given later in another place (see $\beta$-naphthol). Poly-acid phenols may also be obtained from poly-basic sulphonic acids. The formation of m-dioxybenzene or resorcinol from benzenedisulphonic acid is of practical value:

$$C_6H_4(SO_3K)_2 + 2KOH = C_6H_4{<}^{OH}_{OH} + 2K_2SO_3.$$

If an alkali salt of a sulphonic acid mixed with potassium cyanide or potassium ferrocyanide is subjected to dry distillation, the sulphonic acid group is replaced by cyanogen and an acid-nitrile is obtained, e.g.:

$$C_6H_5 \cdot SO_3K + KCN = C_6H_5 \cdot CN + K_2SO_3.$$
Benzonitrile

The sulphonic acids behave toward phosphorus pentachloride like the carbonic acids, with the formation of acid-chlorides:

$$C_6H_5 \cdot SO_3Na + PCl_5 = C_6H_5 \cdot SO_2 \cdot Cl + NaCl + POCl_3.$$

The sulphonchlorides differ from the carbonic acid chlorides, in that they are not decomposed by cold water. In order to separate them from the phosphorus oxychloride, the mixture is generally poured into cold water; after long standing the oxychloride is converted into phosphoric acid, and the acid-chloride insoluble in water is obtained by decanting the water or extracting with ether; or in case it is solid, by filtering. The sulphonchlorides are generally distinguished by a very characteristic odour. They can be distilled in a vacuum only, without decomposition. Treated with ammonia they form sulphonamides, which

# AROMATIC SERIES

crystallise well and are used for the characterisation of the sulphonic acids:

$$C_6H_5 \cdot SO_2 \cdot Cl + NH_3 = C_6H_5 \cdot SO_2 \cdot NH_2 + HCl.$$

In the sulphonamides, in consequence of the strongly negative character of the $X \cdot SO_2$-group, the hydrogen of the amido-group is so easily replaced by metals, that they dissolve in water solutions of the alkalies to form salts of the amide. (Try it.) If a sulphon-chloride is allowed to stand a long time with an aliphatic alcohol, a sulphonic acid ester is formed, *e.g.*:

$$C_6H_5 \cdot SO_2 \cdot Cl + C_2H_5 \cdot OH = C_6H_5 \cdot SO_2 \cdot OC_2H_5 + HCl.$$
Benzenesulphonic ester

If this is now warmed with an alcohol, an aliphatic ether is formed, with the generation of the sulphonic acid, *e.g.*:

$$C_6H_5 \cdot SO_2 \cdot OC_2H_5 + C_2H_5 \cdot OH = C_6H_5 \cdot SO_3H + C_2H_5 \cdot O \cdot C_2H_5.$$

The formation of ether in this case is analogous to the formation of ethyl ether on heating ethyl sulphuric acid with alcohol:

$$SO_2\!\!\begin{array}{l}\diagup OC_2H_5 \\ \diagdown OH\end{array} + C_2H_5 \cdot OH = H_2SO_4 + C_2H_5 \cdot O \cdot C_2H_5.$$

Since this reaction is continuous, and since the benzene sulphonic acid formed in the reaction is a weaker acid than sulphuric acid, and consequently does not carbonise the alcohol like sulphuric acid, the operation may be continued indefinitely. For these reasons recently attempts have been made to employ the aromatic sulphonic acids for the technical preparation of ether. If the sulphonation is effected as above with fuming sulphuric acid, in many cases, besides the sulphonic acid a small quantity of sulphone is formed, *e.g.*:

$$2\,C_6H_6 + SO_3 = SO_2\!\!\begin{array}{l}\diagup C_6H_5 \\ \diagdown C_6H_5\end{array} + H_2O.$$
Diphenylsulphone<br>= Sulphobenzide

For sulphonating purposes, either ordinary concentrated sulphuric acid or the so-called monohydrate or fuming sulphuric acid of various grades is used, according to the conditions. The reaction is conducted with cooling, at the room temperature, or with heating.

To facilitate the elimination of water, phosphorus pentoxide or potassium sulphate may be added to the sulphonating mixture.

In some cases it is of advantage to use chlorsulphuric acid instead of sulphuric acid; the reaction takes place in accordance with the following equation:

$$C_6H_6 + Cl \cdot SO_3H = C_6H_5 \cdot SO_3H + HCl.$$

### 19. REACTION: REDUCTION OF A SULPHONCHLORIDE TO A SULPHINIC ACID OR TO A THIOPHENOL

EXAMPLES: (*a*) **Benzenesulphinic Acid.**[1] (*b*) **Thiophenol**[2]

(*a*) Heat 40 grammes of water to boiling in a 300 c.c. flask provided with a short reflux condenser and a dropping-funnel; add 10 grammes of zinc dust, and without further heating by the flame, gradually allow to flow in from the funnel 15 grammes of benzenesulphonchloride in small portions. After each addition wait until the vigorous reaction accompanied by a hissing sound has moderated. The mixture is then heated a few minutes over a small flame, filtered after cooling from the precipitate of zinc dust and the zinc salt of benzenesulphinic acid, and the precipitate washed several times with water. The insignificant looking gray precipitate is the reaction-product, and not the filtrate, which can be thrown away. The precipitate is then divided into two approximately equal parts; one portion is heated for about ten minutes not quite to boiling with a solution of 10 grammes of dehydrated sodium carbonate in 50 c.c. of water and then filtered with suction. The precipitate remaining on the filter is worthless, while the filtrate contains the sodium benzenesulphinate in solution. This is evaporated to about one-half its original volume, and, after cooling, acidified with dilute sulphuric acid, upon which the free benzenesulphinic acid separates out in colourless crystals; the separation is facilitated by rubbing the sides of the vessel with a glass rod. After filtering, the substance is recrystallised from a little water. Melting-point, 83–84°.

---

[1] B. 9, 1585.      [2] A. 119, 142, B. 10, 940.

After the separation of the free acid and its recrystallisation, the mother-liquors are preserved for use in the preparation of thiophenol. Should the free acid not separate on acidifying the sodium salt, it is extracted several times with ether; this is evaporated, the residue, in case it does not solidify of itself, is rubbed with a glass rod and then recrystallised.

(*b*) In a large flask of at least 1½ litres' capacity place 90 grammes of concentrated hydrochloric acid and 1½ times its volume of water. The mixture is well cooled with ice-water, and treated with a few drops of a solution of copper sulphate and 20 grammes of granulated zinc; to the mixture, which is continuously cooled, generating hydrogen rapidly, gradually add the second half of the zinc benzenesulphinate and the mother-liquors obtained above. As soon as the evolution of hydrogen becomes moderate, add 20 grammes of zinc dust, allow to stand, with cooling, for a half-hour, then add 20 grammes more of zinc dust and heat on the water-bath until the liquid shows a faint acid reaction or none at all. This point is easily recognised in that the zinc dust, which has been carried upwards by the evolution of hydrogen in the acid liquid, sinks completely to the bottom. As soon as this happens, the mixture is allowed to cool somewhat, and is then carefully made strongly acid with much concentrated hydrochloric acid, and the free thiophenol is distilled over with steam. If on the introduction of steam an active foaming with evolution of hydrogen should suddenly take place, the steam is shut off until the foaming has ceased. As soon as no more oil passes over, the distillation is discontinued. Still, before throwing away the non-volatile residue, it is tested for acid. If it does not show an acid reaction, a moderately large amount of concentrated hydrochloric acid is added and it is again distilled with steam. The thiophenol is taken up with ether, and after the evaporation of the latter it is rectified. Boiling-point, 173°.

In the preparation of thiophenol, care is taken that there are no flames in the neighbourhood of the flask in which the reaction is conducted, otherwise there may be an explosion of the mixture of oxygen and hydrogen. Since the thiophenol possesses an extremely unpleasant odour, and the vapours attack the eyes, causing tears, the experiment must not be carried out in the laboratory, but in a side room (hydrogen sulphide room), or in the open air, in the basement, or at least under a hood with a good draught. Further, care must be taken not to allow the substance to come in contact with the skin, since it produces a violent burning.

If zinc dust is allowed to act on a sulphonchloride, the zinc salt of the sulphinic acid is formed:

$$C_6H_5.SO_2.Cl \qquad C_6H_5.SO_2$$
$$C_6H_5.SO_2.Cl + ZnZn = C_6H_5.SO_2 \!\!\!\diagup\!\!\!\!\diagdown Zn + ZnCl_2.$$
$$\text{Zinc benzenesulphinate}$$

The zinc salts thus formed are insoluble in water, and can be easily obtained by filtering off. In order to prepare the free sulphinic acid from a zinc salt, it is first converted into the easily soluble sodium salt by boiling with a sodium carbonate solution; the solution of the sodium salt is concentrated, and the free acid is precipitated with dilute sulphuric acid. The sulphinic acids differ from the sulphonic acids in that they are difficultly soluble in cold water, and can, therefore, be recrystallised from water. On fusing with potassium hydroxide, the sulphinic acids pass over to the hydrocarbons:

$$C_6H_5.SO_2K + KOH = K_2SO_3 + C_6H_6.$$

If they are reduced, a thiophenol is finally obtained, as above:

$$C_6H_5.SO_2H + 4H = C_6H_5.SH + 2H_2O.$$

Instead of using the pure acid for this purpose, it is more convenient to use the crude zinc sulphinate, as above. The reduction does not appear to take place in so simple a manner as that indicated by the equation, but phenyldisulphide is first formed in the acid solution:

$$2\,C_6H_5.SO_2H + 6H = C_6H_5.S.S.C_6H_5 + 4H_2O.$$
$$\text{Phenyldisulphide}$$

If this is warmed with zinc dust, the zinc salt of the thiophenol is formed:

$$C_6H_5.S.S.C_6H_5 + Zn = \begin{array}{c} C_6H_5.S \\ C_6H_5.S \end{array}\!\!\!\!\diagup\!\!\!\!\diagdown Zn,$$

from which the free thiophenol may be separated by hydrochloric acid. The thiophenols are liquids of unpleasant odours; the higher members of the series are solids. Like the mercaptans of the aliphatic series, they form difficultly soluble salts with lead and mercury.

EXPERIMENT: Dissolve mercuric chloride, or lead acetate, in a test-tube with alcohol, by heating; then cool, and filter. If the alcoholic solution is treated with a few drops of thiophenol, a precipitate of the difficultly soluble salt is obtained. The lead salt is yellow, and possesses the composition represented by the formula:

$$(C_6H_5 \cdot S)_2 Pb.$$

In the air, and on treatment with oxidising agents like nitric acid, chromic acid, etc., the thiophenols are oxidised to disulphides:

$$2 C_6H_5 \cdot SH + O = C_6H_5 \cdot S—S \cdot C_6H_5 + H_2O.$$

EXPERIMENT: A few drops of phenyl mercaptan are dissolved in alcohol, treated with some ammonia, and evaporated to dryness on the water-bath in a watch-glass. (Under the hood.) Colourless needles of the disulphide remain. Melting-point, $61°$.

By reduction the disulphides are easily converted back to the thiophenols:

$$C_6H_5 \cdot S—S \cdot C_6H_5 + 2 H = 2 C_6H_5 \cdot SH.$$

Like the phenols the thiophenols have the power of forming ethers, *e.g.*:

$$C_6H_5 \cdot SCH_3 = \text{Thioanisol,}$$
$$C_6H_5 \cdot S \cdot C_6H_5 = \text{Phenylsulphide.}$$

### 20. REACTION: SULPHONATION OF AN AROMATIC HYDROCARBON (II)

EXAMPLE: β-Naphthalenesulphonic Acid

A mixture of 50 grammes of finely pulverised naphthalene and 60 grammes of pure concentrated sulphuric acid is heated in an open flask in an oil-bath for 4 hours to $170$–$180°$. After cooling the solution is poured, with stirring, into 1 litre of water, and the naphthalene not attacked is filtered off; the mixture is neutralised at the boiling temperature in a large dish with a paste of lime, not too thin, prepared by triturating about 70 grammes of dry slaked lime with water. The mixture is filtered while hot as possible through a filter-cloth, which has been previously thoroughly moistened (see page 54) and the precipitate washed with hot water. The filter-cloth is then folded together and thoroughly squeezed out in another dish; the expressed, generally, turbid liquid, after filtering, is united with the main quantity. The solution is then evaporated in a dish over a free flame until a test-portion will solidify to a crystalline paste on rubbing with a glass rod. After the solution has been allowed to stand over night the calcium $\beta$-naphthalenesulphonate is filtered off with suction, washed once with a little water, pressed firmly together with a pestle, and spread out on a porous plate. In order to obtain the sodium salt, it is dissolved in hot water, and the solution gradually treated with a concentrated solution of 50 grammes of crystallised sodium carbonate until a test-portion filtered off no longer gives a precipitate with sodium carbonate. After cooling, the precipitate of calcium carbonate is filtered off with suction, washed with water, and the filtrate evaporated over a free flame until crystals begin to separate from the hot solution. After standing several hours at the ordinary temperature, the crystals are filtered off, and the mother-liquor further concentrated; after long standing, the second crystallisation is filtered off, and the mixture of the two lots of crystals dried on the water-bath. Yield, 60–70 grammes.

Naphthalene is sulphonated on heating with sulphuric acid, in accordance with the following equation:

$$C_{10}H_8 + H_2SO_4 = C_{10}H_7 \cdot SO_3H + H_2O.$$

There is formed not as in the case of benzene, in which the six hydrogen atoms are equivalent, a single sulphonic acid, but a mixture of two isomeric sulphonic acids:

a-Naphthalenesulphonic acid      and      $\beta$-Naphthalenesulphonic acid

According to the temperature at which the sulphonation takes place, more of one than of the other acid is formed; at lower temperatures an excess of the α-acid is obtained, at higher an excess of the β-acid. If the mixture is heated to 100°, a mixture of 4 parts of the α-acid and 1 part of the β-acid is formed, while at 170° a mixture of the 3 parts of the β-acid and 1 part of the α-acid is obtained. In order to separate the sulphonic acids from the excess of sulphuric acid, advantage is taken of the fact that sulphonic acids differ from sulphuric acid in that they form soluble salts of calcium, barium, and lead, as mentioned under benzenesulphonic acid. For the separation of the sulphonic acid from sulphuric acid, the calcium salt is prepared by neutralising the acid mixture with chalk or lime, since it is cheaper than lead carbonate or barium carbonate. This method is followed technically on the large scale as well as in laboratory preparations. Since the calcium salts of the two isomeric sulphonic acids possess a very different solubility in water, — at 10° 1 part of the α-salt dissolves in 16.5 parts of water, and 1 part of the β-salt dissolves in 76 parts of water, — the β-salt, which is more difficultly soluble, and consequently crystallises out first, can be separated by fractional crystallisation from the α-salt which remains in solution. For the conversion into naphthol the calcium salt cannot be used directly; it must first be changed into the sodium salt by treatment with sodium carbonate:

$$(C_{10}H_7 \cdot SO_3)_2Ca + Na_2CO_3 = 2\,C_{10}H_7 \cdot SO_3Na + CaCO_3.$$

In order to remove the last portions of the α-salt, it is advisable not to evaporate the solution of sodium salt directly to dryness, but to allow the more difficultly soluble β-salt to crystallise out, upon which the α-salt remains dissolved in the mother-liquor.

The reactions of the naphthalenesulphonic acids are similar to those given above under benzenesulphonic acid. It is still to be mentioned that the α-acid is converted into the β-acid by heating with concentrated sulphuric acid to almost 200°; a reaction which is explained by the fact that the sulphonic acid decomposes in the small amount of water always present, into naphthalene and sulphuric acid, and that the former is then sulphonated to the β-acid at the higher temperature (200°). The sulphonation of naphthalene to the α- and β-acids is carried out on the large scale in technical operations, since when fused with sodium hydroxide these acids yield naphthols of great importance for the manufacture of dyes. The next preparation deals with this reaction.

## 21. REACTION: CONVERSION OF A SULPHONIC ACID INTO A PHENOL

EXAMPLE: β-Naphthol from Sodium-β-Naphthalene Sulphonate and Sodium Hydroxide [1]

In order to convert sodium-$\beta$-naphthalene sulphonate into $\beta$-naphthol the proportions of the necessary reagents used are:

10 parts sodium-$\beta$-naphthalene sulphonate;
30 parts sodium hydroxide, as pure as possible;
1 part water.

The sodium hydroxide is broken in pieces about a centimetre in length, or the size of a bean, treated with the water, and heated in a nickel crucible (a crucible 11 cm. high and 8 cm. in diameter is a convenient size), with stirring, to 280° (Fig. 67). The stirring is done with a thermometer, the lower end of which is protected by a case of copper or nickel, about 16 cm. long and 8 mm. wide. This is supported by a cork, containing a narrow canal at the side, fitting the case. In order to be able to determine the temperature as exactly as possible, a layer of oil 1 cm. high is placed in the case, in which the bulb of the thermometer is immersed. If the stirring is done with the case, the upper portion is covered with several layers of asbestos board, secured with wire, or a cork is pushed over the case (Fig. 67). Since, on fusion of the sodium hydroxide, a troublesome spattering takes place, the hand is protected by a glove, and the eyes by glasses. As soon as the temperature reaches 280°, the heating is continued with a somewhat smaller flame, and the sodium naphthalene sulphonate is gradually added, with stirring. After each new addition, the temperature falls somewhat; no more of the salt is added, until the temperature again reaches 280°. After all the salt is added, the flame is made somewhat larger, upon which the fusion becomes viscid with evolution of steam and frothing, until finally, at about 310°, the real reaction takes place. After the temperature is held

---

[1] E. Fischer-Kling, Prep. of Organic Compounds, page 54. Z. 1867, 299.

at 310–320° for about 5 minutes, the fusion becomes liquid, and the reaction is complete. The mass is now allowed to cool somewhat, without further stirring, and the upper layer, consisting essentially of sodium naphtholate, as soon as it has attained a pasty consistency, is removed while still hot, with a spatula; the portions adhering to the spatula are dissolved off each time, by immersion in cold water. Should the two layers not separate well, the whole fusion may be dissolved in water. The water solution, diluted to about 1 litre, is then acidified while hot, with a mixture of equal parts of concentrated hydrochloric acid and water (under the hood), and after cooling, the separated $\beta$-naphthol is filtered off, washed with water, and firmly pressed together with a pestle. It may be purified by distillation, or recrystallisation from much hot water. In the former case, the moist preparation is previously fused in a porcelain dish, and after solidifying, the water is removed by decantation, or absorbed by filter-paper. Melting-point, 123°. Boiling-point, 286°. Yield, half the weight of the sulphonate used.

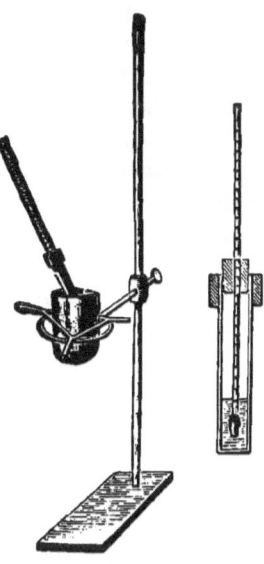

FIG. 67.

As above indicated, in a sodium hydroxide, or potassium hydroxide fusion of a sulphonic acid, besides the phenol, the alkali sulphite is formed, *e.g.*:

$$C_{10}H_7 \cdot SO_3Na + 2\,NaOH = C_{10}H_7 \cdot ONa + Na_2SO_3 + H_2O.$$
Sodium naphtholate

The free phenol is, therefore, not directly obtained on fusion, but the alkali salt of it, from which, after the solution of the fusion in water, the phenol is liberated on acidifying with hydrochloric acid.

The reaction just effected is in practice carried out on the largest scale in iron kettles to which stirring apparatus is attached. $\beta$-naphthol as well as its numerous mono- and poly-sulphonic acid derivatives obtained by treatment with sulphuric acid find extensive application for the manufacture of azo dyes. Further, from the $\beta$-naphthol, $\beta$-naphthylamine is prepared by the action of ammonia under pressure:

$$C_{10}H_7 \cdot OH + NH_3 = C_{10}H_7 \cdot NH_2 + H_2O,$$

which also finds technical use for the manufacture of azo dyes, as such and in the form of its sulphonic acids. $a$-Naphthol is also prepared in the same way by fusion of $a$-sodiumnaphthalene sulphonate with sodium hydroxide, although not in so large quantities as the $\beta$-naphthol.

The phenols, in consequence of the negative character of the aromatic hydrocarbon residue, are weak acids which dissolve in water solutions of the alkalies to form salts. Still the acid nature is so weak that the salts can be decomposed by carbon dioxide; use is frequently made of this property for the purification and separation of phenols.

EXPERIMENT: A mixture of $\beta$-naphthol and benzoïc acid is dissolved in a diluted caustic soda solution, and carbon dioxide passed into it for a long time. $\beta$-Naphthol only separates out; this is filtered off. The filtrate is acidified with concentrated hydrochloric acid upon which the benzoïc acid is precipitated.

The naphthols differ from the phenols of the benzene series, in that their hydroxyl groups are more capable of reaction than those of the phenols, cresols, etc. While, for example, the ether of phenol cannot be prepared from the phenol and corresponding alcohol by abstracting water:

$$(C_6H_5 \cdot OH + CH_3 \cdot OH = C_6H_5 \cdot O \cdot CH_3 + H_2O),$$
Does not take place

but can only be obtained by the action of halogen alkyls on phenol salts:

$$C_6H_5 \cdot ONa + ICH_3 = C_6H_5 \cdot O \cdot CH_3 + NaI.$$

By heating the naphthols with an aliphatic alcohol and sulphuric acid, the ethers are easily prepared:

$$C_{10}H_7 \cdot OH + CH_3 \cdot OH = C_{10}H_7 \cdot O \cdot CH_3 + H_2O.$$
Naphthylmethyl ether

## 22. REACTION: NITRATION OF A PHENOL

EXAMPLE: o- and p-Nitrophenol

Dissolve 80 grammes of sodium nitrate in 200 grammes of water by heating; after cooling, the solution is treated, with stirring, with 100 grammes of concentrated sulphuric acid. To the mixture cooled to 25° contained in a beaker, add drop by drop, from a separating funnel, with frequent stirring, a mixture of 50 grammes of crystallised phenol and 5 grammes of alcohol, melted by warming. During this addition the temperature is kept between 25–30° by immersing the beaker in water. Should the phenol solidify in the separating funnel, it is again melted by a short warming in a large flame. After the reaction-mixture has been allowed to stand for two hours, with frequent stirring, it is treated with double its volume of water; the reaction-product collects as a dark oil at the bottom of the vessel. The principal portion of the water solution is then decanted from the oil, this is washed again with water, and after the addition of $\frac{1}{2}$ litre of water, is distilled with steam until no more o-nitrophenol passes over. Concerning the removal of the o-nitrophenol solidifying in the condenser, see page 36 (temporary removal of the condenser-water).

After cooling, the distillate is filtered, the o-nitrophenol washed with water, pressed out on a porous plate, and dried in a desiccator. Since it is obtained completely pure, it is unnecessary to subject it to any further process of purification. In order to obtain the non-volatile p-nitrophenol remaining in the flask, the mixture is cooled by immersion in cold water, the water solution is filtered from the undissolved portions, and the filtrate boiled for a quarter-hour with 20 grammes of animal charcoal, the water evaporating being replaced by a fresh quantity. The charcoal is then filtered off and the filtrate allowed to stand in a cool place over night, upon which the p-nitrophenol separates out in long, almost colourless needles. The oil still present in the distillation is boiled with a mixture of 1 part by volume of concentrated hydrochloric acid and 2 parts by volume of water, with the addition of animal charcoal, filtered after partial cooling and the filtrate allowed to stand over night. There is thus obtained a second crystallisation. If the crystals which have separated out are still contaminated by the oil, they are recrystallised from dilute hydrochloric acid with the use of animal charcoal.

Melting-point of o-Nitrophenol, $45°$;
Melting-point of p-Nitrophenol, $114°$.

Yield, 30 grammes and 5–10 grammes respectively.

The mon-acid phenols of the benzene series are, in contrast to the corresponding hydrocarbons, very easily nitrated. In the nitration of benzene, in order to facilitate the elimination of the water, concentrated sulphuric acid must be added; whereas the action of concentrated nitric acid alone upon phenol is so energetic, that in this case it must be diluted with water. Upon nitrating phenol, the o- and p-nitrophenols are formed simultaneously, the former of which is volatile with steam:

$$C_6H_5 \cdot OH + NO_2 \cdot OH = C_6H_4{<}{{NO_2}\atop{OH}} + H_2O.$$

o- and p-Nitrophenol

On nitrating the homologues of phenol, the nitro-groups always enter the o- and p-positions to the hydroxyl group, and never the m-position. In order to prepare m-nitrophenol, it is necessary to start from m-nitroaniline; this is diazotised and its diazo-solution boiled with water.

The nitrophenols behave in all respects like the phenols. But by the entrance of the negative nitro-group, the negative character of the phenol is so strengthened that the nitrophenols not only dissolve in alkalies, but also in the alkali carbonates.

EXPERIMENT: Dissolve some o-nitrophenol in a solution of sodium carbonate by warming; the scarlet red sodium salt is formed.

In consequence of this action, the nitrophenols cannot be precipitated from their alkaline solutions by carbon dioxide.

In addition the nitrophenols show the characteristics of the nitro-compounds in general, since they, for example, pass over to amidophenols on energetic reduction, etc.

## 23. REACTION: (a) CHLORINATION OF A SIDE-CHAIN OF A HYDROCARBON. (b) CONVERSION OF A DICHLORIDE INTO AN ALDEHYDE

EXAMPLES: (a) **Benzalchloride from Toluene**
(b) **Benzaldehyde from Benzalchloride**

(a) A 100 c.c. round flask with as wide a neck as possible (Fig. 68) containing 50 grammes of toluene is placed in a well-lighted position, best in the sunlight. The toluene is heated to

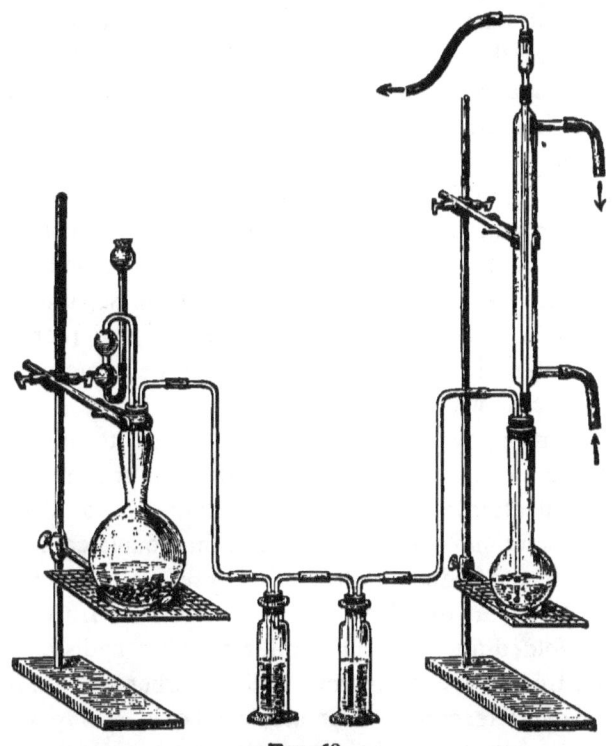

FIG. 68.

boiling and a current of dry chlorine conducted into it until its weight is increased by 40 grammes. In order to be able to judge of the course of the reaction, the flask, with the toluene, is weighed before the experiment. By interrupting the passage of the chlorine from time to time, cooling and weighing the flask, the increase in weight will indicate how far the chlorinating action has proceeded. The length of the operation varies greatly. In summer the reaction is complete in a few hours; during the cloudy days of winter a half or a whole day may be necessary.

(*b*) In order to convert the benzalchloride into benzaldehyde, the crude product thus obtained is treated in a round flask provided with an effective reflux condenser, with 500 c.c. of water and 150 grammes of precipitated calcium carbonate (or floated chalk or finely pulverised marble) and the mixture heated four hours in a hemispherical oil-bath to 130° (thermometer in the oil). Without further heating, steam is passed through the hot contents of the flask until no more oil distils over. For this purpose the apparatus necessary (cork with a glass tube) has been prepared before the heating in the oil-bath.

Before the crude benzaldehyde is subjected to purification, the liquid remaining in the distilling flask is filtered while hot through a folded-filter, and the filtrate acidified with much concentrated hydrochloric acid. On cooling, the benzoïc acid obtained as a by-product in the preparation of benzaldehyde separates out in lustrous leaves. After cooling, it is filtered off and recrystallised from hot water, during which it must not be heated too long, since it is volatile with steam.

The oil passing over with the steam is treated, together with all of the liquid, with a concentrated solution of sodium hydrogen sulphite, until after long shaking the greater part of the oil has passed into solution. Should crystals of the double compound of benzaldehyde and sodium hydrogen sulphite separate out, water is added until they are dissolved. The water solution is then filtered through a folded-filter from the oil remaining undissolved, and the filtrate treated with anhydrous sodium carbonate until it shows a strong alkaline reaction. This alkaline liquid is now subjected to distillation with steam, when perfectly pure benzaldehyde passes over; it is taken up with ether, and, after the evaporation of the ether, is distilled. Boiling-point, 179°.

Under the preparation of brombenzene it has already been mentioned that by the action of chlorine or bromine on aromatic hydrocarbons containing aliphatic side-chains, different products are formed, depending on the temperature at which the action takes place. If, e.g., chlorine acts at lower temperatures on toluene, chlortoluene is formed, the chlorine entering the benzene ring:

$$C_6H_5 \cdot CH_3 + Cl_2 = C_6H_4 \begin{array}{c} CH_3 \\ Cl \end{array} + HCl.$$

At ordinary temperatures      Chlortoluene

If, on the other hand, chlorine is conducted into boiling toluene, the chlorine atom enters the side-chain:

$$C_6H_5 \cdot CH_3 + Cl_2 = C_6H_5 \cdot CH_2Cl + HCl.$$
At boiling temperature      Benzylchloride

If chlorine is conducted into toluene at the boiling temperature for a long time, a second, and finally a third, hydrogen atom of the methyl group is substituted:

$$C_6H_5 \cdot CH_2Cl + Cl_2 = C_6H_5 \cdot CHCl_2 + HCl,$$
Boiling      Benzalchloride

$$C_6H_5 \cdot CHCl_2 + Cl_2 = C_6H_5 \cdot CCl_3 + HCl.$$
Benzotrichloride

The formation of benzotrichloride is the final result of the action of chlorine under these conditions, since the benzotrichloride is not changed even by passing in the chlorine for a longer time.

The replacement of hydrogen by chlorine directly presents the difficulty, unlike the liquid bromine, that weighed quantities cannot be used, and the exact point to which the introduction of chlorine should be continued in order to get a certain definite compound must be determined. This is accomplished if, from time to time, the increase in weight of the substance being chlorinated is determined. Since the conversion of one molecule of toluene to benzylchloride requires an increase of weight equal to the atomic weight of chlorine minus the atomic weight of hydrogen (Cl − H = 34.5), therefore in the preparation of benzylchloride, 100 parts by weight of toluene must take up an additional weight of chlorine = 37.5 parts by weight, and correspondingly in the preparation of benzalchloride or benzotrichloride, the increase in parts by weight must be respectively 2 × 37.5 = 75 and 3 × 37.5 = 112.5.

In most organic chlorinating reactions, besides the main reaction, a side reaction also takes place which, in the above example, results in the conversion of a portion of the toluene to benzalchloride and benzotrichloride, while another portion is only chlorinated to benzyl chloride. Accordingly the reaction-product obtained above consists essentially of benzalchloride mixed with a small quantity of benzyl chloride and benzotrichloride. If the benzalchloride is to be obtained in a pure condition, the mixture must be separated by fractional distillation.

The halogen derivatives of the aromatic hydrocarbons containing the halogen in the side-chain are in part liquids, in part colourless crystallisable solids, which are distinguished from their isomers containing the halogen in the benzene ring, in that their vapours violently attack the mucous membrane of the eyes and nose. Care is taken therefore in the above preparation not to expose the face to the vapours, and further to prevent the chlorination products from coming in contact with the hands.

Concerning their chemical properties, these two isomeric series differ in that the compounds containing the halogen in the side-chain are far more active than those in which the halogen occurs in the benzene ring, as is apparent from the following equations:

$$C_6H_5 \cdot CH_2Cl + NH_3 = C_6H_5 \cdot CH_2 \cdot NH_2 + HCl,$$
Benzylamine

$$C_6H_5 \cdot CH_2Cl + CH_3 \cdot COONa = CH_3 \cdot CO \cdot OCH_2 \cdot C_6H_5 + NaCl,$$
Benzylacetate

$$C_6H_5 \cdot CH_2Cl + KCN = C_6H_5 \cdot CH_2 \cdot CN + KCl.$$
Benzyl cyanide

The chlorides obtained by substituting a side-chain are of importance for making certain preparations, — the aromatic alcohols, aldehydes, and acids. On boiling with water, they decompose, in accordance with the following equations:

(1) $C_6H_5 \cdot CH_2Cl + H_2O = C_6H_5 \cdot CH_2 \cdot OH + HCl,$
Benzyl alcohol

(2) $C_6H_5 \cdot CHCl_2 + H_2O = C_6H_5 \cdot CHO + 2 HCl,$
Benzaldehyde

(3) $C_6H_5 \cdot CCl_3 + 2 H_2O = C_6H_5 \cdot CO \cdot OH + 3 HCl.$
Benzoïc acid

But since, in cases 1 and 2, the hydrochloric acid formed in the reaction acts in the opposite way and regenerates the original chloride, it must be neutralised. This is usually accomplished by the addition of a carbonate, upon which the acid acts with the liberation of carbon dioxide.

In practice, the cheap calcium carbonate (marble dust) is used, and the above method for the preparation of benzaldehyde is, as far as possible, an imitation of the technical process used for obtaining the substance. From benzylchloride, benzaldehyde may also be prepared directly by boiling it with water in the presence of lead nitrate or copper nitrate. From the benzylchloride, benzyl alcohol is first formed, which is oxidised by the nitrate to benzaldehyde.

As above mentioned, the chlorination product obtained consists essentially of benzalchloride, mixed with small amounts of benzylchloride and benzotrichloride. If the mixture is boiled with water, with the addition of calcium carbonate, a mixture consisting mainly of benzaldehyde, besides benzylchloride and benzoïc acid — the latter being converted into the calcium salt by the carbonate — is obtained. If the reaction-mixture is distilled with steam, benzaldehyde, benzyl alcohol, and a small amount of chlorides which have not taken part in the reaction, pass over with the steam, while the calcium benzoate remains in the distillation flask. By acidifying the residue, the free benzoïc acid may be obtained as above. A large proportion of the so-called "benzoïc acid from toluene" is obtained in this way as a by-product in the technical preparation of benzaldehyde. In order to separate the benzaldehyde from benzyl alcohol, the chlorides, and other impurities, advantage is taken of the general property of aldehydes of uniting with acid sodium sulphite to form soluble double compounds. If the distillate is shaken with a solution of sodium hydrogen sulphite, the aldehyde dissolves, while the impurities remain undissolved. These are filtered off, and the sulphite compound of the aldehyde decomposed with sodium carbonate; on a second distillation with steam, the pure aldehyde passes over.

The aromatic aldehydes are in part liquids, in part solids, possessing a pleasant aromatic odour.

They show the characteristic aldehyde reactions, yielding primary alcohols on reduction and carbonic acids on oxidation.

EXPERIMENT: A few drops of benzaldehyde are allowed to stand in a watch-glass in the air. After a long time, crystals of benzoïc acid appear.

$$C_6H_5 \cdot CHO + O = C_6H_5 \cdot CO \cdot OH.$$

That they unite with acid sulphites to form crystalline compounds has been mentioned:

$$C_6H_5 \cdot CHO + HSO_3Na = C_6H_5 \cdot CH \begin{matrix} OH \\ SO_3Na \end{matrix}$$

EXPERIMENT: To $\frac{1}{2}$ c.c. of benzaldehyde add a concentrated solution of sodium hydrogen sulphite, and shake. The mixture solidifies after a short time to a crystalline mass.

Further, the aldehydes react, as mentioned under acetaldehyde, with hydroxylamine and phenyl hydrazine, to form oximes and hydrazones. Aldehydes also condense readily with primary aromatic bases with the elimination of water:

$$C_6H_5 \cdot CHO + C_6H_5 \cdot NH_2 = C_6H_5 \cdot CH = N \cdot C_6H_5 + H_2O.$$
Benzylideneaniline

EXPERIMENT: In a test-tube make a mixture of 1 c.c. of benzaldehyde and 1 c.c. of pure aniline, and warm gently. On cooling, drops of water separate out, and the mixture solidifies, crystals of benzylideneaniline being formed.

An additional number of characteristic aldehyde reactions will be taken up later in practice. Benzaldehyde is prepared technically on the large scale. Its most important application is for the manufacture of the dyes of the Malachite Green series, and of cinnamic acid (see this preparation).

### 24. REACTION: SIMULTANEOUS OXIDATION AND REDUCTION OF AN ALDEHYDE UNDER THE INFLUENCE OF CONCENTRATED POTASSIUM HYDROXIDE

EXAMPLE: **Benzoïc Acid and Benzyl Alcohol from Benzaldehyde**[1]

Treat 20 grammes of benzaldehyde in a stoppered cylinder or a thick-walled vessel with a cold solution of 18 grammes of potassium hydroxide in 12 grammes of water and shake until a permanent emulsion is formed; the mixture is then allowed to stand over night. The vessel is closed by a cork, and not a glass stopper, since at times a glass stopper becomes so firmly fastened that it can be removed only with great difficulty. To the crys-

---

[1] B. 14, 2394.

talline paste (potassium benzoate) separating out, water is added until a clear solution is obtained from which the benzyl alcohol is extracted by repeatedly shaking with ether. After the evaporation of the ether the residue is subjected to distillation; benzyl alcohol passes over at 206°. Yield, about 8 grammes. The benzoïc acid is precipitated from the alkaline solution on acidifying with hydrochloric acid.

While many aliphatic aldehydes (see acetaldehyde) are converted by alkalies into more complex compounds, with higher molecular weights, the so-called aldehyde resins, the aromatic aldehydes under similar conditions react smoothly; two molecules are decomposed by one molecule of potassium hydroxide, one aldehyde molecule being oxidised to the corresponding acid, and the other being reduced to a primary alcohol:

$$2\,C_6H_5.CHO + KOH = C_6H_5.CO.OK + C_6H_5.CH_2.OH.$$

Since the aldehydes are in part easily obtained, the different primary alcohols may be prepared advantageously by this reaction. Thus,

Anisic alcohol p-$C_6H_4\diagdown^{OCH_3}_{CH_2.OH}$ and Cuminic alcohol p-$C_6H_4\diagdown^{C_3H_7}_{CH_2.OH}$

are obtained by treating anisic aldehyde and cuminol, respectively, with alcoholic potash at the ordinary temperature or on heating. m-Nitrobenzylalcohol may also be obtained easily from m-nitrobenzaldehyde and aqueous potash. The aldehyde of furfurane, furfurol, is converted under the same conditions into furfurane alcohol and pyromucic acid:

$$2\ \underset{\text{Furfurol}}{\begin{array}{c}HC{-}\!\!-CH\\ \| \quad\ \ \|\\ HC\diagdown_{O}\!\!\diagup C{-}CHO\end{array}} + H_2O = \underset{\text{Furfurane alcohol}}{C_4H_3O.CH_2.OH} + \underset{\text{Pyromucic acid}}{C_4H_3O.CO.OH}.$$

The primary aromatic alcohols behave in all respects like the corresponding aliphatic alcohols in forming ethers and esters, e.g.:

$\underset{\text{Benzyl ether}}{\begin{array}{c}C_6H_5.CH_2\\ C_6H_5.CH_2\end{array}\!\!\!>\!O}$   $\underset{\text{Benzylmethyl ether}}{\begin{array}{c}C_6H_5.CH_2\\ CH_3\end{array}\!\!\!>\!O}$   $\underset{\text{Aceticbenzyl ester}}{CH_3.CO.OCH_2.C_6H_5}$

S

On oxidation they are converted first into aldehydes and finally into acids:

$$C_6H_5 \cdot CH_2 \cdot OH + O = C_6H_5 \cdot CHO + H_2O,$$
$$C_6H_5 \cdot CH_2 \cdot OH + O_2 = C_6H_5 \cdot COOH + H_2O.$$

## 25. REACTION: CONDENSATION OF AN ALDEHYDE BY POTASSIUM CYANIDE TO A BENZOÏN

EXAMPLE: **Benzoïn from Benzaldehyde**[1]

Mix 10 grammes of benzaldehyde with 20 grammes of alcohol and treat the mixture with a solution of 2 grammes of potassium cyanide and 5 c.c. of water. Boil on the water-bath for one hour (reflux condenser). The hot solution is poured into a beaker and allowed to cool slowly; the crystals separating out are filtered off, washed with alcohol, and dried on the water-bath. For conversion into benzil (see next preparation), they need not be recrystallised. In order to obtain perfectly pure benzoïn, a small portion of the crude product is recrystallised from a little alcohol in a test-tube. Melting point, 134°. Yield, about 90% of the theory.

If an aromatic aldehyde of the type of benzaldehyde is warmed in alcohol solution with potassium cyanide, substances are obtained which possess the same composition, but with double the molecular weight of the aldehyde:

$$2\,C_6H_5 \cdot CHO = \underset{\text{Benzoïn}}{C_6H_5 \cdot CO \cdot \underset{\underset{OH}{|}}{CH} - C_6H_5}$$

Since a small quantity of potassium cyanide has the power to condense large quantities of the aldehyde, the reaction may possibly take place as indicated by the equations below. In the first phase, one molecule of aldehyde reacts with one molecule of potassium cyanide:

(1) $C_6H_5 \cdot CO\boxed{H + CN}K = C_6H_5 \cdot COK + HCN.$
　　　　　　Potassium benzaldehyde

The hydrocyanic acid then reacts with a second molecule of alde-

---

[1] A, 198, 150.

hyde, forming an addition-product, mandelic nitrile, in the second phase:

$$(2)\ C_6H_5 \cdot CHO + HCN = \underset{\text{Mandelic nitrile}}{\underset{OH}{\overset{C_6H_5 \cdot CH-CN}{|}}}.$$

In the third phase a molecule of potassium benzaldehyde acts upon a molecule of the nitrile with the elimination of potassium cyanide, and the formation of the benzoïn:

$$(3)\ C_6H_5 \cdot COK + \underset{OH}{\overset{C_6H_5 \cdot CH \cdot CN}{|}} = \underset{OH}{\overset{C_6H_5 \cdot CO \cdot CH \cdot C_6H_5}{|}} + KCN.$$

The molecule of potassium cyanide required for equation (1) is thus again formed, and may condense two more molecules of aldehyde, etc.

In the same way from anisic aldehyde and cuminol, there is obtained anisoïn and cuminoïn, respectively:

$$2\ p\text{-}C_6H_4\!\!\begin{array}{c}\diagup OCH_3\\ \diagdown CHO\end{array} = CH_3O \cdot C_6H_4 \cdot CO \cdot \underset{OH}{\overset{CH}{|}} \cdot C_6H_4 \cdot OCH_3$$

Anisic aldehyde            Anisoïn

$$2\ p\text{-}C_6H_4\!\!\begin{array}{c}\diagup C_3H_7\\ \diagdown CHO\end{array} = C_3H_7 \cdot C_6H_4 \cdot CO \cdot \underset{OH}{\overset{CH}{|}} - C_6H_4 \cdot C_3H_7.$$

Cuminol            Cuminoïn

With potassium cyanide furfurol yields furoïn:

$$2\ \underset{\text{Furfurol}}{C_4H_3O \cdot CHO} = \underset{\text{Furoïn}}{\underset{OH}{\overset{C_4H_3O \cdot CO \cdot CH \cdot C_4H_3O}{|}}}$$

Benzoïn and its analogues are derivatives of the hydrocarbon dibenzyl, $C_6H_5 \cdot CH_2 \cdot CH_2 \cdot C_6H_5$, and in fact benzoïn on reduction with hydriodic acid is converted into this hydrocarbon.

The benzoïns act, on the one hand, like ketones if the carbonyl group (CO) takes part in the reaction, and, on the other hand, like secondary alcohols if the group CH.OH (the secondary alcohol group) reacts. Thus they have the power to form oximes and hydrazones with hydroxylamine and phenyl hydrazine respectively. If benzoïn is reduced with sodium amalgam, the ketone group is converted into the secondary alcohol group:

$$\underset{\underset{OH}{|}}{C_6H_5 \cdot CO \cdot CH \cdot C_6H_5} + H_2 = \underset{\underset{OH\ \ OH}{|\ \ \ |}}{C_6H_5 \cdot CH-CH \cdot C_6H_5}$$
Hydrobenzoïn

If the reduction is effected by zinc and hydrochloric acid or glacial acetic acid, the carbonyl group is not attacked, but the alcohol group is reduced and desoxybenzoïn is obtained:

$$\underset{\underset{OH}{|}}{C_6H_5 \cdot CO\ \ CH \cdot C_6H_5} + H_2 = C_6H_5 \cdot CO \cdot CH_2 \cdot C_6H_5 + H_2O,$$
Desoxybenzoïn

a compound of especial interest, because in it, as in acetacetic ester, one of the two hydrogen atoms of the methylene group ($CH_2$), in consequence of the acidifying influence of the adjoining negative carbonyl and phenyl groups, may be replaced by sodium; with the sodium compound the same kind of syntheses may be effected as with acetacetic ester:

$$\underset{\underset{Na}{|}}{C_6H_5 \cdot CO \cdot CH \cdot C_6H_5} + IC_2H_5 = \underset{\underset{C_2H_5}{|}}{C_6H_5 \cdot CO \cdot CH-C_6H_5} + NaI.$$

Sodium desoxybenzoïn        Ethyl desoxybenzoïn

Benzoïn, further, acts as an alcohol, the hydroxyl group being capable of reacting with alkyl- and acid-radicals to form ethers and esters. If oxidizing agents act on benzoïn, the alcohol group is oxidized to a ketone group, as is the case with all secondary alcohols:

$$\underset{\underset{OH}{|}}{C_6H_5 \cdot CO \cdot CH \cdot C_6H_5} + O = C_6H_5 \cdot CO \cdot CO \cdot C_6H_5 + H_2O.$$
Dibenzoyl = Benzil

The next preparation will deal with this reaction.

### 26. REACTION: OXIDATION OF A BENZOÏN TO A BENZIL

EXAMPLE: Benzil from Benzoïn

The crude benzoïn obtained in the preceding preparation is finely pulverised after drying, and heated in an open flask, with frequent shaking, with twice its weight of pure concentrated nitric acid, for $1\frac{1}{2}$–2 hours on a rapidly boiling water-bath. When the oxidation is ended, the reaction mixture is poured into cold water;

after the mass solidifies the nitric acid is poured off; it is then washed several times with water, pressed out on a porous plate and crystallised from alcohol. After filtering off the crystals separating out, they are dried in the air on several layers of filter-paper and *not* on the water-bath. Melting-point, 95°. Yield, about 90 % of the theory.

The equation representing the oxidation of benzoïn to benzil has been given under the preceding preparation. The analogues of benzoïn also give, on oxidation, compounds of the benzil series. Thus from anisoïn and cuminoïn, anisil and cuminil respectively are obtained:

$$CH_3O . C_6H_4 . CO . CO . C_6H_4 . OCH_3; \quad C_3H_7 . C_6H_4 . CO . CO . C_6H_4 . C_3H_7.$$
Anisil  Cuminil

Benzil acts like a ketone in that it forms oximes with hydroxylamine. The oximes are of exceptional interest, since our knowledge of the stereochemistry of nitrogen proceeds from them. Benzil forms *two* monoximes and *three* dioximes. The constitution of these compounds will be discussed later, under the preparation of benzophenone-oxime.

On fusion with potassium hydroxide or by long heating with a water solution of potassium hydroxide, benzil undergoes a remarkable change, in that by taking up water it passes over to the so-called benzilic acid:

$$C_6H_5 . CO . CO . C_6H_5 + H_2O = \begin{matrix} C_6H_5 \\ C_6H_5 \end{matrix} \!\! > \!\! \underset{OH}{C} . CO . OH.$$

Diphenylglycolic acid = Benzilic acid

Anisil and cuminil also yield, in a similar way, anisilic and cuminilic acids.

### 27. REACTION: THE ADDITION OF HYDROCYANIC ACID TO AN ALDEHYDE

EXAMPLE : **Mandelic Acid from Benzaldehyde**[1]

(a) *Mandelic Nitrile*

In a flask containing 13 grammes of finely pulverised 100 % potassium cyanide, or an equivalent amount of the purest salt

---

[1] B. 14, 235

available, pour 20 grammes of freshly distilled benzaldehyde, and add to this from a separating funnel, the flask being cooled with ice, a quantity of the most concentrated hydrochloric acid, corresponding to 7 grammes of anhydrous hydrochloric acid (about 20 grammes concentrated acid), drop by drop, with frequent shaking. The reaction-mixture is then allowed to stand, with frequent shaking, for one hour, then poured into about 5 volumes of water, the oil washed with water several times, and finally separated in a dropping-funnel. Owing to the ease with which the nitriles decompose, a further purification is not possible. Yield, almost quantitative.

(b) *Saponification of the Nitrile*

The nitrile is mixed with three times its weight of concentrated hydrochloric acid in a porcelain dish, and evaporated on the sand-bath until crystals begin to separate out on the upper surface of the liquid. Enough water to dissolve the residue is then added, the solution filtered, if necessary, from any undissolved oil; the easily soluble mandelic acid is extracted with ether several times, the latter is evaporated, and the residue warmed in a watch-glass on the water-bath for some time, in order to free it from water. The mandelic acid remaining solidifies on cooling, and can be obtained pure by recrystallisation from benzene. Melting-point, 118°. Yield, about 10–15 grammes.

Hydrocyanic acid unites with aromatic as well as aliphatic aldehydes and ketones with the formation of α-oxyacid nitriles:

$$CH_3 \cdot CHO + HCN = CH_3 \cdot CH \begin{smallmatrix} OH \\ CN \end{smallmatrix},$$

Aldehydecyanhydrine
α-lactic nitrile

$$C_2H_5 \cdot CO \cdot C_2H_5 + HCN = \begin{smallmatrix} C_2H_5 \\ C_2H_5 \end{smallmatrix} C \begin{smallmatrix} OH \\ CN \end{smallmatrix},$$

Diethylketone   Diethylglycolic nitrile

$$C_6H_5 \cdot CHO + HCN = C_6H_5 \cdot CH \begin{smallmatrix} OH \\ CN \end{smallmatrix},$$

Benzaldehyde    Mandelic nitrile

# AROMATIC SERIES

$$C_6H_5 \cdot CO \cdot CH_3 + HCN = \underset{CH_3}{\overset{C_6H_5}{>}}C\underset{CN}{\overset{OH}{<}}$$

Acetophenone          Acetophenonecyanhydrine

This reaction also takes place with more complex compounds containing the carbonyl group:

$$CH_3 \cdot CO \cdot CH_2 \cdot CO \cdot OC_2H_5 + HCN = CH_3 \cdot \underset{OH\ \ CN}{\overset{\wedge}{C}} - CH_2 \cdot CO \cdot OC_2H_5,$$

Acetacetic ester

$$CH_3 \cdot CO \cdot CO \cdot OH + HCN = CH_3 \cdot \underset{OH\ \ CN}{\overset{\wedge}{C}} - CO \cdot OH,$$

Pyroracemic acid

*a*-Cyan-*a*-lactic acid

$$C_6H_5 \cdot CO \cdot CH_2 \cdot OH + HCN = C_6H_5 \cdot \underset{OH\ \ CN}{\overset{\wedge}{C}} - CH_2 \cdot OH.$$

Benzoylcarbinol

The reaction may be effected by digestion with already prepared hydrocyanic acid at ordinary or higher temperatures, but in most cases it is more advantageous to employ nascent hydrocyanic acid as above.

If the oxynitriles are subjected to saponification, for example, by boiling with hydrochloric acid, the free oxyacid is obtained, *e.g.*:

$$C_6H_5 \cdot CH{\overset{OH}{\underset{CN}{<}}} + 2\ H_2O + HCl = C_6H_5 \cdot CH{\overset{OH}{\underset{CO \cdot OH}{<}}} + NH_4Cl.$$

Mandelic nitrile          Mandelic acid

Since the cyanhydrine reaction takes place smoothly in most cases, it is frequently used for the preparation of *a*-oxacids.

Thus in the sugar group the cyanhydrine reaction is of extreme importance, not only for its value in determining constitution, but also for the syntheses of sugars or sugar-like substances containing long chains of carbon atoms.

In reference to the latter one example may be mentioned. If hydrocyanic acid is united with grape sugar, which is an aldehyde, there is first obtained an oxynitrile, which on saponification yields an oxyacid. If this, or rather the inner anhydride (lactone) into which it easily passes, is reduced, the carboxyl group is reduced to an aldehyde group, and there is thus obtained a sugar, containing one more secondary alcohol (CHOH) group than the original grape sugar:

$$
\begin{array}{c}
\text{CHO} \\
| \\
(\text{CH.OH})_4 + \text{HCN} = \\
| \\
\text{CH}_2.\text{OH}
\end{array}
\begin{array}{c}
\text{CN} \\
| \\
\text{CH.OH} \\
| \\
(\text{CH.OH})_4 \\
| \\
\text{CH}_2.\text{OH}
\end{array}
\xrightarrow{\text{saponified}}
\begin{array}{c}
\text{CO.OH} \\
| \\
\text{CH.OH} \\
| \\
(\text{CH.OH})_4 \\
| \\
\text{CH}_2.\text{OH}
\end{array}
\xrightarrow{\text{reduced}}
\begin{array}{c}
\text{CHO} \\
| \\
(\text{CH.OH})_5. \\
| \\
\text{CH}_2.\text{OH}
\end{array}
$$

Aldoheptose

With the substance thus obtained a similar reaction may be carried out, etc.

Mandelic acid belongs to the class of substances containing an asymmetric carbon atom, *i.e.* one which is in combination with four different substituents:

$$
*C\begin{cases} C_6H_5 \\ H \\ OH \\ CO.OH \end{cases}
$$

Like all compounds of this class, it exists in two different space modifications, which bear the same relation to each other as does an object and its image, and owing to their power of revolving the plane of polarisation, are called dextro- and lævo-mandelic acids. The acid obtained in the above synthesis is optically inactive; since, in the synthesis of compounds with an asymmetric carbon atom from inactive substances, an equal number of molecules of the dextro- and lævo-varieties are always obtained, which, in the above case, unite to form the inactive, so-called, para-mandelic acid. But, by different methods, the active acids can be obtained from the inactive modifications. If, *e.g.*, the cinchonine salt of para-mandelic acid is allowed to crystallise, the more difficultly soluble salt of the dextro-acid separates out first, and then, later, the lævo-salt crystallises. By treatment with acids, the free active acids may be obtained.

With the aid of certain micro-organisms, the inactive compounds may be decomposed into their active constituents. If, *e.g.*, the well-known *Penicillium glaucum* is allowed to grow in a solution of ammonium para-mandelate, it destroys the lævo-modification; while another organism, *Saccharomyces, ellipsoïdeus* consumes the dextro-modification, and leaves the other.

## 28. REACTION: PERKIN'S SYNTHESIS OF CINNAMIC ACID [1]

EXAMPLE: Cinnamic Acid from Benzaldehyde and Acetic Acid

A mixture of 20 grammes of benzaldehyde, 30 grammes of acetic anhydride, both freshly distilled, and 10 grammes of anhydrous pulverised sodium acetate (for the preparation, see page 114), is heated in a flask provided with a wide vertical air-condenser about 60 cm. long, for 8 hours, in an oil-bath at 180°. If the experiment cannot be completed in one day, a calcium chloride tube is placed in the upper end of the condenser over night. After the reaction is complete, the hot reaction-product is poured into a large flask; add water, and then distil with steam, until no more benzaldehyde passes over. The quantity of water used here is large enough so that all of the cinnamic acid dissolves except a small portion of an oily impurity. The solution is then boiled a short time, with some animal charcoal, and filtered; on cooling, the cinnamic acid separates out in lustrous leaves. Should it not possess the correct melting-point, it is recrystallised from hot water. Melting-point, 133°. Yield, about 15 grammes.

The reaction involved in the Perkin synthesis takes place in accordance with this equation:

$$C_6H_5 \cdot CHO + CH_3 \cdot CO \cdot ONa = C_6H_5 \cdot CH{=}CH \cdot CO \cdot ONa + H_2O.$$
Sodium cinnamate

The reaction, however, does not take place, as appears from the equation, by the direct union of the aldehyde-oxygen atom with the hydrogen atoms of the methyl group and a combination of the resulting residues, but it proceeds in two phases.

In the first, the sodium acetate unites with the aldehyde, forming sodium phenyl lactate:

$$(1) \quad C_6H_5 \cdot CHO + CH_3 \cdot CO \cdot ONa = \underset{\text{Sodium phenyl lactate}}{C_6H_5 \cdot \underset{\underset{OH}{|}}{CH} \cdot CH_2 \cdot CO \cdot ONa.}$$

---

[1] J. 1877, 789; B. 10, 68; 16, 1436; A. 227, 48.

In the second phase, this salt, under the influence of acetic anhydride, loses water, upon which the sodium cinnamate is formed:

(2) $C_6H_5.CH(OH).CH_2.CO.ONa = C_6H_5.CH{=}CH.CO.ONa + H_2O$.

That sodium acetate, and not the acetic anhydride, condenses with the benzaldehyde, is proved by the following facts: If, instead of sodium acetate, sodium proprionate is used, and this is heated with benzaldehyde and acetic anhydride, cinnamic acid is not obtained, but methyl cinnamic acid:

(1) $C_6H_5.CHO + CH_3.CH_2.CO.ONa = C_6H_5.CH(OH){-}CH(CH_3){-}CO.ONa$,

(2) $C_6H_5.CH(OH){-}CH(CH_3).CO.ONa = C_6H_5.CH{=}C(CH_3){-}CO.ONa + H_2O$.

Sodium phenyl cinnamate

It follows from this that the sodium salt used always takes part in the reaction. In the experiment it is of course necessary that the fusion is not carried out at so high a temperature as in the above example, but only at the heat of the water-bath; at higher temperatures the sodium salt of proprionic acid and acetic anhydride decompose into sodium acetate and proprionic anhydride, so that cinnamic acid is obtained, and therefore, *apparently*, the anhydride reacts with the aldehyde.

The Perkin reaction is capable of numerous modifications, since in place of benzaldehyde, its homologues, its nitro- and oxy-derivatives, etc., may be used. On the other hand, the homologues of sodium acetate may be used as has been pointed out. The condensation in these cases always takes place at the carbon atom adjoining the carboxyl group. Halogen substituted aliphatic acids will also react; thus from benzaldehyde and chloracetic acid, chlorcinnamic acid is obtained:

$C_6H_5.CHO + CH_2Cl.CO.OH = C_6H_5.CH{=}CCl.CO.OH + H_2O$.

In place of the aliphatic homologues of acetic acid the aromatic substituted acetic acids can also be used, *e.g.*:

$C_6H_5.CHO + C_6H_5.CH_2.CO.OH = C_6H_5.CH{=}C(C_6H_5){-}CO.OH$.

Phenyl acetic acid     Phenyl cinnamic acid

These examples are sufficient to show the wide application of the Perkin reaction.

A very similar reaction takes place on heating sodium acetate with the cheaper benzalchloride, instead of benzaldehyde:

$$C_6H_5.CHCl_2 + CH_3.CO.ONa = C_6H_5.CH{=}CH.CO.ONa + 2HCl.$$

Cinnamic acid, its homologues and analogues, behave on the one hand like acids, since they form salts, esters, chlorides, amides, etc. Further, they show the properties of the ethylene series in that they take up by addition the most various kinds of atoms and groups. By the action of nascent hydrogen two atoms of hydrogen are added to the molecule of cinnamic acid with a change from double to single union:

$$C_6H_5.CH{=}CH.CO.OH + H_2 = C_6H_5.CH_2.CH_2.CO.OH.$$
Hydrocinnamic acid

It also combines with chlorine and bromine:

$$C_6H_5.CH{=}CH.CO.OH + Cl_2 = C_6H_5.CHCl.CHCl.CO.OH,$$
Dichlorhydrocinnamic acid

$$C_6H_5.CH{=}CH.CO.OH + Br_2 = C_6H_5.CHBr.CHBr.CO.OH.$$
Dibromhydrocinnamic acid

Further, it unites with hydrochloric, hydrobromic, and hydriodic acids, e.g.:

$$C_6H_5.CH{=}CH.CO.OH + HBr = C_6H_5.CHBr.CH_2.CO.OH.$$
β-bromhydrocinnamic acid

The halogen atom in these cases always unites with the carbon atom not adjoining the carboxyl group.

Hypochlorous acid also unites with cinnamic acid with the formation of phenylchlorlactic acid:

$$C_6H_5.CH{=}CH.CO.OH + ClOH = \begin{array}{c} C_6H_5.CH{-}CHCl.CO.OH. \\ | \\ OH \end{array}$$

The o-nitrocinnamic acid from which indigo is synthetically prepared is of technical importance. If cinnamic acid, or better, an ester of it, is nitrated, a mixture of the o- and p-nitroderivatives is obtained which can be separated into its constituents. If bromine is allowed to act on the o-nitrocinnamic acid, there is obtained:

$$O{-}C_6H_4{\Big<}\begin{array}{l}NO_2 \\ CHBr{-}CHBr.CO.OH\end{array}.$$

If this acid is boiled with alcoholic potash, two molecules of hydrobromic acid are split off as in the preparation of acetylene from ethylene bromide, and o-nitrophenylpropriolic acid is formed, which, with alkaline reducing agents, yields indigo, and is used in indigo printing:

$$o - C_6H_4 \begin{array}{c} NO_2 \\ C \equiv C . CO . OH \end{array}$$

By the decomposition of an alkaloid found with cocaine, a stereoisomeric cinnamic acid (allocinnamic acid) is obtained, which bears the same relation to cinnamic acid that maleïc acid does to fumaric acid:

$$\begin{array}{cc} C_6H_5.C.H & C_6H_5.CH \\ \parallel & \parallel \\ H.C.CO.OH & HO.OC.CH \\ \text{Cinnamic acid} & \text{Allocinnamic acid} \end{array}$$

## 29. REACTION: ADDITION OF HYDROGEN TO AN ETHYLENE DERIVATIVE

EXAMPLE: **Hydrocinnamic Acid from Cinnamic Acid**

In a glass-stoppered cylinder, or a thick-walled preparation glass, treat 10 grammes of cinnamic acid with 75 c.c. of water; add a dilute solution of caustic soda until the acid passes into solution and the liquid is *just alkaline*. It is then treated gradually with about 200 grammes of 2 % sodium amalgam, and heated gently, as soon as this has become liquid, on the water-bath for a short time. The liquid is then decanted from the mercury and acidified, upon which the hydrocinnamic acid separates out as an oil; when cooled with ice-water and rubbed with a glass rod, it solidifies to a crystalline mass. After pressing it out on a porous plate, the acid is recrystallised from water. Since it possesses a low melting-point, it may separate out as an oil on cooling, in which case proceed according to the directions given on page 8. Melting-point, 47°.

The equation for the reaction has been given under cinnamic acid. The same reaction also takes place on heating with hydriodic acid and red phosphorus. The acid does not show any noteworthy reactions.

## 30. REACTION: PREPARATION OF AN AROMATIC ACID-CHLORIDE FROM THE ACID AND PHOSPHORUS PENTACHLORIDE

EXAMPLE: **Benzoyl Chloride from Benzoïc Acid**[1]

Treat 50 grammes of benzoïc acid in a dry $\frac{1}{2}$-litre flask, with 90 grammes of finely pulverised phosphorus pentachloride under the hood; the two are shaken well together, upon which, after a short time, reaction takes place with energetic evolution of hydrochloric acid, and the mass becomes liquid. In order to prevent the vessel, which has become strongly heated by the reaction, from cracking, it is not placed on the cold stone floor of the hood, but on a wooden block or straw ring. After standing a short time, the completely liquid mixture is twice fractionated (under the hood) with the use of a wide, long air condenser, observing the directions given on pages 22 and 23. Boiling-point of benzoyl chloride, 200°. Yield, 90 % of the theory.

The formation of benzoyl chloride takes place in accordance with the following reaction:

$$C_6H_5 \cdot CO \cdot OH + PCl_5 = C_6H_5 \cdot CO \cdot Cl + POCl_3 + HCl.$$

It has been mentioned under acetyl chloride that, for the preparation of the aromatic acid-chlorides, phosphorus pentachloride is generally used. Benzoyl chloride differs from acetyl chloride in that it is more difficultly decomposed by water.

EXPERIMENT: Treat $\frac{1}{2}$ c.c. of benzoyl chloride with 5 c.c. of water, and shake. While acetyl chloride, under these conditions, decomposes violently, the benzoyl chloride is scarcely changed. It is then warmed somewhat. It must be subjected to a longer heating until all the oil has been decomposed.

In other respects, benzoyl chloride is a wholly normal acid-chloride, and what was said under acetyl chloride is applicable to this chloride; only it is possible to prepare aromatic amides by a different method from that used for the preparation of acetamide.

EXPERIMENT: In a porcelain dish, 15 grammes of finely pulverised ammonium carbonate is treated with 5 grammes of benzoyl

---

[1] A. 3, 262.

chloride; they are intimately mixed with a glass rod and heated on the water-bath until the odour of the acid-chloride has vanished. The mixture is then diluted with water, filtered and washed with water, and crystallised from water. Melting-point of benzamide, 128°.

$$C_6H_5 \cdot CO \cdot Cl + NH_3 = C_6H_5 \cdot CO \cdot NH_2 + HCl.$$

Since the aromatic amides are generally insoluble in water, they are usually prepared by the method just given, and not, as in the case of acetamide, by heating the ammonium salt of the acid.

### 31. REACTION: THE SCHOTTEN-BAUMANN REACTION FOR THE RECOGNITION OF COMPOUNDS CONTAINING THE AMIDO-, IMIDO-, OR HYDROXYL-GROUP.

EXAMPLE: **Benzoïcphenyl Ester from Phenol and Benzoylchloride.**[1]

Dissolve a small quantity of crystallised phenol (about $\frac{1}{2}$ gramme) in 5 c.c. of water in a test-tube and add $\frac{1}{2}$ c.c. of benzoyl chloride; make the solution alkaline with a solution of caustic soda and, with shaking, heat gently a short time over a free flame. If the reaction-mixture is cooled by water and then shaken and the sides of the tube rubbed with a glass rod, the oil separating out solidifies to colourless crystals, which are filtered off with suction, washed with water, pressed out on a porous plate and recrystallised from a little alcohol. Melting-point, 68–69°.

As already mentioned under acetyl chloride, acid-chlorides react with alcohols, phenols, primary and secondary amines, the chlorine atom uniting with the hydrogen of the hydroxyl-, amido-, or imido-group, with the elimination of hydrochloric acid, while the residues combine to form an ester or a substituted amide. The value of the Schotten-Baumann reaction depends on the fact that this reaction is so essentially facilitated by the presence of sodium hydroxide or potassium hydroxide, that even in the presence of water the decomposition takes place, which in the absence of alkalies is not possible:

$$C_6H_5 \cdot OH + C_6H_5 \cdot CO \cdot Cl + NaOH = C_6H_5 \cdot O \cdot OC \cdot C_6H_5 + NaCl + H_2O.$$

[1] B. 19, 3218; 21, 2744; 23, 2962; 17, 2545.

The reaction is of great importance, especially for the recognition and characterisation of soluble compounds containing the groups mentioned above. It is obvious that if it is desired to test even small quantities of those compounds, the most difficultly soluble acid derivatives of them must be prepared. The benzoyl derivatives are particularly well adapted to this purpose. A few examples may render this statement clearer: If a water solution of a poly-acid aliphatic alcohol, *e.g.*, glycerol, or of the various sugars, from which the dissolved substance will only separate with difficulty, is treated with benzoyl chloride and alkali, a benzoate is formed, which is generally insoluble in water, and which can be recognised by its melting-point. For the recognition of primary and secondary amines the method of procedure is the same. Thus, *e.g.*, it is not difficult to convert aniline (one drop dissolved in water) by the above method to benzanilide, which can be recognised by its melting-point, $161°$. (Try the experiment.)

$$C_6H_5.NH_2 + C_6H_5.CO.Cl + NaOH = C_6H_5.NH.CO.C_6H_5 + H_2O + NaCl.$$

The soluble amido-phenols, di- and poly-amines are also converted into difficultly soluble benzoyl derivatives:

$$C_6H_4\diagup\!\!\!\!{}^{NH_2}_{OH} + 2\,C_6H_5.CO.Cl = C_6H_4\diagup\!\!\!\!{}^{NH.CO.C_6H_5}_{O.OC.C_6H_5} + 2\,HCl,$$

$$C_6H_4\diagup\!\!\!\!{}^{NH_2}_{NH_2} + 2\,C_6H_5.CO.Cl = C_6H_4\diagup\!\!\!\!{}^{NH.CO.C_6H_5}_{NH.CO.C_6H_5} + 2\,HCl.$$

In place of benzoyl chloride, other chlorides, *e.g.*, phenylacetyl chloride, or benzenesulphon chloride, can be used, which act in a similar way. Acetyl derivatives may also be prepared in the presence of alkalies in water solution, only in this case acetic anhydride and not the easily decomposed acetyl chloride is used. At times the reaction takes place better by using potassium hydroxide in place of sodium hydroxide.

**32. REACTION:** (*a*) FRIEDEL AND CRAFTS' KETONE SYNTHESIS [1]
(*b*) PREPARATION OF AN OXIME
(*c*) BECKMANN'S TRANSFORMATION OF AN OXIME

EXAMPLE: **Benzophenone from Benzoylchloride, Benzene and Aluminium Chloride**

(*a*) To a mixture of 30 grammes of benzene, 30 grammes of benzoyl chloride, and 100 c.c. (= 130 grammes) of carbon disulphide in a dry flask, add, in the course of about 10 minutes, with frequent shaking, 30 grammes of freshly prepared and finely pulverised aluminium chloride, which is weighed in a dry test-tube closed by a cork. The flask is then connected with a long reflux condenser, and heated on a gently boiling water-bath until only small amounts of hydrochloric acid are evolved: this will require about 2–3 hours. The carbon disulphide is then distilled off, and the residue, while still warm, is carefully poured into a large flask containing 300 c.c. of water and small pieces of ice. The residue adhering to the walls of the first flask is treated with water, and the water added to the main quantity. After the reaction-mixture has been treated with 10 c.c. of concentrated hydrochloric acid, steam is passed into it for about a quarter-hour. The residue remaining in the flask is, after cooling, extracted with ether, the ethereal solution washed several times with water, filtered, and shaken up with dilute caustic soda solution. After drying with calcium chloride, the ether is evaporated, and the residue distilled from a fractionating flask, the side-tube of which is as near as possible to the bulb. Boiling-point, 297°. Yield, about 30 grammes.

(*b*) A solution of 2 grammes of benzophenone in 15 c.c. of alcohol is, with cooling, treated with a cold solution of 2.5 grammes of hydroxylamine hydrochloride in 5 c.c. of water, and 6 grammes of caustic potash in 6 grammes of water; the mixture is heated two hours on the water-bath, with a reflux condenser. Then add 50 c.c. of water, and filter off, if necessary, any un-

---
[1] A. ch. [6] 1, 518.

changed ketone which balls together very easily on shaking; acidify the filtrate slightly with dilute sulphuric acid, and recrystallise the free oxime from alcohol. Melting-point, 140°.

(c) A weighed amount of the oxime is dissolved in some anhydrous, alcohol-free ether, at the ordinary temperature, and gradually treated with $1\frac{1}{2}$ times its weight of finely pulverised phosphorus pentachloride. The ether is then distilled off, the residue, with cooling, is treated with water, and the precipitate separating out is recrystallised from alcohol. Melting-point, 163°.

(a) If an aromatic or an aliphatic acid-chloride is allowed to act on an aromatic hydrocarbon in the presence of an anhydrous aluminium chloride, one of the benzene-hydrogen atoms will be replaced by an acid radical, a *ketone* being formed:

$$C_6H_6 + C_6H_5.CO.Cl = C_6H_5.CO.C_6H_5 + HCl,$$
Diphenyl ketone<br>=Benzophenone

$$C_6H_6 + CH_3.CO.Cl = C_6H_5.CO.CH_3 + HCl.$$
Phenylmethyl ketone<br>=Acetophenone

The reaction may be varied if (1) in place of benzene a homologue is used:

$$C_6H_5.CH_3 + C_6H_5.CO.Cl = p\text{-}C_6H_4\diagup\!\!\!\begin{array}{l}CH_3\\CO.CH_3\end{array} + HCl.$$
Toluene          Phenyltolyl ketone

In cases of this kind, the acid-radical always enters the para-position to the alkyl radical. If this is already occupied, it then goes to the ortho-position. (2) In place of hydrocarbons, phenol-ethers, which react with extreme ease, can be used:

$$C_6H_5.OCH_3 + C_6H_5.CO.Cl = C_6H_4\diagup\!\!\!\begin{array}{l}OCH_3\\CO.C_6H_5\end{array} + HCl.$$
Anisol          Anisylphenyl ketone

Concerning the entrance of the acid-radical, the statements made above are also true for this case. (3) In place of benzoyl chloride or acetyl chloride, their homologues can be used:

$$C_6H_6 + C_6H_4\diagup\!\!\!\begin{array}{l}CH_3\\CO.Cl\end{array} = C_6H_5.CO.C_6H_4.CH_3 + HCl,$$
Toluyl chloride

T

$$C_6H_6 + CH_3 \cdot CH_2 \cdot CO \cdot Cl = C_6H_5 \cdot CO \cdot CH_2 \cdot CH_3 + HCl,$$

$$C_6H_6 + C_6H_5 \cdot CH_2 \cdot CO \cdot Cl = C_6H_5 \cdot CO \cdot CH_2 \cdot C_6H_5 + HCl.$$
Phenylacetyl chloride      Phenylbenzyl ketone = desoxybenzoïn

In this way, starting from o- or m-toluic acid, the o- or m-tolylphenyl ketone can be prepared; it cannot be obtained by the action of benzoyl chloride on toluene. (4) Substituted acid-chlorides like brombenzoyl chloride, nitrobenzoyl chloride, etc., can be used, and thus halogen or nitroketones are obtained:

$$C_6H_6 + C_6H_4\!\!<^{Br}_{CO \cdot Cl} = C_6H_5 \cdot CO \cdot C_6H_4 \cdot Br + HCl,$$
Brombenzoyl chloride

$$C_6H_6 + C_6H_4\!\!<^{NO_2}_{CO \cdot Cl} = C_6H_5 \cdot CO \cdot C_6H_4 \cdot NO_2 + HCl.$$
Nitrobenzoyl chloride

(5) Finally, the chlorides of dibasic acids react with the formation of diketones or ketonic acids:

$$\begin{array}{l} CH_2\!-\!CO \cdot Cl \\ |\\ CH_2\!-\!CO \cdot Cl \end{array} + 2\,C_6H_6 = \begin{array}{l} CH_2 \cdot CO \cdot C_6H_5 \\ |\\ CH_2 \cdot CO \cdot C_6H_5 \end{array} + 2\,HCl,$$
Succinic chloride

$$\text{m- and p-}C_6H_4\!\!<^{CO \cdot Cl}_{CO \cdot Cl} + 2\,C_6H_6 = C_6H_4\!\!<^{CO \cdot C_6H_5}_{CO \cdot C_6H_5} + 2\,HCl,$$
Iso- and tere-phthalyl chloride

$$CO\!\!<^{Cl}_{Cl} + 2\,C_6H_6 = C_6H_5 \cdot CO \cdot C_6H_5 + 2\,HCl.$$
Phosgene            Benzophenone

In these reactions if but one chlorine atom should react, the chlorides of the three following acids would be obtained:

$$\begin{array}{l} CH_2 \cdot CO \cdot C_6H_5 \\ |\\ CH_2 \cdot CO \cdot OH \end{array}, \quad C_6H_4\!\!<^{CO \cdot C_6H_5}_{CO \cdot OH}, \quad C_6H_5 \cdot CO \cdot OH.$$
Benzoylproprionic acid    Benzoylbenzoic acid    Benzoïc acid

AROMATIC SERIES 275

From the chloride of phthalic acid phthalophenone is formed, important on account of its relation to the fluoresceïn dyes:

$$C_6H_4\diagup\genfrac{}{}{0pt}{}{CCl_2}{CO}\diagdown + 2\,C_6H_6 = C_6H_4\diagup\genfrac{}{}{0pt}{}{C\diagup\genfrac{}{}{0pt}{}{C_6H_5}{C_6H_5}}{CO}\diagdown O + 2\,HCl.$$

Phthalophenone

Michler's ketone, tetramethyldiamidobenzophenone is of technical importance; it is obtained from dimethyl aniline and phosgene, and is used in the preparation of dyes of the fuchsine series (see Crystal Violet):

$$2\,C_6H_5\cdot N(CH_3)_2 + COCl_2 = CO\diagup\genfrac{}{}{0pt}{}{C_6H_4\cdot N(CH_3)_2}{C_6H_4\cdot N(CH_3)_2} + 2\,HCl.$$

The Friedel-Crafts reaction can also be used for the preparation of the homologous aromatic hydrocarbons, since in place of the acid-chloride, halogen alkyls may be caused to act on the hydrocarbons:[1]

$$C_6H_6 + C_2H_5Br = C_6H_5\cdot C_2H_5 + HBr,$$

$$C_6H_5\cdot CH_3 + CH_3Cl = C_6H_4\diagup\genfrac{}{}{0pt}{}{CH_3}{CH_3} + HCl.$$

Toluene            Xylene

But in this connection the reaction is in many cases, and indeed in the simplest case, not of equal importance with its application for the ketone syntheses, for three reasons: First, the product of the reaction is a hydrocarbon which can again react; thus it is often difficult to limit the reaction to the desired point. For example, in the action of methyl chloride on toluene, not only is one hydrogen atom substituted, with the formation of dimethyl benzene, but varying quantities of tri-, tetra-, penta-, and hexa-methyl benzene are also formed. A second disadvantage is this: In the different series a mixture of isomers is obtained; in the above example, e.g., not only one of the three dimethyl benzenes, but a mixture of the o-, m-, and p-varieties is formed, which cannot be separated like the homologues by fractional distillation. The reaction is still further complicated in that the aluminium chloride partially splits off the alkyl groups:

$$C_6H_5\cdot CH_3 + HCl = C_6H_6 + CH_3Cl.$$

---

[1] B. 14, 2627.

Since the lower homologues thus formed again react synthetically with the halogen alkyls, and the halogen alkyls on elimination also take part in the reaction, mixtures often difficult to separate are formed. In some favourable cases the reaction is of use in the preparation of the homologues of benzene. The reaction is also applicable to aromatic chlorides which contain the halogen in the side-chain:

$$C_6H_5.CH_2.Cl + C_6H_6 = C_6H_5.CH_2.C_6H_5 + HCl,$$
Benzyl chloride             Diphenyl methane

$$NO_2.C_6H_4.CH_2.Cl + C_6H_6 = NO_2.C_6H_4.CH_2.C_6H_5 + HCl.$$
Nitrodiphenyl methane

As the chlorides of dibasic acids yield diketones, the alkylene chlorides or bromides, as well as tri- and tetra-halogen substituted hydrocarbons, can react with several hydrocarbon molecules, e.g.:

$$2\,C_6H_6 + CH_2Br-CH_2Br = C_6H_5.CH_2.CH_2.C_6H_5 + 2\,HBr,$$
Dibenzyl = Diphenyl ethane

$$3\,C_6H_6 + CHCl_3 = CH.(C_6H_5)_3 + 3\,HCl,$$
Chloroform       Triphenyl methane

$$4\,C_6H_6 + CHBr_2-CHBr_2 = \underset{\underset{C_6H_5}{|}}{\overset{\overset{C_6H_5}{|}}{CH}} - \underset{\underset{C_6H_5}{|}}{\overset{\overset{C_6H_5}{|}}{CH}} + 4\,HBr.$$
Acetylene tetrabromide      Tetraphenyl ethane

In the latter reaction, anthracene is also formed, according to the equation:

$$C_6H_4\boxed{H_2+}\!\!\begin{array}{c}Br\\ |\\ CH\\ |\\ CH\\ |\\ Br\end{array}\!\!\begin{array}{c}Br\\ |\\ \\ \\ \\ |\\ Br\end{array}\!\!\boxed{+H_2}C_6H_4 = C_6H_4\!\!<\!\!\begin{array}{c}CH\\ |\\ CH\end{array}\!\!>\!\!C_6H_4 + 4\,HBr.$$
Anthracene

For the synthesis of aromatic acids the Friedel-Crafts reaction is also of value, although the acids themselves are not directly obtained, but derivatives of them, which upon saponification yield the free acid, e.g.:

$$C_6H_6 + Cl.CO.NH_2 = C_6H_5.CO.NH_2 + HCl,$$
Urea chloride       Benzamide

# AROMATIC SERIES

$$\underset{\text{Phenyl cyanate}}{C_6H_6 + C_6H_5 \cdot NCO} = \underset{\text{Benzanilide}}{C_6H_5 \cdot CO \cdot NH \cdot C_6H_5,}$$

$$\underset{\text{Phenyl mustard oil}}{C_6H_6 + C_6H_5 \cdot NCS} = \underset{\text{Thiobenzanilide}}{C_6H_5 \cdot CS \cdot NH \cdot C_6H_5.}$$

The last two reactions are to be considered as cases of the normal Friedel-Crafts reaction, since the cyanate and mustard oil unite in the first phase with hydrochlorid acid, forming an acid-chloride, which then reacts with the hydrocarbon with elimination of hydrochloric acid, *e.g.*:

$$C_6H_5 \cdot NCO + HCl = CO \begin{matrix} \nearrow NH \cdot C_6H_5 \\ \searrow Cl \end{matrix}$$
Phenyl carbamine chloride

If one considers that in the modifications, in place of the hydrocarbons, ethers, mono- and poly-acid phenols, naphthalene, thiophene, diphenyl, naphthol-ethers, and many other compounds can be used, the great value of the Friedel-Crafts reaction will be readily understood.

Concerning the role which aluminium chloride plays in the reaction, it is still not perfectly clear; certain it is that hydrocarbons as well as phenol-ethers unite with it to form double compounds which are of assistance in causing the reaction to take place.

(*b*) By the action of hydroxylamine on aldehydes and ketones, oximes [1] (aldoximes, ketoximes) are formed in accordance with the following typical reactions:

$$C_6H_5 \cdot CHO + NH_2 \cdot OH = \underset{\text{Benzaldoxime}}{C_6H_5 \cdot CH = N \cdot OH + H_2O,}$$

$$C_6H_5 \cdot CO \cdot C_6H_5 + NH_2 \cdot OH = \begin{matrix} C_6H_5 \cdot C \cdot C_6H_5 \\ \| \\ N \\ | \\ OH \end{matrix} + H_2O.$$
Benzophenone oxime

Oximes may be obtained by three methods: (1) The alcoholic solution of the aldehyde or ketone may be treated, generally, with a concentrated water solution of hydroxylamine hydrochloride and the mixture allowed to stand at the ordinary temperature, or it may be heated in a flask provided with a reflux condenser, or in a bomb-tube. An addition of a few drops of concentrated hydrochloric acid often

---

[1] B. 15, 1324.

expedites the reaction. (2) The formation of oximes may be brought about by the use of free hydroxylamine obtained by treating its hydrochloride with the theoretical amount of a solution of sodium carbonate. (3) Oximes may in many cases be very easily obtained, if, as above, for one carbonyl-group three molecules of hydroxylamine hydrochloride and nine molecules of potassium hydroxide are used; in the presence of a large excess of hydroxylamine in a strongly alkaline solution, generally a very smooth decomposition takes place. Since the oximes possess a weak acid character, under these conditions the alkali salt of the oxime is first obtained, *e.g.*:

$$C_6H_5 \cdot \underset{\underset{OK}{|}}{\overset{\|}{\underset{N}{C}}} \cdot C_6H_5$$

from which the free oxime is liberated by treating it with an acid.

Of especial significance for the stereo-chemistry of nitrogen are the oximes of aldehydes as well as those of the unsymmetrical ketones. By the action of hydroxylamine on benzaldehyde, *e.g.*, there is formed not only a single oxime, but a mixture of *two* stereo-isomers. This is also true when oximes are formed from many unsymmetrical ketones. The existence of these isomers is explained by the assumption that the three valencies of nitrogen do not lie in a plane, but that they extend into space, proceeding from a point like the three edges of a regular triangular pyramid.[1] Since, *e.g.*, in the formation of benzaldoxime, the hydroxyl-group of the hydroxylamine is vicinal to either the phenyl-group or hydrogen atom, the two following stereo-isomers are possible:

$$C_6H_5 \cdot \overset{\|}{\underset{HO-N}{C}} \cdot H \qquad \text{and} \qquad C_6H_5 \cdot \overset{\|}{\underset{N-OH}{C}} \cdot H$$

OH vicinal to $C_6H_5$          OH vicinal to H

The stereoisomeric forms of an unsymmetrical ketone are, according to this conception, to be expressed by the following formulæ, *e.g.*:

$$BrC_6H_4 \cdot \overset{\|}{\underset{HO-N}{C}} \cdot C_6H_5 \qquad \text{and} \qquad BrC_6H_4 \cdot \overset{\|}{\underset{N-OH}{C}} \cdot C_6H_5$$

OH vicinal to $C_6H_4Br$          OH vicinal to $C_6H_5$

---

[1] B. 23, II, 1243.

AROMATIC SERIES 279

With symmetrical ketones it is obviously immaterial upon which of the two similar sides the hydroxyl-group finds itself, so that here only one oxime is possible.

In this place the two mono-oximes and three dioximes of benzil may be referred to again. These compounds gave the impetus to the investigations[1] of this class of compounds. They are explained by the following space-formulæ:

$$\begin{array}{cc} C_6H_5.C.CO.C_6H_5 & C_6H_5.C.CO.C_6H_5 \\ \parallel & \text{and} \quad \parallel \\ HO-N & N-OH \end{array},$$

$$\begin{array}{cc} C_6H_5.C.C.C_6H_5 & C_6H_5.C\!-\!-\!-\!-\!C.C_6H_5 \\ \parallel \quad \parallel & \parallel \quad \parallel \\ HO-N \;\; N-OH \end{array},\;\; \begin{array}{c} HO-N \;\; HO-N \end{array},\;\; \begin{array}{c} C_6H_5.C\!-\!-\!-\!-\!C.C_6H_5 \\ \parallel \quad \parallel \\ N-OH \;\; HO-N \end{array}.$$

Not all aldehydes and unsymmetrical ketones yield two oximes. In many cases one form is so unstable (labile) that only the other stabile modification exists. It may be briefly pointed out that, according to another view, it is not the position of the *hydroxyl-group*, but that of the *hydrogen* atom which gives rise to the different forms. The two benzaldoximes, *e.g.*, would then be formulated thus:

$$\begin{array}{cc} C_6H_5C.H & C_6H_5.C.H \\ \parallel & \text{and} \quad \parallel \\ N & N \\ | & | \\ H-O & O-H \end{array}$$

From a third side it is maintained, finally, that the isomerism of the oximes depends not on the space relations but on structural differences.

(c) If phosphorus pentachloride is allowed to act on an oxime, it is transformed into an anilide,[2] *e.g.*:

$$\begin{array}{c} C_6H_5.C.C_6H_5 \\ \parallel \\ N \\ | \\ OH \end{array} = C_6H_5.NH.CO.C_6H_5.$$

Benzophenone oxime     Benzanilide

This so-called Beckmann transformation has been of great significance for the explanation of the constitution of the isomeric oximes. If, *e.g.*, phosphorus pentachloride is allowed to act on both of the above

---

[1] B. 16, 503; 21, 784, 1304, 3510; 22, 532, 564, 1985, 1996.
[2] B. 19, 988; 20, 1507 and 2580; A. 252, 1.

formulated stereoisomeric oximes of the brombenzophenone, the same compounds are not obtained from both, but two different ones, which, as follows from their saponification products, correspond on the one hand to the benzoyl derivative of bromaniline, and, on the other, to the brombenzoyl derivative of aniline:

$$C_6H_5 \cdot CO \cdot NH \cdot C_6H_4 \cdot Br \text{ and } BrC_6H_4 \cdot CO \cdot NH \cdot C_6H_5.$$

The transformation takes place, probably, in the following way: If phosphorus pentachloride is allowed to act on an oxime, the hydroxyl-group is replaced by chlorine:

$$\underset{\underset{OH}{|}}{\overset{C_6H_5 \cdot C \cdot C_6H_5}{\|}} + PCl_5 = \underset{\underset{Cl}{|}}{\overset{C_6H_5 \cdot C \cdot C_6H_5}{\|}} + HCl + POCl_3.$$

But a compound of this kind, in which chlorine is united with nitrogen, is unstable, and it is immediately transformed into a more stable imido-chloride, the chlorine atom being replaced by a phenyl-group:

$$\underset{\underset{Cl}{|}}{\overset{C_6H_5 \cdot C \cdot C_6H_5}{\|}} \longrightarrow \underset{\underset{C_6H_5}{|}}{\overset{C_6H_5 \cdot C \cdot Cl}{\|}}$$

Imidochloride of Benzanilide
(Compare p. 127)

If this is now treated with water, benzanilide is formed, in accordance with the following equation:

$$C_6H_5 \cdot C \cdot Cl = N \cdot C_6H_5 + H_2O = C_6H_5 \cdot CO \cdot NH \cdot C_6H_5 + HCl.$$

If the oxime of brombenzophenone, formulated above, is subjected to a similar reaction, the unstable chlorides are first obtained:

$$\underset{Cl-N}{\overset{Br \cdot C_6H_4 \cdot C \cdot C_6H_5}{\|}} \text{ and } \underset{N-Cl}{\overset{Br \cdot C_6H_4 \cdot C \cdot C_6H_5}{\|}}.$$

Cl vicinal to $C_6H_4Br$    Cl vicinal to $C_6H_5$

If the most probable assumption is now made, that the chlorine atom gives up its position to the vicinal hydrocarbon radical, there are formed:

AROMATIC SERIES 281

$$\begin{array}{cc} \text{Cl.C.C}_6\text{H}_5 & \text{Br.C}_6\text{H}_4\text{.C.Cl} \\ \| & \text{and} \quad \| \\ \text{Br.C}_6\text{H}_4\text{.N} & \text{N.C}_6\text{H}_5 \end{array},$$

from which, by treatment with water, there are obtained:

Br.C$_6$H$_4$.NH.CO.C$_6$H$_5$ and Br.C$_6$H$_4$.CO.NH.C$_6$H$_5$.
Benzoyl bromanilide                Brombenzoyl anilide

Upon saponification, these yield:

Br.C$_6$H$_4$.NH$_2$ + C$_6$H$_5$.CO.OH and Br.C$_6$H$_4$.CO.OH + C$_6$H$_5$NH$_2$.
Bromaniline  Benzoïc acid        Brombenzoïc acid aniline

That hydrocarbon radical which in the oxime was vicinal to the hydroxyl-group, is, therefore, on saponification of the polymerised product, obtained in the form of a primary amine. In this way, the constitution of the stereoisomeric oximes of the unsymmetrical ketones is determined.

### 33. REACTION: REDUCTION OF A KETONE TO A HYDROCARBON

EXAMPLE: **Diphenyl Methane from Benzophenone** [1]

A mixture of 10 grammes of benzophenone, 12 grammes of hydriodic acid (boiling-point, 127°), and 2 grammes of red phosphorus is heated in a sealed tube for 6 hours, at 180°. The reaction-mixture is then treated with ether, poured into a small evaporating funnel, and shaken up with water several times. The ethereal solution is filtered through a small folded filter, the ether evaporated, and the residue distilled. Boiling-point, 263°. On cooling, the diphenyl methane solidifies to crystals which melt at 27°. Yield, almost quantitative.

Hydriodic acid, especially at high temperatures, is an extremely energetic reducing agent, which can be used to effect reduction when, as in the above case, another reducing agent, *e.g.*, a metal and acid, could not be employed. The reducing action depends on the following decomposition:

$$2\,\text{HI} = \text{H}_2 + \text{I}_2.$$

---

[1] B. 7, 1624.

The above reaction takes place in accordance with the following equation:

$$C_6H_5 \cdot CO \cdot C_6H_5 + 4\,HI = \underset{\text{Diphenyl methane}}{C_6H_5 \cdot CH_2 \cdot C_6H_5} + H_2O + 2\,I_2.$$

With the aid of hydriodic acid, not only ketones but also acids may be reduced to the hydrocarbon from which they are derived, *e.g.*:

$$\underset{\text{Benzoic acid}}{C_6H_5 \cdot CO \cdot OH} + 6\,HI = \underset{\text{Toluene}}{C_6H_5 \cdot CH_3} + 2\,H_2O + 3\,I_2,$$

$$\underset{\text{Stearic acid}}{C_{17}H_{35} \cdot CO \cdot OH} + 6\,HI = \underset{\text{Octodecane}}{C_{18}H_{38}} + 2\,H_2O + 3\,I_2.$$

Alcohols, iodides, etc., can also be reduced to their final reduction products, the hydrocarbons, *e.g.*:

$$\underset{\text{Ethyl iodide}}{C_2H_5I} + HI = \underset{\text{Ethane}}{C_2H_6} + I_2,$$

$$\underset{\text{Glycerol}}{\begin{array}{c}CH_2 \cdot OH \\ | \\ CH \cdot OH \\ | \\ CH_2 \cdot OH\end{array}} + 6\,HI = \underset{\text{Propane}}{\begin{array}{c}CH_3 \\ | \\ CH_2 \\ | \\ CH_3\end{array}} + 3\,H_2O + 3\,I_2.$$

By heating with hydriodic acid, the unsaturated compounds take up hydrogen, *e.g.*:

$$C_6H_6 + 6\,HI = \underset{\text{Hexahydrobenzene}}{C_6H_{12}} + 3\,I_2.$$

The effect of hydriodic acid is increased by the addition of red phosphorus. Under these conditions, during the course of the reaction, the liberated iodine unites with the phosphorus to form phosphorus tri-iodide:

$$3\,I + P = PI_3,$$

which with the water present again decomposes to form hydriodic acid:

$$PI_3 + 3\,H_2O = 3\,HI + \underset{\text{Phosphorous acid}}{P(OH)_3}.$$

A definite amount of hydriodic acid can thus, provided a sufficient quantity of phosphorus is present, act as a continuous reducing agent.

## 34. REACTION: SAPONIFICATION OF AN ACID-NITRILE

EXAMPLE: Toluic Acid from Tolyl Nitrile[1]

The p-tolyl nitrile obtained in Reaction 9 is heated with slightly diluted sulphuric acid on the sand-bath in a round flask with reflux condenser until crystals of toluic acid appear in the condenser. For each gramme of the nitrile a mixture of 6 grammes of concentrated sulphuric acid with 2 grammes of water is used. After cooling it is diluted with water, the acid separating out is filtered off and washed several times with water. A small portion is dissolved in a little alcohol, and hot water added until the solution just becomes turbid; it is then boiled some time with animal charcoal. On cooling, the pure acid is obtained. Melting-point, 177°. Yield, 80–90 % of the theory.

By saponification in a narrow sense is understood the splitting up of an acid-ester into an alcohol and acid. It is, however, used in a wider sense to indicate the conversion of acid-derivatives, like nitriles, amides, substituted amides, e.g., anilides, into acids of the same name. Saponification may be conducted either in an alkaline or an acid solution. Thus, for instance, acetamide reacts on heating with a solution of caustic potash or caustic soda with the formation of the alkali salt of acetic acid and the evolution of ammonia. Nitriles and esters may frequently be saponified by water solutions of the alkalies. Further, alcoholic caustic potash or caustic soda can be used for a similar purpose; it may here be recalled that in Preparation 11 of the fatty series, the ethyl malonic ester was saponified by this method. Finally, saponification may be effected by heating with a sodium carbonate solution under pressure; this method is especially well adapted for difficultly saponifiable amides or anilides.

In order to effect saponification in acid solution, the substance to be saponified is heated with either hydrochloric acid or sulphuric acid in varying degrees of dilution, e.g.:

$$C_6H_4\!\!\begin{array}{c}\diagup CH_3 \\ \diagdown CN\end{array} + 2\,H_2O = C_6H_4\!\!\begin{array}{c}\diagup CH_3 \\ \diagdown CO.OH\end{array} + NH_3.$$

p-Tolyl nitrile                    p-Toluic acid

---

[1] A. 258, 10.

The decomposition of the ethers of phenols is also designated as saponification. Such decomposition cannot be effected by the methods hitherto given. Hydriodic acid is used which, when heated with phenol-ethers, decomposes them into the phenol and the alkyl iodide:

$$C_6H_5 \cdot OCH_3 + HI = C_6H_5 \cdot OH + CH_3I.$$
Anisol

Anhydrous aluminium chloride may be used here with great advantage; upon heating, it acts on the phenol-ether in the manner indicated by the following equation:

$$3\,C_6H_5 \cdot OCH_3 + AlCl_3 = (C_6H_5 \cdot O)_3Al + 3\,CH_3Cl.$$
Aluminium salt<br>of phenol

If a phenol salt is treated with an acid, the free phenol will separate out. This method presents the advantage that it may be applied to substances containing, in addition to the phenol-ether radical, a reducible carbonyl group, which, if treated with hydriodic acid, would be changed.

### 35. REACTION: OXIDATION OF THE SIDE-CHAIN OF AN AROMATIC COMPOUND

EXAMPLE: **Terephthalic Acid from p-Toluic Acid**

Dissolve 5 grammes of the crude toluic acid obtained in Reaction 34 in a solution of 3 grammes of sodium hydroxide in 250 c.c. of water; heat in a porcelain dish on the water-bath, and gradually treat with a solution of 12 grammes of finely powdered potassium permanganate in 250 c.c. of water until, after long boiling, the red colour of the permanganate no longer vanishes. Alcohol is then added until the liquid is colourless, and, after cooling, the manganese dioxide separating out is filtered off; this is washed with hot water, and the filtrate, heated to boiling, is acidified with concentrated hydrochloric acid. After cooling, the terephthalic acid is filtered off, washed with water, and dried on the water-bath. Yield, 90 % of the theory. Terephthalic acid is insoluble in water. On heating, it sublimes without melting.

AROMATIC SERIES 285

It is a common property of aliphatic side-chains, united with the benzene nucleus, to pass over to a carboxyl-group on oxidation. A methyl-group requires 3 atoms of oxygen for oxidation:

$$C_6H_5 \cdot CH_3 + 3\,O = C_6H_5 \cdot CO \cdot OH + H_2O.$$
Toluene       Benzoic acid

If several side-chains are present in a compound, either all or a portion of them may be converted into carboxyl-groups:

$$C_6H_4\!\!<\!\!{CH_3 \atop CH_3} \text{ gives } \nearrow C_6H_4\!\!<\!\!{CH_3 \atop CO.OH},$$
$$\searrow C_6H_4\!\!<\!\!{CO.OH \atop CO.OH}.$$

$$C_6H_3\!\!<\!\!{CH_3 \atop CH_3 \atop CH_3} \text{ gives } \longrightarrow$$

$$\nearrow C_6H_3\!\!<\!\!{CH_3 \atop CH_3 \atop CO.OH},$$
$$C_6H_3\!\!<\!\!{CH_3 \atop CO.OH \atop CO.OH},$$
$$\searrow C_6H_3\!\!<\!\!{CO.OH \atop CO.OH \atop CO.OH}.$$

If a side-chain contains several carbon atoms, in many cases only the methyl-group at the end of the chain can be oxidised, e.g.:

$$X \cdot CH_2 \cdot CH_3 + 3\,o = X \cdot CH_2 \cdot CO \cdot OH + H_2O.$$

But by an energetic oxidation all the carbon atoms, with the exception of the last, are split off, e.g.:

$$C_6H_5 \cdot CH_2 \cdot CH_3 + 3\,O_2 = C_6H_5 \cdot CO \cdot OH + CO_2 + 2\,H_2O.$$
Ethyl benzene

The basicity of the acid derived from the oxidation of a hydrocarbon accordingly gives an indication concerning the number of side-chains of the hydrocarbon. Derivatives of hydrocarbons are also capable of similar reaction, e.g.:

$$C_6H_4\!\!<\!\!{CH_3 \atop Cl} + 3\,O = C_6H_4\!\!<\!\!{CO.OH \atop Cl} + H_2O.$$
Chlortoluene       Chlorbenzoic acid

$$C_6H_4{\scriptsize\begin{matrix}CH_3\\NO_2\end{matrix}} + 3O = C_6H_4{\scriptsize\begin{matrix}CO.OH\\NO_2\end{matrix}} + H_2O.$$

Nitrotoluene   Nitrobenzoïc acid

The reaction carried out above, takes place in accordance with this equation:

$$C_6H_4{\scriptsize\begin{matrix}CH_3\\CO.OH\end{matrix}} + 3O = C_6H_4{\scriptsize\begin{matrix}CO.OH\\CO.OH\end{matrix}} + H_2O.$$

Amines and phenols cannot be directly oxidised in most cases, but an indirect method must be employed, by which the former are converted into an acid derivative, and the latter into an ester. If, *e.g.*, it is desired to convert p-toluidine into p-amidobenzoïc acid, the base is first acetylated, and the acetatoluide is then oxidised:

$$C_6H_4{\scriptsize\begin{matrix}CH_3\\NH.CO.CH_3\end{matrix}} + 3O = C_6H_4{\scriptsize\begin{matrix}CO.OH\\NH.CO.CH_3\end{matrix}} + H_2O.$$

The acid thus obtained is then saponified, and the desired amidobenzoïc acid is formed:

$$C_6H_4{\scriptsize\begin{matrix}CO.OH\\NH.CO.CH_3\end{matrix}} + H_2O = C_6H_4{\scriptsize\begin{matrix}NH_2\\CO.OH\end{matrix}} + CH_3.CO.OH.$$

If it is desired, on the other hand, to oxidise a phenol, *e.g.*, cresol, $C_6H_4{\scriptsize\begin{matrix}CH_3\\OH\end{matrix}}$, the sulphuric acid- or phosphoric acid-ester of it is first prepared and oxidised; the reaction-product is then saponified. As oxidising agent, dilute nitric acid (1 vol. conc. nitric acid to 2 vol. water [1]), chromic acid or potassium permanganate is used. The mildest effect is obtained with the nitric acid, which is therefore used when all the side-chains are not to be oxidised, but only a portion of them, *e.g.*:

$$C_6H_4{\scriptsize\begin{matrix}CH_3\\CH_3\end{matrix}} \longrightarrow C_6H_4{\scriptsize\begin{matrix}CH_3\\CO.OH\end{matrix}}$$

Nitric acid is also used in other cases, where, as frequently happens with other derivatives, other oxidising agents totally destroy the substance.

---

[1] A. 133, 41; 137, 302.

Chromic acid in the form of its anhydride generally, dissolved in glacial acetic acid, or as a water solution of potassium dichromate or sodium dichromate acidified with dilute sulphuric acid, can also be used as an oxidising agent, not only in the case in hand, but also for the oxidation of alcohols, ketones, etc. In oxidation reactions, two molecules of chromic anhydride ($CrO_3$) give three atoms of oxygen:

$$2\,CrO_3 = C_2O_3 + 3\,O.$$

For the oxidation of aromatic hydrocarbons,[1] experience has shown that a good oxidising mixture is 40 parts of potassium dichromate, 55 parts of concentrated sulphuric acid, diluted with twice its volume of water.

With potassium permanganate,[2] oxidation can be effected either in alkaline or in acid solution. In the first case, manganese dioxide is deposited:

$$2\,KMnO_4 + H_2O = 3\,O + 2\,MnO_2 + 2\,KOH.$$

Two molecules of potassium permanganate yield, therefore, in alkaline solution, three atoms of oxygen.

In acid solution (sulphuric acid), no manganese dioxide separates out, since it is dissolved by the sulphuric acid, with evolution of oxygen, to form manganous sulphate:

$$2\,KMnO_4 + 3\,H_2SO_4 = 5\,O + K_2SO_4 + MnSO_4 + 3\,H_2O.$$

Two molecules of the permanganate in acid solution, therefore, yield 5 atoms of available oxygen.

In oxidising with potassium permanganate, a 2-5% solution is generally used. An excess of the permanganate can be removed by the addition of alcohol or sulphurous acid.

### 36. REACTION: KOLBE'S SYNTHESIS OF OXYACIDS

EXAMPLE: **Salicylic Acid from Sodium Phenolate and Carbon Dioxide** [3]

Dissolve $12\frac{1}{2}$ grammes of chemically pure sodium hydroxide in 20 c.c. of water in a porcelain dish, or better a nickel dish, and with stirring, treat gradually with 30 grammes of crystallised phe-

---

[1] A. 133, 41; 137, 302.   [2] B. 7, 1057.   [3] J. pr. [2] 10, 89; 27, 39; 31, 397.

nol. The greatest portion of the water is then evaporated by heating over a free flame, the mass being continually stirred. As soon as a crystalline film forms on the surface of the liquid, the heating is continued with a luminous flame, which is not placed directly under the dish, but is kept in constant motion. In order to fasten the dish, a pair of crucible tongs is clamped in a vertical position, and the dish supported between its jaws. There is first obtained a caked, bright-coloured mass which is crushed from time to time, and heated until it has changed to a dusty, dry, almost colourless powder; this is allowed to cool in a desiccator and then placed in a tubulated retort of about 200 c.c. capacity. The retort is then immersed as far as possible in an oil-bath (Fig. 69). After the air has been forced out by dry hydrogen, it is heated a half-hour in a current of hydrogen (the end of the delivery tube is 1 cm. above the upper surface of the sodium phenolate), in order to remove the last traces of water, to $140°$; it is then allowed to cool to $110°$, and the hydrogen is now replaced by a current of dry carbon dioxide; this is passed into the retort for an hour. The temperature is then gradually raised ($20°$ per hour) during the course of four hours, while a not too rapid current is passed in, to $190°$. The mixture is finally heated 1–2 hours at $200°$. After cooling, the phenol in the neck of the retort is melted by the application of a flame to the outside, the dusty, fine powder is poured into a large beaker, the retort is washed out several times with water, and the salicylic acid precipitated with much concentrated hydrochloric acid. After the reaction-mixture has been cooled with ice-water a long time, and the sides of the vessel

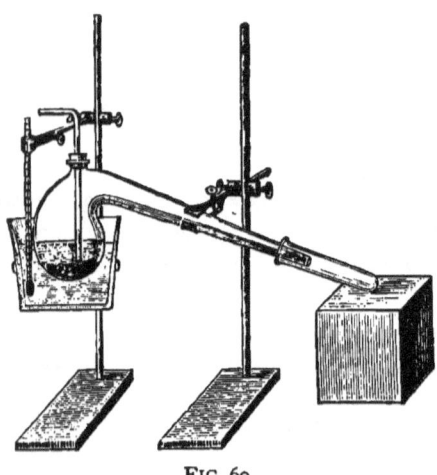

FIG. 69.

rubbed with a glass rod, the crude salicylic acid is filtered off, washed with a little water, and pressed out on a porous plate. From water it crystallises in needles, generally slightly coloured, which melt at 156°. Yield, very varying (5–10 grammes).

The preparation of salicylic acid does not always take place successfully the first time. The success of the experiment depends particularly on the condition of the sodium phenolate, which must be perfectly dry. If it "cakes" in the stream of hydrogen, there is great probability that the experiment will be unsuccessful.

The operation should be so arranged that the sodium phenolate is prepared toward evening, so that it may be allowed to stand in a sulphuric acid desiccator over night. The drying in the current of hydrogen is begun immediately the next morning.

The synthesis is named after its discoverer, Kolbe. It takes place in three phases. In the first, the carbon dioxide is added to the sodium phenolate, which forms sodium phenyl carbonate:

(I.) $C_6H_5.ONa + CO_2 = C_6H_5.O.CO_2Na$.

In the above experiment this reaction is completed during the heating up to 110° for one hour. In the second phase, the sodium phenyl carbonate is transformed into the so-called neutral sodium salicylate:

(II.) $C_6H_5.O.CO_2Na = C_6H_4\!\!\begin{array}{c}\diagup OH \\ \diagdown CO_2Na\end{array}$,

while in the last phase a molecule of this salt reacts with a molecule of unchanged sodium phenolate in the following way:

(III.) $C_6H_4\!\!\begin{array}{c}\diagup OH \\ \diagdown CO.ONa\end{array} + C_6H_5.ONa = C_6H_4\!\!\begin{array}{c}\diagup ONa \\ \diagdown CO.ONa\end{array} + C_6H_5.OH$.

These two latter reactions take place during the gradual heating up to 200°. Only one-half of the phenol, therefore, is converted into salicylic acid, the second half being obtained unchanged.

A modification of the Kolbe synthesis which permits the immediate conversion of all the phenol into salicylic acid is known as Schmitt's Synthesis. According to this method, as in the other, the sodium phenyl carbonate is first prepared; this is then further heated in an autoclave under pressure to 140°, upon which it is completely transformed into sodium salicylate according to Equation II. Instead of preparing the sodium phenyl carbonate with gaseous carbon dioxide, the sodium phenolate may be mixed directly with liquid or solid carbon dioxide in the autoclave.

The Kolbe synthesis is capable of very common application, since from each mon-acid phenol, a carbonic acid may be obtained in the same way as that used above. The carboxyl-group under these conditions primarily seeks the ortho-position to the hydroxyl-group. The derivatives of phenols, *e.g.* the three chlorphenols, yield chlorinated salicylic acids. With acid-ethers of poly-acid phenols which still contain a free hydroxyl-group, as, *e.g.*, guaiacol, $C_6H_4\langle{}^{OCH_3}_{OH}$, this reaction likewise takes place. From the two naphthols $C_{10}H_7.OH$ the oxynaphthoic acids $C_{10}H_6\langle{}^{OH}_{CO.OH}$, can be obtained.

If in the Kolbe reaction instead of sodium phenolate, potassium phenolate is used, the para-oxybenzoïc acid is obtained, and not the ortho-acid. The potassium phenolate, like the sodium phenolate, first absorbs carbon dioxide, and the potassium phenyl carbonate thus formed, heated in carbon dioxide up to 150°, also yields salicylic acid; but if the temperature is increased, an increasingly larger quantity of the para-acid is obtained, until finally at 220° the potassium para-oxybenzoate is the only product.

The addition of carbon dioxide is effected with greater ease in poly-acid phenols. With these compounds the reaction begins if the phenol is boiled in a water-solution of ammonium carbonate or potassium hydrogen carbonate, *e.g.*:

$$C_6H_4\langle{}^{OH}_{OH} + HO.CO.OK = C_6H_3\langle{}^{OH}_{CO.OK}{}^{-OH} + H_2O$$

Salicylic acid is prepared technically on the large scale. Since it is an excellent antiseptic, it finds extensive application in preventing fermentation, for the preservation of meat, for the disinfection of wounds. It may easily be recognised, since its water solution gives a violet colour with ferric chloride; in this action it differs from the para- and meta-modifications. It is volatile with steam; for this reason it must not be boiled too long in an open vessel when it is to be recrystallised. All ortho-oxycarbonic acids show this property; the meta- and para-isomers are not volatile with steam. If salicylic acid is heated strongly, it decomposes into carbon dioxide and phenol:

$$C_6H_4\diagup^{CO.OH}_{\diagdown OH} = C_6H_5 \cdot OH + CO_2.$$

The para-oxycarbonic acids show the same property while the meta-acids are stable.


**37. REACTION: PREPARATION OF A DYE OF THE MALACHITE GREEN SERIES**

EXAMPLE: **Malachite Green from Benzaldehyde and Dimethylaniline** [1]

(*a*) *Preparation of the Leuco-base.* — A mixture of 50 grammes of dimethylaniline and 20 grammes of benzaldehyde is heated in a porcelain dish, with frequent stirring, on the water-bath, for 4 hours, with 20 grammes of zinc chloride, which has been previously fused in a porcelain dish, and pulverised, after cooling. (See p. 295.) This viscous mass, which cannot be poured directly out of the dish, is melted by covering it with hot water, and heating it at the same time, on the water-bath; while hot, it is transferred to a ½-litre flask. Steam is conducted into it, until no drops of it pass over. There is thus obtained the non-volatile leuco-base of the dye, in the form of a viscous mass, which adheres firmly to the walls of the distilling flask. After the liquid is cold, the water is poured off, the base adhering to the sides of the flask is washed with water several times, and then dissolved in the flask with alcohol, on the water-bath. After filtering, the solution is allowed to stand over night in a cool place, upon which the base separates out in colourless crystals; these are filtered off, washed with alcohol, and dried in the air, on several layers of filter-paper. By concentrating the mother-liquor, a second crystallisation may be obtained. Should the base not crystallise, but separate out in an oily condition, which frequently happens after

---

[1] A. 206, 83; 217, 250.

a short standing of the filtered solution, this is due to the fact that an insufficient amount of alcohol has been used. In this case, more alcohol is added, and the mixture heated until the oil is dissolved.

(*b*) *Oxidation of the Leuco-base.*—Dissolve 10 parts, by weight, of the completely dry leuco-base by heating in a quantity of dilute hydrochloric acid corresponding to 2.7 parts, by weight, of anhydrous hydrochloric acid. For the purpose, dilute pure concentrated hydrochloric acid with double its volume of water, determine the specific gravity of the diluted acid, and refer to a table to find out how much anhydrous acid the solution contains, and from this calculate how much of the solution must be taken in order to get the required amount of the anhydrous acid (2.7 grammes). The colourless solution of the leuco-base is then diluted in a large flask with 800 parts, by weight, of water, and treated with 10 parts of 40 % acetic acid (sp. gr. 1.0523), prepared by gradually diluting glacial acetic acid with water; the mixture is well cooled by throwing in pieces of ice; then, with frequent stirring, gradually add (during 5 min.) a quantity of freshly prepared lead peroxide paste corresponding to 7.5 grammes of pure lead peroxide. The peroxide is weighed off in a beaker, and treated with a quantity of water sufficient to form a very thin paste. The residue remaining on the beaker after the first emptying is washed out with water. After the addition of the peroxide, the reaction-mixture is allowed to stand five minutes, with frequent shaking; then add a solution of 10 parts of Glauber's salt and 50 parts of water; the solution is then filtered off through a folded filter from the precipitated lead sulphate and chloride. The filtrate is treated with a filtered solution of 8 parts of zinc chloride dissolved in as small a quantity of water as possible; then a saturated sodium chloride solution is added, until all the dye is precipitated. This is easily recognised by bringing a drop of the solution, with a glass rod on a piece of filter-paper; a bluish green precipitate surrounded by a circle of a still fainter bright green colour will be formed. The precipitated dye is filtered off with suction, washed with a little saturated sodium chloride solution, and pressed out on a porous plate. In order to purify it further, it may be dissolved again in water; and from the filtered solution, after cooling, it is again thrown out by sodium chloride.

The reaction just carried out, discovered by Otto Fischer in 1877, is also used in the large scale for the preparation of Malachite Green, or Bitter Almond Green. In the formation of the leuco-base, the following reaction takes place:

$$C_6H_5 \cdot CH \boxed{O + \begin{matrix} H \\ H \end{matrix}} \begin{matrix} \cdot C_6H_4 \cdot N(CH_3)_2 \\ \cdot C_6H_4 \cdot N(CH_3)_2 \end{matrix} = C \begin{matrix} C_6H_5 \\ C_6H_4 \cdot N(CH_3)_2 \\ C_6H_4 \cdot N(CH_3)_2 \\ H \end{matrix} + H_2O.$$

Tetramethyldiamidotriphenylmethane
= Leuco-base of Malachite Green

The substance thus obtained is not a dye, but the reduction product of the real dye, which, on oxidation, passes over to the dye. Formerly, the dye formation was believed to take place in accordance with the following equation:

$$C \begin{matrix} C_6H_5 \\ C_6H_4 \cdot N(CH_3)_2 \\ C_6H_4 \cdot N(CH_3)_2 \\ H \end{matrix} \boxed{\cdot H | Cl \atop + O} = C \begin{matrix} C_6H_5 \\ C_6H_4 \cdot N(CH_3)_2 \\ C_6H_4 \cdot N(CH_3)_2 Cl \end{matrix} + H_2O.$$

The union, effected by the oxidation between the pentavalent nitrogen atom of the dimethyl aniline residue and the common methane-carbon atom, was considered to be the condition which determined the nature of the dye. At present, the view that the latter is determined by the presence of the quinone-like secondary benzene residue is generally accepted, and the formula of the dye-salt is written thus:

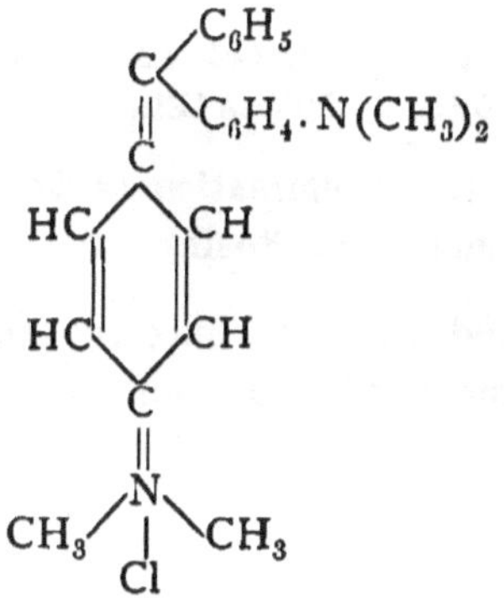

= Quinoïde formula of Malachite Green.

Further, it may be pointed out in this place that, in the formation of the leuco-base, the hydrogen atom in the para-position to the dimethyl-amido groups $N(CH_3)_2$ unites with the aldehyde oxygen atom to form water. The salt of the formula given above is difficult to separate from its solution. But if zinc chloride is added, a double salt of the same colour is formed:

$$3(C_{23}H_{25}N_2Cl) + 2\,ZnCl_2 + H_2O$$

which may be separated from its water solution by common salt; it comes into the market as a dye.

Malachite Green may also be made by a second method, which was discovered by Döbner: it consists in heating benzotrichloride with dimethyl aniline in the presence of zinc chloride:

$$C\!\!<\!\!\begin{array}{l}C_6H_5\\ \boxed{\begin{array}{ll}Cl & H\\ Cl + H\end{array}}\cdot\begin{array}{l}C_6H_4\cdot N(CH_3)_2\\ C_6H_4\cdot N(CH_3)_2\end{array}\\ Cl\end{array} = C\!\!<\!\!\begin{array}{l}C_6H_5\\ C_6H_4\cdot N(CH_3)_2\\ C_6H_4\cdot N(CH_3)_2\\ Cl\end{array} + 2\,HCl.$$

Since the chlorine atom remaining over migrates toward a nitrogen atom, the dyestuff salt is directly formed by the transformation. Still, since the preparation of pure benzotrichloride on the large scale is difficult, this method, which was formerly used, has been abandoned, and the dye is now prepared exclusively by Fischer's method.

Malachite Green is a representative of a whole series of dyes, — the Malachite Green Series. If, instead of dimethyl aniline, diethyl aniline is used, an analogous substance, which bears the name of Brilliant Green, is formed. In place of benzaldehyde, substituted benzaldehydes, etc., can be used. The dyes of the Bitter Almond Series colour only the animal fibres, silk and wool, directly. Vegetable fibre (cotton) is not coloured unless it has been previously mordanted.

### 38. REACTION: CONDENSATION OF PHTHALIC ANHYDRIDE WITH A PHENOL TO FORM A PHTHALEÏN

EXAMPLE: (*a*) Fluoresceïn.[1]  (*b*) Bromination of Fluoresceïn with the Formation of Eosin

(*a*) In a mortar grind up and intimately mix 15 grammes of phthalic anhydride with 22 grammes of resorcinol, and heat in an

---

[1] A. 183, 1.

oil-bath to 180° (Fig. 70). As a vessel for heating the mixture, the "extract of beef" jars, which are generally glazed inside, are well adapted to the purpose; they can be obtained readily at a small cost, and may be used several times for the same fusion. It is suspended by its projecting edge from a triangle into the oil-bath. To the fused mass add, with stirring (glass rod), in the course of 10 minutes, 7 grammes of pulverised zinc chloride. This is prepared in the following way: 10 grammes of the commercial salt, which always contains water, is carefully heated to fusion over a free flame in a porcelain dish. After the mass has been kept in a fused condition for a few minutes, it is allowed to cool, and the solidified substance is removed from the dish with a knife and pulverised. After adding 7 grammes of the anhydrous salt thus obtained, the temperature is increased to 210°, and the heating continued until the liquid, which gradually thickens, becomes solid, for which about 1-2 hours is required. The cold, friable melt is removed from the crucible with a sharp instrument (it is best to use a chisel), finely pulverised, and boiled 10 minutes in a porcelain dish with 200 c.c. of water and 10 c.c. of concentrated hydrochloric acid. This causes the solution of the substance which did not enter into the reaction; the addition of hydrochloric acid is necessary to dissolve the zinc oxide and basic zinc chloride. The fluoresceïn is filtered from the solution, washed with water until the filtrate no longer gives an acid reaction; it is then dried on the water-bath. Yield, almost quantitative.

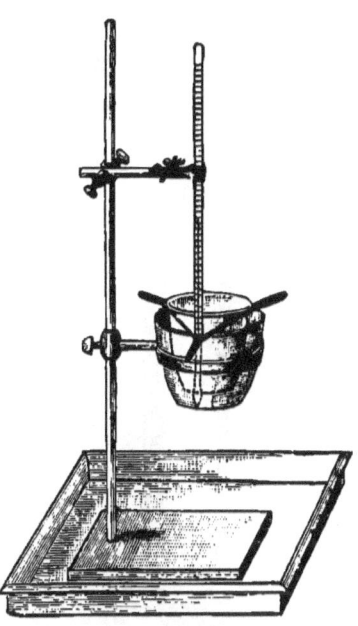

FIG. 70.

(*b*) Over 15 grammes of fluoresceïn in a flask, pour 60 grammes of alcohol (about 95 %), add, with frequent shaking, 33 grammes of bromine, drop by drop, from a separating funnel. This should require about a quarter-hour. In place of a separating-funnel, it is advisable, as in all cases of bromination, to use a burette, by which the troublesome weighing of bromine is obviated. Since the specific gravity of bromine at moderate temperatures is very nearly 3, it is only necessary to divide the required weight by 3, in order to find the number of cubic centimetres corresponding to the weight. Of the numerous kinds of burettes, the one best adapted to this

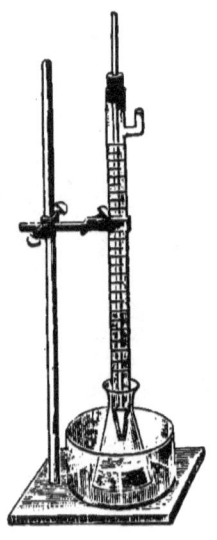

FIG. 71.

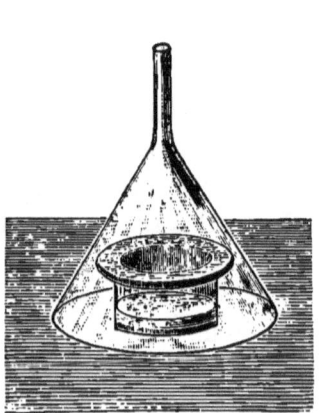

FIG. 72.

purpose is the Winckler form; since it possesses no cock, it can be inserted into the body of a flask with a not too narrow neck, and by this manipulation the disagreeable bromine vapours may be avoided (Fig. 71). In the above case, 11 c.c. of bromine is necessary. On the addition of bromine, it is observed that the quantity of fluoresceïn insoluble in alcohol steadily decreases, and that when about one-half of the bromine has been added, a clear, dark, reddish-brown solution is formed. This is due to the fact that the dibromide is first formed, which is easily soluble in alcohol. On the further addition of bromine, the tetra-bromide is formed, which, since it is difficultly soluble in alcohol, separates out in the form of brick-red leaflets. After all the bromine has been added, the reaction-mixture is allowed to stand for 2 hours, the precipitate is filtered off, washed several times with alcohol, and dried on the water-bath. The product thus obtained is a compound of 1 molecule of eosin and 1 molecule of alcohol. In order to obtain pure eosin from it, the substance is heated a half-hour in an air-bath at 110° : during the heating, its colour becomes brighter. Since eosin is insoluble in water, the soluble potassium-, sodium-, or ammonium-salt is prepared on the large scale for dyeing.

*Ammonium Eosin.* — Over a flat-bottom crystallising dish, $\frac{1}{8}$ filled with a concentrated ammonia solution, place a filter, of paper as strong as possible. Upon this is spread the eosin acid, in a layer about $\frac{1}{2}$ cm. thick, and the whole is covered with a funnel (Fig. 72). The bright-red crystals of the free eosin acid very soon assume a darker colour, and, after about three hours, it is completely converted into the ammonium salt, which forms dark-red crystals with a greenish lustre. The end of the reaction is easily recognized, by testing a small portion with water. If it dissolves, the conversion is complete.

On the large scale, this reaction is carried out in wooden chests containing a number of frames covered with coarse linen, arranged like drawers. After the eosin is spread out on the linen in thin layers, dry ammonia evolved from ammonium chloride and lime is passed into the chest, until a test-portion of the substance will completely dissolve.

*Sodium Eosin.* — Grind 6 grammes of eosin with 1 gramme of dehydrated sodium carbonate, and in a not too small beaker moisten it with a little alcohol; after the addition of 5 c.c. of water, heat on the water-bath until the evolution of carbon dioxide ceases. To the water solution of sodium eosin thus obtained, add 20 grammes of alcohol, heat to boiling, and filter the hot solution. On cooling, the soluble sodium salt separates out in the form of splendid, brownish-red needles of a metallic lustre. As is the case with many dyes, the crystallisation requires a long time; one day, at least, is necessary.

Phthalic anhydride and phenols can react with each other in two different ways. (1) An *equal* number of molecules of each can condense, the oxygen atom of the anhydride, which unites the carbonyl groups, can combine with two ring-hydrogen atoms of the phenol to form one molecule of water; this action results in the formation of an anthraquinone derivative:

$$C_6H_4\diagup_{CO}^{CO}\!\!\!\!>\boxed{O+H_2}\cdot C_6H_3\cdot OH = C_6H_4\diagup_{CO}^{CO}\!\!\!\!>C_6H_3\cdot OH$$
　　　　　　　　Phenol　　　　　　　Oxyanthraquinone

Or (2) *one* molecule of the anhydride can react with *two* molecules of the phenol in such a way that one of the two carbonyl-oxygen atoms of the former combines with one ring-hydrogen of the two phenol molecules to form a so-called phthaleïn:

$$C_6H_4\diagup_{CO}^{C\boxed{O+H}\cdot C_6H_4\cdot OH}_{\phantom{C}O}\!\!\!\overset{\mid H\mid\cdot C_6H_4\cdot OH}{} = \begin{matrix}HO\phantom{xx}OH\\ |\phantom{xxx}|\\ C_6H_4\phantom{x}C_6H_4\\ \diagdown\!\!\!/\\ C\\ C_6H_4\diagup^{\phantom{x}}\!\!\!\!>O\\ CO\end{matrix}$$

　　　　　　　　　　　　　　　Phenolphthaleïn=
　　　　　　　　　　　　　　　dioxyphthalophenone

For the knowledge concerning this class of compounds, to which belong numerous important dyestuffs, we are indebted to the investigations of A. Baeyer (1871). Phthalophenone is considered to be the mother-substance of the group:

$$C_6H_4\diagup_{CO}^{C\diagup_{C_6H_5}^{C_6H_5}}\!\!\!\!>O$$

which, as already stated, is obtained from phthalyl chloride and benzene in the presence of aluminium chloride. If one conceives that the mother-substance can take up one molecule of water, a hypothetical mono-carbonic acid of triphenyl carbinol would result:

$$C\!\!\diagup_{OH}^{\diagup C_6H_5}_{\diagdown C_6H_4\cdot CO\cdot OH,}^{C_6H_5}$$

the formula of which shows very clearly the connection between the phthaleïns and the triphenyl methane derivatives.

If, as expressed by the above equation, phthalic anhydride is allowed to act on phenol, phenolphthaleïn is obtained, a substance of acid properties, colourless in the free condition; its salts are red. It is used as an indicator in volumetric analysis.

By the action of phthalic anhydride on resorcinol, the formation of a *tetraoxyphthalophenone* would naturally be expected; but fluoresceïn, containing the constituents of one molecule of water less than this, is obtained, an inner anhydride formation taking place between the two hydroxyl groups:

$$C_6H_4\begin{array}{c}CO\\CO\end{array}O + \begin{array}{c}H.C_6H_3\\H.C_6H_3\end{array}\begin{array}{c}OH\\OH\\OH\\OH\end{array} = C_6H_4\begin{array}{c}C\\CO\end{array}\begin{array}{c}C_6H_3\\C_6H_3\end{array}O\begin{array}{c}OH\\OH\end{array} + 2\,H_2O.$$

Fluoresceïn

Fluoresceïn is technically prepared on the large scale by the method given above. While phenolphthaleïn, in spite of the intense colour of its salts, is not a dye, in that it does not colour fibres, fluoresceïn is a true dye which colours animal fibres a fast yellow. But it is not manufactured as a dye, since it has been replaced by other dyes that give as beautiful colours and are cheaper. A number of its halogen- and nitro-substitution products have valuable colouring properties, and are prepared from it. The simplest dye of this kind is eosin or tetra-brom-fluoresceïn, discovered in 1874 by Caro. The four bromine atoms are equally divided between the two resorcinol residues:

$$C_6H_4\begin{array}{c}C\\CO\end{array}\begin{array}{c}C_6HBr_2\\C_6HBr_2\end{array}O\begin{array}{c}HO\\OH\end{array}$$

which follows from the fact that eosin in fusion with potassium hydroxide yields di-bromresorcinol besides phthalic acid. Instead of phthalic anhydride, the di- and tetra-chlor-substitution products are fused with resorcinol on the large scale; and so there is obtained in the phthalic acid residue, the di- and tetra-chlorfluoresceïns from which halogen substitution products, nitro-derivatives, ethers, etc., are prepared on the large scale (Phloxine, Rose Bengal).

Besides fluoresceïn there is practically only one other phthaleïn prepared technically, Galleïn. This is done by heating phthalic anhydride with the $v$-trioxybenzene — pyrogallol. In this case the same anhydride formation takes place as in the preparation of fluoresceïn, but there is a simultaneous splitting off of the hydrogen atoms of two hydroxyl groups resulting in a peroxide union:

$$C_6H_4 \begin{array}{c} CO \\ >O \\ C-C_6H_2 \\ \diagdown \\ C_6H_2 \end{array} \begin{array}{c} OH \\ -O \\ >O \\ -O \\ OH \end{array} = \text{Galleïn}.$$

From galleïn, a derivative of anthracene, cœruleïn, a new dye, is obtained by heating with sulphuric acid. Since 1887 the phthaleïns have been on the market under the name of rhodamines, which are prepared in a manner similar to that of fluoresceïn, except that instead of resorcinol, m-amidophenol, or amidophenols substituted by alkyls in the amido-group, are used:

$$C_6H_4 \begin{array}{c} CO \\ >O \\ CO \end{array} + \boxed{\begin{array}{c} H.\ C_6H_3 \diagup^{NH_2} \\ \boxed{OH} \\ \boxed{O\ H} \\ H.\ C_6H_3 \diagdown_{NH_2} \end{array}} = \begin{array}{c} H_2N \quad NH_2 \\ O \\ C_6H_3 \diamond C_6H_3 \\ C \\ C_6H_4 \diamond O \\ CO \end{array} + 2\,H_2O.$$

Simplest Rhodamine

The rhodamine on the market is the tetra-ethyl derivative of this mother-substance.

## 39. REACTION: CONDENSATION OF MICHLER'S KETONE WITH AN AMINE TO A DYE OF THE FUCHSINE SERIES

EXAMPLE: **Crystal Violet from Michler's Ketone and Dimethyl Aniline**

A mixture of 25 grammes of dimethyl aniline, 10 grammes of Michler's ketone (this is on the market), and 10 grammes of phosphorus oxychloride, is heated in an open, dry flask, 5 hours, on an actively boiling water-bath. The blue-coloured mass is then poured into water, made alkaline with a solution of caustic soda, and treated with steam until no drops of the unattacked dimethyl aniline pass over. After cooling, the solidified colour-base remaining in the distillation flask is filtered from the alkaline solution, washed with water, and boiled with a mixture of 1 litre of water and 5 grammes of concentrated hydrochloric acid. The blue solution is filtered while hot from the colour-base, which remains undissolved; the latter is boiled again with a fresh quantity of dilute hydrochloric acid; this operation is repeated until the substance has been almost entirely dissolved. After cooling, the solution of the dye is treated with finely pulverised salt (stirring) until the dye is precipitated. It is then filtered with suction, pressed out on a porous plate, and crystallised from a little water. On cooling, the Crystal Violet separates out in coarse crystals of a greenish colour; these are filtered off and dried in the air on filter-paper.

If Michler's ketone is heated with an amine in the presence of a condensation agent (phosphorus oxychloride, $POCl_3$), addition takes place, in accordance with the following equation:

$$\underset{\text{Michler's ketone}}{\begin{matrix} C_6H_4 \cdot N(CH_3)_2 \\ | \\ CO \\ | \\ C_6H_4 \cdot N(CH_3)_2 \end{matrix}} + H \cdot C_6H_4 \cdot N(CH_3)_2 = \underset{\substack{\text{Hexamethylpararosaniline =} \\ \text{Colour-base of Crystal Violet}}}{C \begin{matrix} \diagup C_6H_4 \cdot N(CH_3)_2 \\ - C_6H_4 \cdot N(CH_3)_2 \\ \diagdown C_6H_4 \cdot N(CH_3)_2 \\ \phantom{\diagdown} OH \end{matrix}}$$

If this is dissolved in hydrochloric acid, one molecule of this is added, and, as in the formation of Malachite Green, the elimination of a molecule of water immediately takes place and the dye is formed:

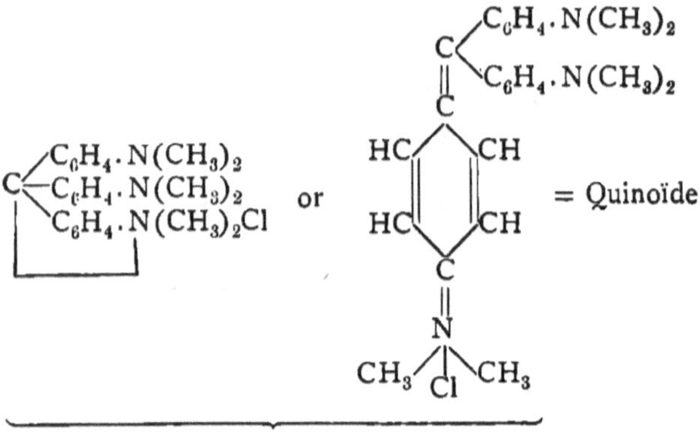

$$C\!\!\begin{array}{l}\diagup C_6H_4\cdot N(CH_3)_2 \\ \diagdown C_6H_4\cdot N(CH_3)_2 \\ \phantom{\diagdown}C_6H_4\cdot N(CH_3)_2Cl\end{array}$$

or

[quinoid formula structure with central ring, HC=CH groups, and =N-C(CH$_3$)(Cl)(CH$_3$) substituent; C attached to two $C_6H_4\cdot N(CH_3)_2$ groups]

= Quinoïde formula.

Crystal Violet

It is a derivative of parafuchsine:

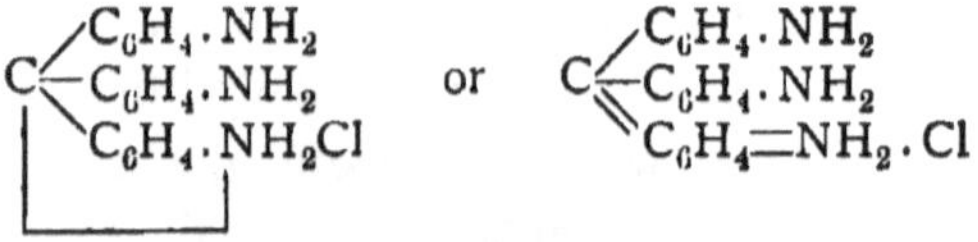

$$C\!\!\begin{array}{l}\diagup C_6H_4\cdot NH_2 \\ \diagdown C_6H_4\cdot NH_2 \\ \phantom{\diagdown}C_6H_4\cdot NH_2Cl\end{array} \quad \text{or} \quad C\!\!\begin{array}{l}\diagup C_6H_4\cdot NH_2 \\ \diagdown C_6H_4\cdot NH_2 \\ \phantom{\diagdown}C_6H_4\!=\!NH_2\cdot Cl\end{array}$$

indeed, it may be considered as a hexamethyl parafuchsine. It is prepared technically in the same way, and forms the principal constituent of the Methyl Violet obtained by the oxidation of dimethyl aniline.

Dyes can also be prepared in the same way by the combination of other amines with Michler's ketone, of which it is only possible to mention here Victoria Blue and Night Blue.

**40. REACTION: CONDENSATION OF PHTHALIC ANHYDRIDE WITH A PHENOL TO AN ANTHRAQUINONE DERIVATIVE**

EXAMPLE: **Quinizarin from Phthalic Anhydride and Hydroquinone** [1]

A mixture of 5 grammes of pure hydroquinone and 20 grammes of phthalic anhydride is heated in an open flask with a mixture of 100 grammes of pure concentrated sulphuric acid and 10

---

[1] B, 6, 506; 8, 152; A. 212, 10.

grammes of water for 3 hours to 170–180°, and finally for 1 hour at 190–200°. The directions as to time and temperature must be followed as exactly as possible. The hot solution is poured, with stirring, into about 400 c.c. of water in a porcelain dish, heated just to boiling, and filtered hot with the aid of a Büchner funnel. The residue remaining on the filter is again boiled out with water and filtered while hot. In order to separate the quinizarin from carbonaceous decomposition products, the precipitate is boiled with 200 c.c. of glacial acetic acid, filtered hot with suction, the filtrate poured into a beaker, and, while hot, treated with its own volume of hot water. The residue remaining on the filter is again boiled up with 100 c.c. glacial acetic acid, and, after filtering, treated as above. On cooling of the diluted acetic acid solution, the crude quinizarin separating out is filtered off, washed with water several times, dried first on the water-bath, and finally in an air-bath at 120°. Since it is difficult to obtain it pure by crystallisation, after drying it is distilled from a small retort of difficultly fusible glass, and is driven over as rapidly as possible with a large flame. A beaker is used as a receiver. After the distillate in the receiver and that in the neck of the retort (this is broken) has been finely pulverised, it is crystallised from glacial acetic acid, from which, on cooling, the quinizarin separates out in the form of large, orange-yellow leaves; these are filtered off and washed with glacial acetic acid, which is steadily diluted with water, until finally only pure water is used.

Under the preparation of fluoresceïn, it has already been mentioned that phthalic anhydride condenses with phenols in certain proportions, to form derivatives of anthraquinone. The reaction just effected takes place in accordance with the following equation:

$$C_6H_4\diagup^{CO}_{\diagdown CO}\diagdown\boxed{O+H_2\cdot}\,C_6H_2\cdot(OH)_2 = C_6H_4\diagup^{CO}_{\diagdown CO}\diagdown C_6H_4(OH)_2 + H_2O.$$

Quinizarin

In an analogous way, mono-acid- as well as poly-acid phenols, combine with phthalic anhydride. It is of theoretical importance that from pyrocatechol (o-dioxybenzene), besides a second isomer, alizarin is obtained, showing that the two hydroxyl-groups in alizarin are in the ortho-position to each other. Of practical significance is the above reaction for the preparation of anthragallol, which is obtained on the large scale by heating pyrogallol with phthalic anhydride:

$$C_6H_4\left\langle\begin{matrix}CO\\CO\end{matrix}\right\rangle\boxed{O + H_2}.C_6H.(OH)_3 = C_6H_4\left\langle\begin{matrix}CO\\CO\end{matrix}\right\rangle C_6H.(OH)_3 + H_2O.$$

Pyrogallol — Trioxyanthraquinone = Anthragallol

It may be mentioned briefly that by the condensation of benzoïc acid with oxybenzoïc acids, similar compounds are also obtained:

$$C_6H_4\left\langle\begin{matrix}CO\boxed{OH\quad H}\\ + \\ \boxed{H\quad HO}CO\end{matrix}\right\rangle C_6H.(OH)_3 = C_6H_4\left\langle\begin{matrix}CO\\CO\end{matrix}\right\rangle C_6H.(OH)_3 + 2\,H_2O.$$

Benzoïc acid — Gallic acid — Anthragallol

Quinizarin dissolves, like oxyanthraquinones, in alkalies with a violet colouration. (Try it.) Since it does not contain the hydroxyl-groups in the vicinal $\alpha$-$\beta$-positions, it will not form dyes with metallic-salt mordants. This will be explained further under Alizarin.

## 41. REACTION: ALIZARIN FROM SODIUM β-ANTHRAQUINONE-MONOSULPHONATE[1]

In an autoclave or an iron pipe with a cap which can be screwed on (see page 60), heat a mixture of 10 parts commercial sodium anthraquinonemonosulphonate, 30 parts of sodium hydroxide, 1.8 parts of finely pulverised potassium chlorate, with 40 parts of water, for 20 hours to 170°. After cooling, the melt is boiled out with water several times, and acidified at the boiling-point of the solution in a large dish with concentrated hydrochloric acid. The alizarin separating out is then filtered off according to the quantity, either with suction or with the aid of a filter-press, washed with water, pressed out on a porous plate, and dried in an air-bath at 120°. In order to obtain it completely pure, it is distilled rapidly from a small retort, and is

[1] A. Spl. 7, 300; B. 3, 359; 9, 281.

crystallised from glacial acetic acid, or in large quantities from nitrobenzene.

The sodium hydroxide fusion of the sodium anthraquinonemonosulphonate is an abnormal reaction to the extent that besides the replacement of the sulphonic acid group by hydroxyl, a hydrogen atom is also oxidised to a hydroxyl-group:

$$C_6H_4\begin{picture}(0,0)\end{picture}\genfrac{}{}{0pt}{}{CO}{CO}C_6H_2\genfrac{}{}{0pt}{}{H}{SO_3Na} + 3\,NaOH + O$$

$$= C_6H_4\genfrac{}{}{0pt}{}{CO}{CO}C_6H_2\genfrac{}{}{0pt}{}{ONa}{ONa} + Na_2SO_3 + 2\,H_2O.$$

The tendency to the formation of alizarin is so great that even without the addition of an oxidising agent (potassium chlorate), it is formed with the evolution of hydrogen. Formerly the oxygen of the air was used as the oxidising agent, the reaction being effected in air.

In order to prepare alizarin on the large scale, anthracene is the starting-point; this is obtained from the highest-boiling fractions of coal tar (anthracene oil). It is oxidised by chromic acid to anthraquinone (see below), and this on heating with sulphuric acid is converted into the monosulphuric acid. The separation of this latter compound is greatly facilitated by the fact that it forms a sodium salt difficultly soluble in water, which, on account of its silvery appearance, is called "Silver salt." If the sulphonation mixture is diluted with water and neutralised with sodium carbonate, the sodium anthraquinonemonosulphonate is precipitated directly, which thus obviates the necessity of removing the excess of sulphuric acid beforehand. On the large scale the alizarin fusion is conducted exactly as on the small scale, except that autoclaves, with stirring attachments, are used. The constitutional formula of alizarin is:

[Structural formula of alizarin showing anthraquinone ring system with OH groups at α and β positions]

The salts are intensely coloured. The red aluminium salt, the violet ferric salt, and the granite-brown chromic salt are especially important in dyeing. With alizarin and all its related compounds the dyeing is effected by mordanting the fibre with a salt of one of the

x

three oxides just mentioned; the thus prepared fibre is heated with a thin dilute water-paste of the free insoluble dye, whereby salts are formed on the fibre (Lakes).

Of the numerous di- and poly-oxyanthraquinones only those are actual dyes which contain, like alizarin, two hydroxyl-groups in the vicinal $\alpha$-$\beta$-position, *i.e.* the derivatives of alizarin (Rule of Liebermann and Kostanecki). The above prepared quinizarin dissolves in alkalies with a violet colouration, but with metallic oxides it forms no salts on fibres.

From two disulphonic acids of anthraquinone, two trioxyanthraquinones, flavo- and anthra-purpurin, are prepared in a manner analogous to that by which alizarin is obtained from the monosulphonic acid.

From alizarin there can be prepared, further, by nitration, the $\alpha$- or $\beta$-nitro-alizarin, and from this, by reduction, the corresponding amido-alizarin. From $\beta$-nitro- and amido-alizarin, by heating with glycerol and sulphuric acid, the important Alizarin Blue is obtained. Further, by the action of fuming sulphuric acid on alizarin there is obtained a tetraoxyanthraquinone (Bordeaux), etc.

## 42. REACTION: ZINC DUST DISTILLATION

EXAMPLE: **Anthracene from Alizarin or Quinizarin**

To a paste prepared by rubbing up 100 grammes of zinc dust with 30 c.c. of water, add pieces of porous pumice stone of a size that will conveniently pass into a combustion tube, and stir them around so that they become covered with the zinc dust paste. They are removed from the paste with pincers, heated in a porcelain dish over a free flame (in constant motion) until the water is evaporated. A combustion tube of hard glass 60–70 cm. long is drawn out at one end to a narrow tube, the narrowed end is closed by a loose plug of asbestos, and a layer of zinc dust 5 cm. long is placed next to the plug; then follows a mixture of $\frac{1}{2}$–1 gramme of alizarin or quinizarin with 10 grammes of zinc dust, and finally, a layer of pumice-zinc dust 30 cm. long. After a canal has been formed over the zinc dust, by placing the tube in a horizontal position and tapping it, the tube is transferred to a combustion furnace inclined at an oblique angle, and dry hydrogen is passed through the tube without heating. In order to test whether the air has been completely expelled from the tube, the open end is closed by a cork bearing a small glass tube to which is attached a piece of rubber tubing; the gas being evolved is conducted into a soap solution, and the bubbles formed are ignited, during which the greatest care must be taken to keep the flame from coming in contact with the gas issuing from the rubber tubing, otherwise a serious explosion may result. If an explosion accompanied by a report takes place when the bubbles are ignited, the air has not been completely removed, but if they burn quietly, then only pure hydrogen is present.[1] When this is the case, the current of gas is diminished so that only two bubbles per second pass through the wash-bottle; the pumice-zinc dust is then heated with small flames, these are increased in size gradually, and finally, the tiles being placed in position, it is heated as strongly as possible; then the rear layer of 5 cm. of zinc dust is similarly heated, and as soon as this glows, as in the nitrogen determination, the mixture of the substance and zinc dust is gradually heated. The anthracene formed condenses to crystals in the forward cool part of the tube. After the reaction is complete, while the tube is allowed to cool, a moderately rapid current of hydrogen is passed through it; the forward part of the tube containing the anthracene is broken off and the substance removed with a small spatula; it is purified by sublimation in a suitable apparatus (see pages 14 and 15).

The sublimed anthracene is dissolved by heating in a test-tube with a little glacial acetic acid; after a moment, it is treated with double its weight of chromic anhydride, and heated a short time to boiling. The solution is then diluted with several times its volume of water, the anthraquinone separating out is filtered off, washed with some dilute sulphuric acid, then with water, and is finally crystallised in a test-tube from a little glacial acetic acid. Long colourless needles of anthraquinone, which melt at $277°$, are thus obtained.

---

[1] As described under Acetylene, the test may also be made by filling a test-tube with the gas over water, and applying a match to the mouth of the tube.

Zinc dust is, especially at high temperatures, an excellent reducing agent (Baeyer, A. 140, 295), which can be used for the reduction of almost all aromatic compounds derived from hydrocarbons containing oxygen, *e.g.*:

$$C_6H_5 \cdot OH + Zn = C_6H_6 + ZnO,$$
Phenol — Benzene

$$C_{10}H_7 \cdot OH + Zn = C_{10}H_8 + ZnO.$$
Naphthol — Naphthalene

Also ketone-oxygen, as the above example shows, can be replaced by hydrogen. The reaction given under Alizarin possesses an historical interest, since, by means of it, Gräbe and Liebermann, in 1868, discovered that alizarin, which had been previously obtained from madder root, was a derivative of anthracene, and could be prepared synthetically from it. (B. 1, 43.)

## III. PYRIDINE OR QUINOLINE SERIES

### 1. REACTION: THE PYRIDINE SYNTHESIS OF HANTZSCH[1]

EXAMPLE: **Collidine = Trimethylpyridine**

*Dihydrocollidinedicarbonic Acid Ester.* — A mixture of 25 grammes of acetacetic ester and 8 grammes of aldehyde ammonia is heated in a small beaker in a wire-gauze, about 3 minutes, to 100–110°, the mixture being stirred with the thermometer. The warm reaction-mixture is then treated with double its volume of dilute hydrochloric acid, and stirred without further heating vigorously, until the liquid mass solidifies. It is then thoroughly triturated in a mortar, filtered, washed with water, and dried, either by pressing out, or by warming on the water-bath. For the further working up of the collidinedicarbonic acid ester, the crude product can be directly used. In order to obtain the dihydroester in a crystallised condition, 2 grammes of the crude product is dissolved in a small quantity of alcohol in a test-tube, by heat, and allowed

---

[1] A. 215, 1.

to cool slowly. Colourless tablets with a bluish fluorescence are thus obtained. Melting-point, 131°.

*Collidinedicarbonic Acid Ester.* — The crude dihydroester is treated in a small flask with an equal weight of alcohol; complete solution does not take place. Into the mixture cooled by water pass nitrous fumes (Fig. 73), until the dihydroester goes into solution, and a test-portion dissolves to a clear solution in dilute hydrochloric acid. The alcohol is then evaporated by heating on the water-bath, the thick residue is treated with a sodium carbonate solution to alkaline reaction; the oil separating out is taken up with ether. After the ethereal solution has been dried by a small piece of potassium hydroxide, or potash, the ether is evaporated, and the residue subjected to distillation; on account of the high boiling-point of the ester, a fractionating flask is selected, having the condensation tube as near as possible to the bulb. The fraction passing over between 290–310° can be used for the following experiment:

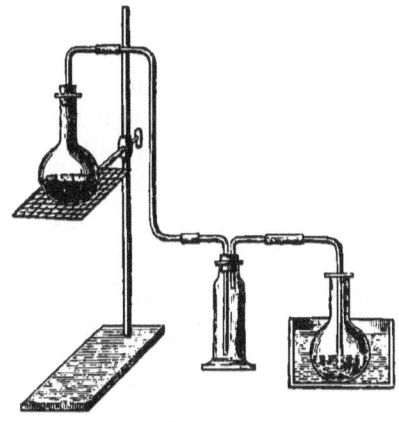

FIG. 73.

*Potassium Collidine Dicarbonate.* — The saponification of the ester is effected by boiling with alcoholic potash, prepared in the following manner: Finely pulverised potassium hydroxide (2 parts to 1 part of ester) is moderately heated in a flask on a wire-gauze with 3 times its weight of absolute alcohol, until the greater portion has passed into solution. The alcoholic solution is then poured off from the portion remaining undissolved, treated with the ester to be saponified, and heated 4–5 hours on a rapidly boiling water-bath (with reflux condenser); the potassium salt separates out in crusts. The alcoholic liquid is then poured off from the salt, and the latter washed on the filter with alcohol and finally with ether.

*Collidine.* — The dried potassium salt is intimately mixed in a mortar with double its weight of slaked lime, and placed in one end of a hard glass tube (about 2 mm. wide and 55 cm. long). In order to prevent the mixture from being carried over into the receiver on heating, a plug of asbestos is placed in the tube in front of it. After a canal has been made by tapping, the tube is connected with an adapter bent downwards, by means of a cork or asbestos paper; it is then transferred to a combustion furnace, the front end of which is somewhat elevated and warmed throughout its entire length with small flames, beginning at the closed end. The flames are steadily increased in size until, with the tiles in position, the tube is heated as strongly as possible. The collidine passing over is taken up with ether, dried with potassium hydroxide, and, after the evaporation of the ether, is subjected to distillation. Boiling-point, 172°.

On heating acetacetic ester with aldehyde-ammonia, the following reaction takes place (see A. 215, 8):

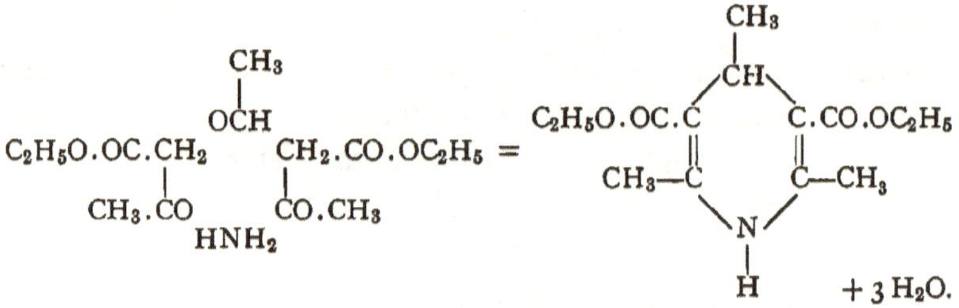

$$C_2H_5O.OC.CH_2 \overset{\overset{\displaystyle CH_3}{|}}{\underset{\underset{\displaystyle CH_3.CO}{|}}{OCH}} \quad \overset{}{\underset{\underset{\displaystyle CO.CH_3}{|}}{CH_2.CO.OC_2H_5}} = $$

Dihydrocollidinedicarbonicethyl ester $+ 3 H_2O$.

The reaction may be modified by using other aldehydes instead of acetaldehyde; thus there is obtained from benzaldehyde, acetacetic ester, and ammonia, the dihydrophenyllutidinedicarbonic ester:

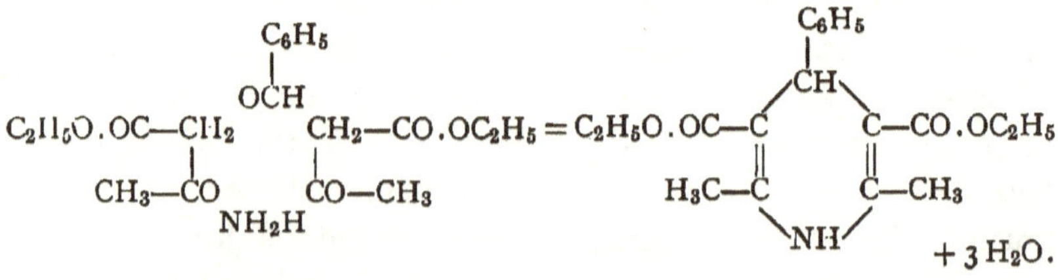

$$C_2H_5O.OC-CH_2 \overset{\overset{\displaystyle C_6H_5}{|}}{\underset{\underset{\displaystyle CH_3-CO}{|}}{OCH}} \quad \overset{}{\underset{\underset{\displaystyle CO-CH_3}{|}}{CH_2-CO.OC_2H_5}} = $$

$+ 3 H_2O$.

With proprionic aldehyde, butyraldehyde, valeraldehyde, œnanthol, myristic aldehyde, nitrobenzaldehyde, phenylacetaldehyde, furfurol, and others, the reaction can be carried out. All the compounds obtained contain the methyl groups of the two acetacetic ester molecules, but the third side-chain is different, depending upon the nature of the aldehyde employed.

By passing nitrous fumes into an alcoholic solution of the dihydro-ester, two hydrogen atoms, and those particular hydrogen atoms in combination with carbon and nitrogen in the methenyl- and imido-groups, respectively, will be oxidised off, and there is formed a derivative of pyridine, containing no ring hydrogen. While the dihydro-esters possess no basic properties, the pyridine derivative dissolves in acid. Therefore, by treating the solution with hydrochloric acid, it can be determined whether any unchanged dihydroester (insoluble in acid) is present.

Concerning the saponification of esters, refer to what was said under Reaction 34.

The splitting off of carbon dioxide from a carbonic acid, or a salt of a carbonic acid, is generally designated as a "pyro-reaction." For this kind of action a calcium salt is most frequently used; this is mixed with slaked lime and subjected to distillation, $e.g.$:

$$C_6H_5 \cdot \boxed{COOca + caO}\, H = C_6H_6 + CaCO_3.$$
Calcium benzoate
(ca=½ Ca)

In poly-basic acids, all the carboxyl groups can be replaced by hydrogen. In this way an acid may be transformed into the hydrocarbon from which it was derived. In the above case, the potassium salt may be used instead of the calcium salt.

### 2. REACTION: SKRAUP'S QUINOLINE SYNTHESIS

EXAMPLE: **Quinoline**

In a flask of about 1½ litres capacity containing a mixture of 24 grammes of nitrobenzene, 38 grammes of aniline, and 120 grammes of glycerol, add, with stirring, 100 grammes of concentrated sulphuric acid. The flask is then connected with a long, wide reflux condenser, and heated on the sand-bath. As soon as the reaction begins, which is recognised by the sudden evolution of bubbles of vapour ascending through the liquid, the flame is removed, and the energetic reaction is allowed to complete itself without further heating from without. When the reaction-mixture has become quiet, it is again heated for three hours on the sand-bath, diluted with water, and from the acid liquid the unchanged nitrobenzene is removed with steam. As soon as no drops of oil pass over, the distillation with steam is discontinued. The liquid remaining in the distillation flask is allowed to cool somewhat, and then made alkaline with concentrated caustic soda solution, upon which the liberated quinoline, mixed with the unchanged aniline, is distilled over with steam. Since these substances cannot be separated by fractional distillation, their separation must be effected by a chemical method. For this purpose the distillate (oil and water solution) is treated with dilute sulphuric acid until all oil is dissolved, and to the cold solution a solution of sodium nitrite is added until a drop of the liquid will cause a blue spot on potassium iodide-starch paper. The aniline (primary amine) is converted into diazobenzenesulphate, while the tertiary quinoline remains unchanged. The mixture is heated for some time on the water-bath, by which, as in Reaction 7, the diazo-sulphate is converted into phenol. The liquid is again made alkaline, upon which the phenol goes into solution, while the quinoline is liberated. The mixture is now distilled with steam, and the quinoline is obtained in a pure condition: it is taken up with ether, the ether evaporated, and the residue distilled. Boiling-point, $237°$. Yield, 40–45 grammes. (See Wiener Monatshefte 2, 141.)

Quinoline is formed in the above reaction, according to the following equation:

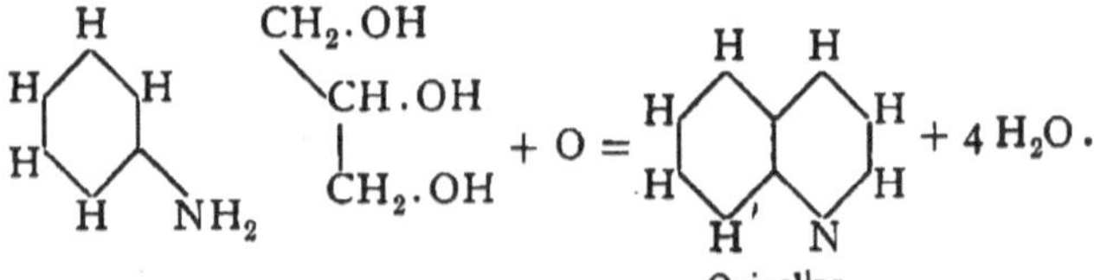

$$\text{C}_6\text{H}_5\text{NH}_2 + \begin{matrix} \text{CH}_2.\text{OH} \\ \text{CH}.\text{OH} \\ | \\ \text{CH}_2.\text{OH} \end{matrix} + \text{O} = \text{Quinoline} + 4\,\text{H}_2\text{O}.$$

The oxygen necessary for the reaction is taken from the nitrobenzene, which is hereby reduced in a manner that is not wholly clear. It is possible that the reaction may take place in this way: first, acroleïn is formed from glycerol, under the influence of sulphuric acid:

$$\begin{array}{ccc} CH_2.OH & CH_2 \\ | & \| \\ CH.OH & = CH & +2\,H_2O. \\ | & | \\ CH_2.OH & CHO \end{array}$$

Like all aldehydes, this condenses with aniline to form acroleïn-aniline.

$$C_6H_5.NH_2 + CHO.CH{=}CH_2 = C_6H_5.N{=}CH{-}CH{=}CH_2 + H_2O.$$

While this, under the influence of the oxidising action of the nitro-compound, loses two atoms of hydrogen, and thus quinoline is formed:

Quinoline

The Skraup reaction is capable of a very many-sided application. If, instead of aniline, its homologues are used, methyl-, dimethyl-aniline, etc., quinoline is obtained. Also halogen-, nitro-, substituted amines, etc., yield halogen-, nitro-, substituted quinolines, etc. Amidocarbonic acids, amidosulphonic acids, amidophenols, yield carbonic acid-, sulphonic acid- or oxy-derivatives of quinoline. The reaction is also applicable to the corresponding amido-compounds of the naphthalene series. By starting from the diamines, two new pyridine rings, connected with the benzene ring, are formed; in this way the so-called phenanthrolines, etc., are obtained.

Of technical and historical interest is the discovery which was made by Prudhomme in the year 1877, that $\beta$-nitroalizarin, on heating with glycerol and sulphuric acid, yields a blue dye, Alizarin Blue. This gave the impetus to Skraup's synthesis. To Gräbe's investigations we are indebted for the knowledge of the process by which, as above, a quinoline synthesis is effected in the following way:

Nitroalizarin →

Residue of the glycerol added = Alizarin Blue

## IV. INORGANIC PART

### 1. CHLORINE

A flask is one-third filled with manganese dioxide (pyrolusite) in pieces the size of filberts; to this is added a quantity of concentrated hydrochloric acid which is just sufficient to cover it. On heating the mixture on a wire-gauze with a free flame, a regular current of chlorine is generated; this is passed through two wash-bottles containing water and concentrated sulphuric acid respectively; the water retains any hydrochloric acid which is carried along with the gas, and the sulphuric acid dries it. (See Figs. 68 and 79.) A piece of thin asbestos-paper is placed on the wire-gauze, as is always done on heating large flasks, by which the danger of breaking is essentially diminished. A very regular current of chlorine can also be obtained from finely pulverised potassium dichromate and crude concentrated hydrochloric acid by heating the mixture on the water-bath. To 1 litre of hydrochloric acid, use 180–200 grammes of pulverised potassium dichromate.

### 2. HYDROCHLORIC ACID

Gaseous hydrochloric acid, which is frequently needed for the preparation of acid-esters, is generated most conveniently in a

# HYDROCHLORIC ACID

Kipp apparatus charged with fused ammonium chloride in pieces as large as possible, and concentrated sulphuric acid. The operation is conducted in the same way as that for the generation of carbon dioxide or hydrogen from a Kipp apparatus.

If the apparatus is not available, the acid can be generated very conveniently in the following manner:

In concentrated hydrochloric acid contained in a suction flask allow to flow from a separating funnel concentrated sulphuric acid, drop by drop (Fig. 74). The hydrochloric acid evolved is dried by passing it through concentrated sulphuric acid contained in a

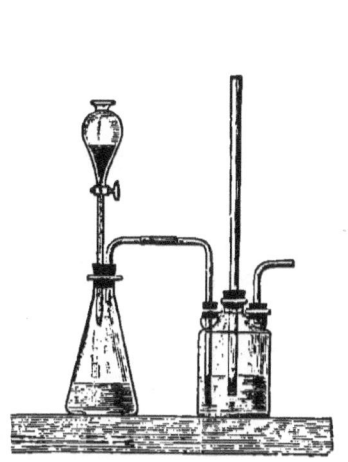

FIG. 74.

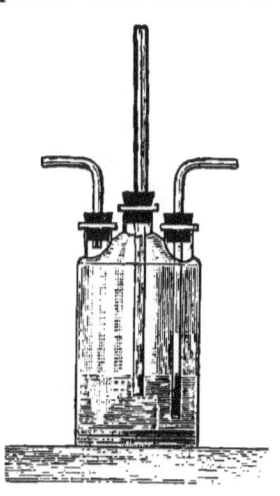

FIG. 75.

safety wash-bottle (Fig. 75); this latter is always used, since otherwise, with an irregular gas current, the liquid to be saturated may be easily drawn back into the wash-bottle and then into the generating mixture. In place of a Woulff-flask with three tubulures, a single-neck wash-bottle may be converted into a safety-bottle as follows (see Fig. 76): Into a two-hole cork place a straight tube as wide as possible; through this insert a narrow delivery-tube, bent at a right angle, which reaches almost to the bottom of the bottle.

The liquid to be saturated cannot flow back into the wash-bottle with this arrangement, since in case there should be a tendency to do so, air would enter the suction-flask through the space between the delivery tube and the wider tube, thus relieving the pressure. If a wash-bottle having a side-tube is available, it can also be converted into a safety-tube (see Fig. 77).

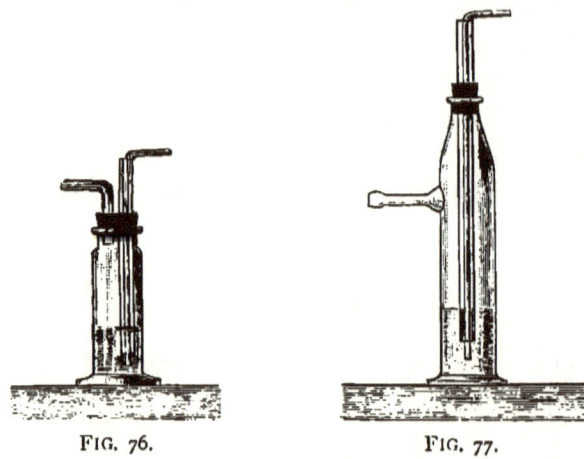

FIG. 76.  FIG. 77.

Hydrochloric acid gas may also be obtained by warming 10 parts of sodium chloride with a cold mixture of 3 parts of water and 18 parts of concentrated sulphuric acid.

### 3. HYDROBROMIC ACID (see Brombenzene)

The hydrobromic acid obtained as a by-product in the bromination reactions is purified by distilling it from a fractionating flask. Water first passes over until finally the temperature remains constant at 126°, when a 48% acid goes over; this is collected.

### 4. HYDRIODIC ACID

To 44 grammes of iodine (not pulverised) contained in a small *round* flask of about 100 c.c. capacity (Fig. 78), gradually add 4 grammes of yellow phosphorus divided into about 8 pieces; these are dried just before transferring them to the flask, by pressing between layers of blotting-paper. The first piece of phosphorus added unites with the iodine with an active evolution of heat and light. When the first action is ended, after shaking the contents of the flask, which soon become liquid, the second piece is added. The reaction still proceeds with evident energy, although it is less intense than when the first piece was added. Care is taken to place the phosphorus as nearly as possible in the middle of the flask, and not to allow it to fall on the walls, since otherwise the flask may be easily broken. When all of the phosphorus is added, a fused, dark mass of phosphorus triiodide is obtained which becomes solid on cooling. The hydriodic acid prepared from this by warming with water, must be passed over red phosphorus in order to free it from iodine which is carried along with it. Proceed as follows: 5 grammes of red phosphorus is rubbed up to a paste with 2 c.c. of a water solution of hydriodic acid, or in case this is not available, with as little water as possible (1 c.c. at the most). In this is placed glass beads, or bits of broken glass, which on stirring around in the mixture, become covered with the paste. They are then transferred to a U-tube. In order to prepare a water solution of hydriodic acid, the gas issuing from the U-tube is passed into 45 c.c. of water (see Fig. 78). The glass tube is not immersed in the water, but its end must be 1 cm. above the surface; otherwise, in consequence of the great affinity of water for hydriodic acid, under certain conditions, the water may be drawn back.

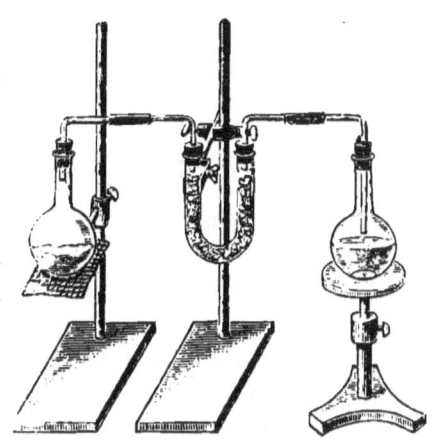

FIG. 78.

The hydriodic acid is now obtained by treating the *completely cooled* phosphorus triiodide with 6 grammes of water and warming with a *very small* luminous flame. The contents of the flask steadily become clearer, while in the other flask the heavy layer of hydriodic acid sinks to the bottom. The heating is continued until only a clear, colourless liquid remains in the generating flask.

In order to obtain a concentrated solution of hydriodic acid, the liquid in the receiver is distilled. At first a few cubic centimetres of water pass over at 100°, then the temperature rises in a short time to 125°; the concentrated acid passing over up to 130° is collected separately. This boils for the most part at 127°.

This experiment teaches much concerning the chemistry of phosphorus and iodine. First, it shows that iodine and phosphorus unite directly with a vigorous reaction, to form phosphorus triiodide:

$$P + 3I = PI_3.$$

The iodide then decomposes with water, to form hydriodic acid, which is evolved, while the phosphorous acid ($H_3PO_3$) remains in the flask:

$$P\boxed{I_3 + 3H} \cdot OH = 3HI + PH_3O_3.$$

The gaseous hydriodic acid is an intensely fuming substance, which may be easily shown by removing the cork from the receiver containing the aqueous acid, for a moment. Hydriodic acid is absorbed by water with great avidity. The acid, boiling constantly at 127°, contains approximately 50% of anhydrous hydriodic acid.

In this experiment it is observed that the connecting tubes of the apparatus, especially those between the generating flask and the U-tube become coated with crystals of a diamond-like brilliancy. These are crystals of phosphonium iodide, $PH_4I$, which is formed by the decomposition of phosphorous acid.

It is a common property of all the lower oxidation products of phosphorus, to pass over to the highest oxidation product — phosphoric acid, with the evolution of phosphine on heating. With phosphorous acid, the reaction takes place as follows:

$$4 PH_3O_3 = 3 H_3PO_4 + PH_3.$$

The phosphine thus formed unites, since it possesses weak basic properties, with hydriodic acid, to form phosphonium iodide:

$$PH_3 + HI = PH_4I.$$

Since this may easily clog the connecting tubes, the tubes selected are as wide as possible. On cleaning the tubes with water, this reacts with the phosphonium iodide with the evolution of phosphine, a gas with a garlic-like odour, and which in this case is not spontaneously inflammable. The phosphonium iodide decomposes with water into its components, in accordance with this equation :

$$PH_4I + H_2O = PH_3 + HI.$$

This reaction, as is well known, is employed for preparing pure phosphine which is not spontaneously inflammable.

### 5. AMMONIA

Gaseous ammonia is prepared most conveniently by heating the *most concentrated* ammonia solution in a flask over a wire-gauze with a small flame. In order to dry the gas, it is passed through a drying tower filled with soda-lime. (See Fig. 57.)

### 6. NITROUS ACID

For the preparation of gaseous nitrous acid, arsenious acid, broken into pieces the size of a pea, is treated with nitric acid, sp. gr. 1.3, and heated gently on a wire-gauze with a free flame (under the hood). In order to condense the nitric acid carried along with the gases, an empty wash-bottle, cooled by cold water, is employed. (See Fig. 73.)

### 7. PHOSPHORUS TRICHLORIDE

Under water, in a porcelain mortar, cut 80 grammes of yellow phosphorus, with a knife or chisel, into pieces which will conveniently pass into the tubulure of a 400 c.c. retort. After the air in the retort has been displaced by dry carbon dioxide (Fig. 79), each single piece of phosphorus is taken from the water by pincers, and dried quickly by pressing it between several layers of filter-paper, and immediately placed in the retort, care being taken to prevent it from becoming ignited by friction in the opening of the tubulure. As soon as all the phosphorus has been transferred

320 SPECIAL PART

to the retort, the tubulure is connected with a delivery tube which must move easily in the cork, and a moderately rapid current of

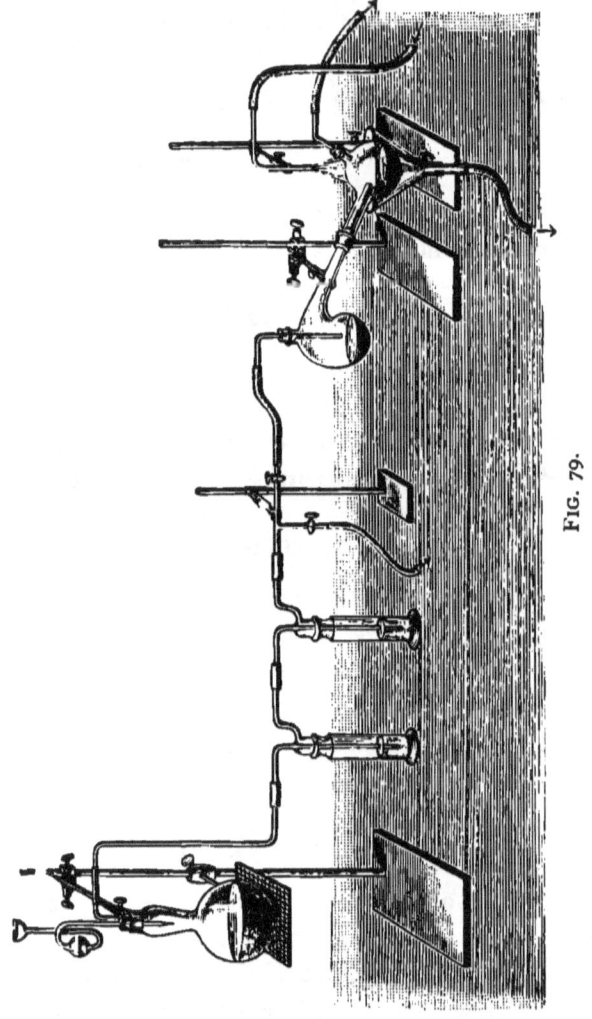

FIG. 79.

dry chlorine passed over the phosphorus; phosphorus chloride is thus formed with evolution of heat and light. If crystals of phosphorus pentachloride should collect in the neck of the retort, the delivery tube is pushed somewhat farther into the retort. If, on the other hand, phosphorus distils to the upper part of the retort, the tube is somewhat raised. The phosphorus trichloride condensing·in the receiver is distilled from a dry fractionating flask. Boiling-point, 74°. Yield, 250–280 grammes.

### 8. PHOSPHORUS OXYCHLORIDE[1]

To 100 grammes of phosphorus trichloride, contained in a large tubulated retort connected with a condenser, add gradually, in small portions of about 2–3 grammes, 32 grammes of finely pulverised potassium chlorate. After each addition, wait until the liquid bubbles up, before adding a new quantity. If, on the addition of the first portion, no reaction takes place, it is started by a gentle warming. During the addition, no liquid should distil into the receiver, but if this does happen, it is poured back into the retort. After all of the chlorate has been added, the phosphorus oxychloride formed is distilled, by heating the retort in an oil-bath, to 130°, or with a luminous flame. A suction-flask is used as a receiver; this is firmly connected with the end of the condenser, by means of a cork. The distillate is rectified from a fractionating flask provided with a thermometer. Boiling-point, 110°. Yield, 100–110 grammes.

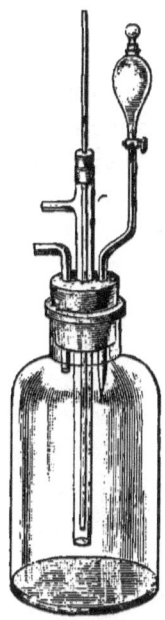

FIG. 80.

### 9. PHOSPHORUS PENTACHLORIDE

Through the upper delivery tube of an apparatus similar to that represented in Fig. 80, a stream of dry chlorine is admitted, which

---

[1] J. pr. Ch. 1883, [2] Vol. 28, 382.

passes out of the lower, right-angled tube. From time to time, several cubic centimetres of phosphorus trichloride are allowed to flow into the bottle from a separating funnel; upon which the trichloride unites with the chlorine to form the solid pentachloride. Since this operation can be repeated, as soon as it is evident that the union is completed, any desired quantity of phosphorus pentachloride can be prepared. Should the delivery tube become stopped up, it is cleared by the glass rod with which the apparatus is provided. As the quantity of the pentachloride formed increases, the tube is correspondingly raised. Yield, quantitative.

### 10. SULPHUROUS ACID

Gaseous sulphurous acid is generated in an apparatus similar to the one represented in Fig. 74, by adding to a concentrated water solution of sodium hydrogen sulphite a cold mixture of equal parts, by volume, of water and concentrated sulphuric acid, drop by drop. The generating flask is shaken frequently, to keep the contents from separating into layers.

### 11. SODIUM

(*a*) *To cut Sodium.* — In order to divide sodium into small portions, it can be cut into scales with a knife, or pressed out into a wire with a sodium-press. To cut it into scales, an apparatus similar to that represented in Fig. 81 is convenient. After both sides of the knife and the front part of the table have been coated with a thin layer of vaseline, a long stick of the metal to be cut, the end of which is wrapped in filter-paper, in order that it may be handled, is placed on the table so that it projects somewhat over the front end; it is then cut with a short stroke of the knife. On the front part of the lower platform is placed a small dish filled with ether or ligroïn, into which the scales fall. When

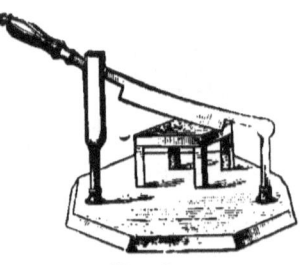

FIG. 81.

using the knife, two points are to be especially observed. The eye is never placed in front of the knife, but always behind it, so that the fingers holding the sodium can always be seen. Only in this way can a wound be prevented. Further, the *cross-section of the piece of sodium must not be too large*, otherwise the metal adheres to the knife. Quadratic scales, the edge of which must not, at most, be more than 5–6 mm. long, are cut. With a little practice, large quantities of the metal can be cut in very thin scales in a short time.

The sodium residues are not thrown into water nor into waste-jars, but are dropped into alcohol contained in a beaker or flask.

(*b*) *Sodium Amalgam.* — Sodium scales, about the size of a 20-cent piece, are pressed to the bottom of mercury contained in a porcelain mortar, in rather rapid succession, by means of a short, moderately thick glass rod, drawn out to a point and bent at a short right angle. The scales are speared on the glass rod (under the hood; eyes protected by spectacles; hands, with gloves).

The mercury may also be warmed in a porcelain casserole on the water-bath (60–70°), and, without further heating, small pieces of sodium, the size of a half bean, are thrust to the bottom of the vessel with the aid of a glass rod.

## 12. ALUMINIUM CHLORIDE

A tube as wide as possible, diameter $1\frac{1}{2}$–2 cm., of hard glass drawn out to a narrow tube, is at one end connected by means of a cork with a wide-neck so-called "salt bottle" (Fig. 82). The cork with which this is closed is supplied with a second, smaller hole, bearing a delivery tube of at least 8 mm. diameter, extending to the centre of the receiver. The tube is half filled (half of its cross-section) with aluminium shavings, which have been previously freed from oil by boiling with alcohol and then dried in an air-bath at 120°; an asbestos plug is placed at each end of the layer. A rapid current of hydrochloric acid gas, most conveniently obtained from a Kipp apparatus charged with fused ammonium chloride and concentrated sulphuric acid, is passed

through the apparatus. Care must be taken that the drying flasl
containing sulphuric acid is not too small, since the acid foam
easily. As soon as the air is driven out of the apparatus, — thi
has been accomplished when the gas evolved is completel:
absorbed by water (a piece of rubber tubing is attached to th
tube, and the gas tested from time to time by immersing the en(
of the tubing in water in a beaker), — the tube is heated in a com
bustion furnace throughout its entire length, at first with smal
flames, which are gradually increased (Fig. 82). When the flame
have reached a certain size, white vapours of aluminium chloride
condensing in the receiver, are noticed. The reaction is ende(
as soon as the aluminium, except for a small, dark-coloured resi

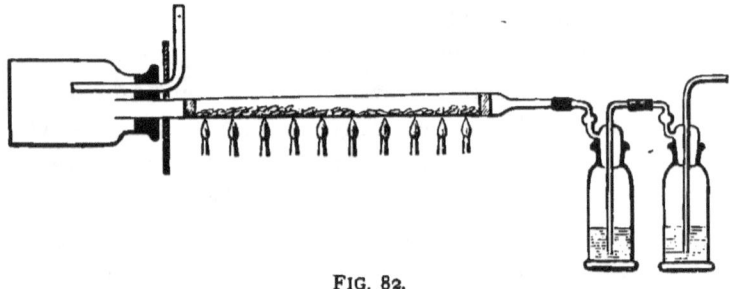

FIG. 82.

due, disappears. For the success of the preparation, the followin;
points are particularly observed : (1) All parts of the apparatu
must be perfectly dry. (2) The air must be removed as com
pletely as possible, since, otherwise, an explosion of oxygen an(
hydrogen may take place. (3) The portion of the tube extend
ing beyond the furnace must be as short as possible, to preven
the aluminium chloride from condensing in it, which results in ;
stopping up of the apparatus. In order that the cork may no
burn, it is protected by an asbestos plate, provided with a circula
hole in the centre. (4) The aluminium must not be heated t(
melting. If this should happen at any particular point, the flame:
must be immediately lowered. (5) The hydrochloric acid cur
rent must be *extremely* rapid. One should not be able to coun
single bubbles of the gas, but they should follow one anothe:

uninterruptedly. The evolution of a small quantity of a smoky vapour from the outlet-tube will always occur, but the greatest part of the aluminium chloride is condensed even if the hydrochloric acid *rushes through* the wash-bottles. Should the first experiment be unsuccessful, in consequence of a stoppage of the tube, the method for connecting this will readily suggest itself.

The aluminium chloride condensing in the receiver is preserved in well-closed bottles, or best, in a desiccator.

### 13. LEAD PEROXIDE

In a large porcelain dish dissolve, with heat, 50 grammes of lead acetate in 250 c.c. of water, and treat with a solution of bleaching-powder, prepared by shaking 100 grammes of bleaching-powder with $1\frac{1}{2}$ litres of water and filtering, and heat not quite to boiling, until the precipitate, bright at first, becomes deep dark brown. A small test-portion is then filtered *hot*, and the filtrate treated with the bleaching-powder solution and heated to boiling; if a dark brown precipitate is formed, more of the bleaching-powder solution is added to the main quantity, and it is heated until a test gives no precipitate with the bleaching-powder solution. The main quantity of the liquid is separated from the heavy precipitate by decantation; the latter is washed several times with water (decantation), and then filtered with suction; the precipitate is washed repeatedly with water. The lead peroxide is not dried, but is preserved in a *closed* vessel in the form of a thick paste.

*Value Determination.* — In order to determine the value of the paste, a weighed portion is heated with hydrochloric acid, the chlorine evolved is passed into a solution of potassium iodide, and the liberated iodine is titrated with a $\frac{N}{10}$ solution of sodium thiosulphate (refer to a text-book on Volumetric Analysis). The determination, carried out as follows, is sufficiently accurate for preparation work: On an analytical balance weigh off exactly 6.2 grammes of pure, crystallised sodium thiosulphate; this is dissolved in enough cold water to make the volume of the solution just 250 c.c. In a small flask weigh off 0.5–1 gramme of the peroxide paste; treat this (with cooling) with a mixture of equal volumes of concentrated hydrochloric acid and water; the flask is immediately connected with a delivery tube, and this is inserted in an inverted retort, the neck of which has been expanded to a bulb, and which contains a solution of four grammes of potassium iodide in water. When heat is applied to the flask, chlorine is generated, which liberates iodine from the potassium iodide. After the end of the heating, care is taken that the potassium iodide solution is not drawn back into the flask. The contents of the retort are then poured into a beaker and treated with the thiosulphate solution from a burette until the yellow colour of the iodine just disappears. Since a molecule of the peroxide liberates two atoms of iodine, a cubic centimetre of the thiosulphate solution corresponds to $\frac{0.0239}{2} = .012$ grammes pure lead peroxide.

# INDEX

Acetaldehyde, 136.
Acetamide, 124.
Acetacetic ester, 148.
Acetonitrile, 128.
Acetic ester, 130.
Acetic anhydride, 120.
Acetyl chloride, 114.
Acetylene, 165.
Acetylene tetrabromide, 167.
Alcohol, test for, 155.
Aldehyde, 136.
Aldehyde-ammonia, 137.
Alizarin, 304.
Aluminium chloride, 323.
Amidoazobenzene, 220.
Amidodimethyl aniline, 213.
Ammonia, 319.
Ammonium eosin, 297.
Aniline, 176.
Animal charcoal, 43.
Anthracene, 306.
Anthraquinone, 307.
Autoclaves, 60.
Azobenzene, 183.
Azo dyes, 211.
Azoxybenzene, 183.

Beckmann's reaction, 272.
Benzal chloride, 251.
Benzaldehyde, 251.
Benzamide, 269.
Benzene from aniline, 193.
Benzene from phenylhydrazine, 206.

Benzenesulphinic acid, 240.
Benzenesulphon amide, 235.
Benzenesulphon chloride, 235.
Benzenesulphonic acid, 235.
Benzidine, 188.
Benzil, 260.
Benzoïc acid, 256.
Benzoïcphenylester, 270
Benzoïn, 258.
Benzophenone, 272.
Benzophenone oxime, 272.
Benzotrichloride, 253.
Benzoyl chloride, 269.
Benzyl alcohol, 256.
Benzyl chloride, 253.
Bitter almond-oil green, 293.
Boiling-point, corrections of, 30.
Bomb-furnace, 58.
Bomb-tubes, 55.
Brombenzene, 226.
Bromethane, 105.
Bromine carrier, 228.
Bromine, determination of, 71.
" Bumping," 30.
Butyric acid, 154.

Carbon, determination of, 92.
Chloracetic acid, 132.
Chlorine, 314.
Chlorine, determination of, 71.
Cinnamic acid, 265.
Cleaning the hands, 67.
Cleaning vessels, 66.

# INDEX

Collidine, 308.
Collidinedicarbonic ester, 309.
Congo-paper, 213.
Crystallisation, 1.
Crystal violet, 301.

Decolourising, 43.
Diazoamidobenzene, 217.
Diazobenzeneimide, 195.
Diazobenzeneperbromide, 195.
Diazotisation, 193.
Dibrombenzene, 226.
Dihydrocollidinedicarbonic ester, 308.
Dinitrobenzene, 173.
Diphenylmethane, 281.
Diphenylthiourea, 190.
Disazo dyes, 216.
Distillation, 16.
Distillation with steam, 35.
Drying, 45.
Drying agents, 46.
Drying, of vessels, 66.

Eosin, 294.
Ether, pure, 232.
Ethyl benzene, 231.
Ethyl bromide, 105.
Ethylene, 160.
Ethylene bromide, 171.
Ethyl iodide, 106.
Ethyl malonic acid, 154.
Ethyl malonic ester, 154.
Extraction with ether, 39.

Filter press, 52.
Filtration, 49–54.
Fittig's Synthesis, 231.
Fluoresceïn, 294.
Friedel-Crafts' Reaction, 272.
Fuchsine-paper, 213.

Halogens, determinations of, 71.
Heating under pressure, 55.

Helianthine, 211.
Hydrazobenzene, 184.
Hydrazones, 209.
Hydriodic acid, 316.
Hydrobromic acid, 316.
Hydrochloric acid, 314.
Hydrocinnamic acid, 268.
Hydrogen, determination of, 92.
Hydroquinone, 225.

Iodine chloride, 134.
Iodine, determination of, 71.
Iodobenzene, 201.
Iodoethane, 106.
Iodosobenzene, 202.
Isonitrile reaction, 182.

Lead peroxide, 325.

Malachite green, 291.
Malonic ester, 154.
Mandelic acid, 262.
Mandelic acid nitrile, 261.
Melting-point, determination of, 62.
Methyl amine, 144.
Methylene blue, 217.
Monobrombenzene, 226.
Monochloracetic acid, 132.

Naphthalenesulphonic acid ($\beta$), 24.
Naphthol ($\beta$), 246.
Nitroaniline, 176.
Nitrobenzene, 172.
Nitrogen, determination of, 80.
Nitrophenol (o and p), 249.
Nitrous acid, 319.

Opening bomb tubes, 58.
Osazones, 209.

Perkin's reaction, 265.
Phenol from aniline, 200.
Phenyldisulphide, 242.

Phenylhydrazine, 206.
Phenyliodide, 201.
Phenyliodide chloride, 202.
Phenyliodite, 203.
Phenyl mercaptan, 242.
Phenyl mustard oil, 190.
Phosphorus oxychloride, 321.
Phosphorus pentachloride, 321.
Phosphorus trichloride, 319.
Pipette, capillary, 40.
Potassium-iodide-starch-paper, 197.
Pressure flasks, 60.
Pukall's cells, 52.
Pyro-reaction, 311.

Qualitative tests for carbon, hydrogen, nitrogen, sulphur, chlorine, bromine, iodine, 68.
Quantitative determination of carbon and hydrogen, 92.
Quantitative determination of halogens, 71.
Quantitative determination of nitrogen, 80.
Quantitative determination of sulphur, 76.
Quinizarin, 302.
Quinoline, 311.
Quinone, 221.

Reduction of an azo dye, 211.

Safety wash-bottle, 314.
Salicylic acid, 287.
"Salting out," 42.
Sandmeyer's reaction, 205.

Schotten-Baumann's reaction, 270.
Sealing of bomb-tubes, 55.
Separation of liquids, 39.
Sodium, 322.
Sodium acetate, anhydrous, 120.
Sodium amalgam, 323.
Sodium eosin, 297.
Specific gravity, 169.
Steam distillation, 35.
Sublimation, 14.
Sulphanilic acid, 192.
Sulphobenzide, 235.
Sulphur, determination of, 76.
Sulphurous acid, 322.

Tarry matter, removal of, 43.
Terephthalic acid, 284.
Testing thermometers, 65.
Tests for carbon, 68.
Tests for halogen, 70.
Tests for hydrogen, 68.
Tests for nitrogen, 68.
Tests for sulphur, 69.
Thermometer, tests of, 65.
Thiocarbanilide, 189.
Thiophenol, 230.
Toluic acid, 283.
Tolyl nitrile, 204.
Trimethylpyridine, 308.
Triphenylguanidine, 190.

Vacuum distillation, 25.
Vinyl bromide, 167.

Zinc dust distillation, 306.

# ABBREVIATIONS

A.     = Liebig's Annalen der Chemie.
A. ch. = Annales de chimie et de physique.
B.     = Berliner Berichte.
Bl.    = Bulletin de la société chimique de Paris.
J.     = Jahresbericht über die Fortschritte der Chemie.
J. pr. = Journal für praktische Chemie.
P.     = Poggendorff's Annalen.
R.     = Journal der russischen chemischen Gesellschaft.
Z.     = Zeitschrift für Chemie.

THE MACMILLAN COMPANY'S
# ELEMENTARY TEXT-BOOKS ON CHEMISTRY.

**COHEN.** — The Owens College Course of Practical Organic Chemistry. 18mo. 70 cents.

"It is with great pleasure that we announce the appearance of this useful little work, in which the author has cut out a new path of his own, by the exclusively practical character of the lessons and by the style he has adopted." — *Chemical Trade Journal.*

**FISHER.** — A Class-Book of Elementary Chemistry. With 60 Illustrations. 12mo. $1.10.

**HEWITT and POPE.** — Elementary Practical Chemistry, Inorganic and Organic. By J. T. HEWITT, Ph.D., F.C.S., and F. G. POPE. 18mo. Limp cloth. 25 cents.

**JONES.** — The Owens College Junior Course of Practical Chemistry. Illustrated. 18mo. 70 cents.

Questions on Chemistry. A Series of Problems and Exercises in Inorganic and Organic Chemistry. 18mo. 75 cents.

**MUIR and CARNEGIE.** — Practical Chemistry. A Course of Laboratory Work. With numerous Illustrations. 80 cents.

**MUIR and SLATER.** — Elementary Chemistry. $1.25.

**RAMSAY.** — Experimental Proofs of Chemical Theory for Beginners. 70 cents.

**RICHARDSON.** — Laboratory Manual and Principles of Chemistry for Beginners. By GEORGE M. RICHARDSON, Associate Professor of Chemistry in the Leland Stanford Junior University. 16mo. $1.10.

**ROSCOE.** — Lessons in Elementary Chemistry, Inorganic and Organic. New Edition, revised and enlarged, with Illustrations. $1.25.

**ROSCOE and LUNT.** — Inorganic Chemistry for Beginners. By Sir HENRY ROSCOE, F.R.S., assisted by JOSEPH LUNT, B.Sc. Globe 8vo. 75 cents.

**THORPE and TAIT.** — A Series of Chemical Problems. With Key. By T. E. THORPE, B.Sc., Ph.D., F.R.S. New Edition, revised and enlarged by W. TAIT. 16mo. 65 cents.

**TURPIN.** — Works by G. S. TURPIN, M.A., D.Sc.

Lessons in Organic Chemistry. Part I. Elementary. 16mo. 75 cents.

Practical Inorganic Chemistry. 16mo. 60 cents.

**WALKER and DOBBIN.** — Chemical Theory for Beginners. 16mo. 70 cents.

PUBLISHED BY

## THE MACMILLAN COMPANY,
66 FIFTH AVENUE, NEW YORK.

# Important Works of Reference on Chemistry.

**BENEDIKT.** — **Chemical Analysis of Oils, Fats, and Waxes**, and of the Commercial Products derived therefrom. From the German of Professor R. BENEDIKT. Revised and Enlarged by Dr. J. LEWKOWITSCH, Consulting Chemist. 8vo. $7.00.

"As many of our readers are aware, the work of Prof. Dr. Benedikt on this subject (in German) has been considered an authority in this field, and the fact that its author himself requested Dr. Lewkowitsch to translate and adapt it to the English-speaking public, with such alterations as the progress of time demands, is an indication that the work now before us is one of a merit not altogether usual in technical books." — *American Soap Journal and Perfume Gazette.*

"It is rarely that one finds the work of the translator so excellently performed. The part of the work devoted to Quantitative Analysis is excellently written, disclosing at once that the author is thoroughly familiar with the work. The latest researches are carefully quoted and criticised, the criticisms being usually strengthened by results obtained in his own laboratory." — *Science.*

**GAMGEE.** — **A Text-Book of the Physiological Chemistry of the Animal Body.** Illustrated. 8vo.

Vol. I. **The Proteids.** $4.50.

"This book ought to be thoroughly studied by every practitioner who is anxious to bring himself fully abreast of the times in the treatment of disease, and we heartily recommend it to all." — *American Medico-Surgical Bulletin.*

Vol. II. **The Physiological Chemistry of Digestion.** $4.50.

"This is a second volume of a great work, the first of which appeared in 1880. This second volume is devoted to the Physiological Chemistry of Digestion, an investigation that covers an account of the agents, processes, and results covered in the conversion of food into the assimilable fluid. The discussion embraces a consideration of the Chemical Constitution of the Alimentary Juices' taking an active part in the process of digestion, both general and special, with full illustrations from the like processes in the lower animals. . . . Too much cannot be said in commendation." — *Medical Herald.*

**LASSAR-COHN.** — **A Laboratory Manual of Organic Chemistry.** A Compendium of Laboratory Methods for the use of Chemists, Physicians, and Pharmacists. By Dr. LASSAR-COHN, Professor of Chemistry in the University of Königsberg. Translated, with the Author's sanction, from the Second German Edition by ALEXANDER SMITH, B.Sc., Ph.D., Assistant Professor of General Chemistry in the University of Chicago. 12mo. $2.25.

**MENSCHUTKIN.** — **Analytical Chemistry.** By N. MENSCHUTKIN, Professor in the University of St. Petersburg. Translated from the Third German Edition, under the Supervision of the Author, by JAMES LOCKE. 8vo. $4.00.

"The work of Professor Menschutkin has for many years been known and favorably regarded by analytical chemists, who will welcome this translation. In part first, Qualitative Analysis, there are three sections, successively treating the metals, and the metalloids, and giving preliminary operations. Under the head of 'General Reactions' the corresponding compounds of all the metals of a group are studied, and also the conditions necessary for separation of one group from another. The 'Special Reactions' include such as are necessary to detect each individual metal." — *The School Review.*

---

PUBLISHED BY

## THE MACMILLAN COMPANY,
### 66 FIFTH AVENUE, NEW YORK.

# Important Works of Reference on Chemistry.

**NERNST.** — **Theoretical Chemistry from the Standpoint of Avogadro's Rule and Thermodyamics.** By Prof. WALTER NERNST, Ph.D., of the University of Göttingen. Translated by Prof. CHARLES SKEELE PALMER, Ph.D., of the University of Colorado. With twenty-six Woodcuts and two Appendices. 8vo. $5.00.

" English-speaking chemists who interest themselves in the theoretical side of their science owe a debt of gratitude to Professor Palmer for so perfect a translation of this able and valuable work. We know of no other book within which a student can find the subject of theoretical chemistry so thoroughly and exhaustively treated from the inorganic standpoint. Its surprising lucidity, where the topic permits of lucidity, makes it as good an elementary treatise as that of Tilden, while at the same time it is so exhaustive in detail and depth that we deem it superior to Ostwald's, Meyer's, or Mendeleeff's works on the same subject." — *Popular Science News.*

" Dr. Nernst everywhere speaks with the authority of a master of his subject. So that his book, notwithstanding the treatises of Ostwald and others on Physical Chemistry, seems to us, in the excellence of its arrangement, the clearness of its style, and the thoroughness of its subject-matter, to be the best book of its kind which has yet appeared. Dr. Palmer deserves especial thanks for putting the book so admirably into its English dress. Typographically, also, the book is a credit to its publishers." — *American Journal of Science.*

**OSTWALD.** — **Manual of Physico-Chemical Measurements.** By WILHELM OSTWALD, Professor of Chemistry in the University of Leipzig. Translated, with the Author's sanction, by JAMES WALKER, D.Sc., Ph.D., Professor of Chemistry in University College, Dundee. 8vo. $2.25.

" Both purpose and contents distinguish this book from the existing handbooks for the physical laboratory, and it introduces the student to a region of experimental work which has already proved of great fruitfulness to chemistry. No one is better qualified than Professor Ostwald for the preparation of such a work. He has been fortunate in his translator, who has given us an excellent piece of work, in pleasing contrast to much that is inflicted on a long-suffering public." — *Evening Post.*

**Outlines of General Chemistry (Physical and Theoretical).** By Prof. W. OSTWALD. Translated by JAMES WALKER, D.Sc., Ph.D. 8vo. $3.50.

**The Scientific Foundations of Analytical Chemistry.** Treated in an Elementary Manner by WILHELM OSTWALD, Ph.D., Professor of Chemistry in the University of Leipzig. Translated with the Author's sanction by GEORGE MCGOWAN, Ph.D. 12mo. $1.60.

**SCHORLEMMER.** — **The Rise and Development of Organic Chemistry.** By CARL SCHORLEMMER, LL.D., F.R.S. Revised Edition, edited by ARTHUR SMITHELLS. 12mo. $1.60.

**SCHULTZ** (G.) and **JULIUS** (P.). — **Systematic Survey of the Organic Colouring Matters.** Translated and Edited, with extensive additions, by ARTHUR G. GREEN, F.I.C., F.C.S. Imp. 8vo. $5.00.

PUBLISHED BY

## THE MACMILLAN COMPANY,
### 66 FIFTH AVENUE, NEW YORK.

A DICTIONARY

OF

# CHEMICAL SOLUBILITIES (INORGANIC).

BY

### ARTHUR MESSINGER COMEY, Ph.D.,

*Formerly Professor of Chemistry, Tufts College; Director of Dept. of Chemistry, Summer School of Harvard University; etc.*

Cloth.    8vo.    pp. xx., 515.    Price, $5.00, *net.*

"One of the most valuable additions to the library of the practical chemist. . . . The work has been thoroughly done and is extraordinarily comprehensive. . . . The entire volume is a credit to the author, and will prove an invaluable aid to the chemist, manufacturer, pharmacist, and student."
— From *Merck's Market Report.*

"An indispensable adjunct to the chemical library. . . . That it has been brought up to a recent date gives it a high value."
—·From the *Scientific American.*

"More than thirty years have elapsed since the appearance of Storer's Dictionary of Solubilities, and Professor Comey has rendered a valuable service to chemists in writing a modern work upon this subject in such an excellent manner."
— From the *American Journal of Science.*

"The work is invaluable, occupies alone a field of great importance, is worth many times its price, and should be in every library that makes any pretence in the direction of chemistry or pharmacy."
— JOHN U. LLOYD in *The Spatula.*

"No student should be without it. A voice of gratitude from the chemists greets Professor Comey for his great work."
— Prof. OTIS C. JOHNSON, *University of Michigan.*

"It is a most valuable work, and will be welcomed by all chemists."
— Prof. C. Y. CHANDLER, *School of Mines, Columbia University.*

Similar comments have been made by Professors of Chemistry in many of the leading schools of the United States, from the Massachusetts Institute of Technology to the University of California.

PUBLISHED BY

## THE MACMILLAN COMPANY,
### 66 FIFTH AVENUE, NEW YORK.

www.ingramcontent.com/pod-product-compliance
Lightning Source LLC
Chambersburg PA
CBHW030324240426
43673CB00040B/1265